Plant Diseases

INFECTION, DAMAGE AND LOSS

Plant Diseases

INFECTION, DAMAGE AND LOSS

Edited for the
British Society for Plant Pathology
by
R.K.S. WOOD FRS
Imperial College of Science and
Technology, London
and
G.J. JELLIS
Plant Breeding Institute
Trumpington, Cambridge

BLACKWELL SCIENTIFIC PUBLICATIONS
OXFORD LONDON EDINBURGH
BOSTON PALO ALTO MELBOURNE

Editorial offices:
Osney Mead, Oxford, OX2 OEL
8 John Street, London, WC1N 2ES
9 Forrest Road, Edinburgh, EH1 2QH
52 Beacon Street, Boston
Massachusetts 02108, USA
706 Cowper Street, Palo Alto
California 94301, USA
99 Barry Street, Carlton
Victoria 3053, Australia

First published 1984

Set by Setrite Typesetters Ltd,
Hong Kong
Printed and bound by
the Alden Press, Oxford

DISTRIBUTORS

USA and Canada
Blackwell Scientific Publications Inc
P O Box 50009, Palo Alto
California 94303

Australia
Blackwell Scientific Book Distributors
31 Advantage Road, Highett
Victoria 3190

British Library
Cataloguing in Publication Data
Plant diseases.
1. Plant diseases
I. Wood, R.K.S. II. Jellis, G.J.
III. British Society for Plant Pathology
581.2 SB731

ISBN 0-632-01126-2

Contents

Section 3
Infection and host damage

Section 4
Damage and loss

Preface

This book is based on a symposium on the theme 'Plant Diseases: Infection, Damage and Loss' held at the University of Surrey in December 1982 under the auspices of the British Society for Plant Pathology. It was felt that there was a place for a symposium, and a book based on it, which surveyed the effects of pathogens on plants. This would be of special value to senior undergraduates, postgraduate students and to scientists who may not be plant pathologists.

Certain conditions were imposed by the need to cope with a broad survey within a book of reasonable size. Thus the diversity of topics meant many papers and that each had to be short. Authors were therefore encouraged to be selective in their approach, provided that this allowed the main points of their subject to be made.

The sequence of chapters conforms with the title apart from a few inconsistencies which were difficult to avoid. The first chapters are on substances from pathogens that damage plants and on how they do so. There follow accounts of the effects of pathogens on the main physiological processes of plants and of the damage to substance and structure that result in different types of disease. The last section is about damage to and loss of various types of economically important crops, including forest trees.

We benefitted greatly by reading the many chapters in this book which are outside our special interests. We trust that many others will choose to do so and benefit as we did.

On behalf of the Society we thank the contributors for their cooperation in the preparation of this volume. We also thank B.C. Clifford for his help as programme organizer and M.J. Howe who compiled the index.

R.K.S. WOOD
G.J. JELLIS
British Society for Plant Pathology

Section 1
How pathogens cause disease

1 Role of toxins in plant disease

J.G. TURNER
School of Biological Sciences, University of East Anglia,
Norwich NR4 7TJ, UK

Introduction

Toxins play a significant role in a number of important diseases of plants caused by fungi and bacteria. For example, the destructive effects of the disease of maize (corn) known as southern corn leaf blight are caused by a toxin produced by the pathogen *Helminthosporium maydis* race T and an epidemic of this disease in the USA during 1970 caused greater economic losses than can be attributed to any other plant disease in recorded history. Other examples of diseases in which toxins are involved are described below and in other reviews (Rudolph, 1976; Yoder, 1980; Scheffer & Briggs, 1981). The aim of this chapter is to present an account of the various ways in which toxins damage plants and the role of these toxins in the infection process and in disease development.

Definitions and classifications

For this paper a toxin is a compound that is produced by a pathogen in its host and which causes part or all of the symptoms of the disease. This definition conveys the commonly accepted meaning of the word toxin (Sykes, 1976) and is similar to the definition adopted by plant pathologists in a recent treatise (Durbin, 1981). However, others have used the word in a broader sense to include toxic compounds not necessarily involved in plant disease but which are produced *in vitro* by plant pathogens (Rudolph, 1976; Durbin, 1972). Historically, plant pathologists have excluded enzymes and plant growth regulators from their definitions of a toxin. This is a pity because it creates an unnecessary division in our consideration of the mechanisms of plant disease; for example, fusicoccin is a toxin that is considered in this chapter, but it has many auxin-like properties and its effect on the plant is probably best discussed in relation to the effects of the auxins and gibberellins that are produced by certain other pathogens in the diseases they cause.

Toxins are of various chemical types and include peptides, glycoproteins, polysaccharides, organic acids, fatty acids and derivatives, polyketides and terpenoids. Most classification schemes for toxins, however, are based on the biological activity of these compounds rather than on chemical structure. In the literature, for example, there are references to the terms pathotoxin (Wheeler & Luke, 1963) and

Wood R.K.S. & Jellis G.J. (1984) *Plant Diseases: infection, damage and loss.* Blackwell Scientific Publications, Oxford.

vivotoxin (Dimond & Waggoner, 1953), both of which define the toxin in terms of the extent to which it is known to be involved in the disease. These terms have not been widely used. Toxins have also been classified as primary or secondary determinants of disease (Scheffer & Pringle, 1967) and as virulence or pathogenicity factors (Yoder, 1980). A more useful, and widely accepted, distinction has been made between host-specific and non-specific toxins. Host-specific toxins are selectively toxic to hosts that are susceptible to pathogens producing these toxins whereas non-specific toxins also affect plants that are not susceptible to the pathogen.

Although these classifications may help to summarize information about a toxin, it is far more important to examine critically the evidence that the compound in question is, in fact, a toxin.

Evidence for a toxin

This is usually obtained in two stages. In the first stage, it is demonstrated that sterile filtrates from cultures of a particular pathogen can reproduce part or all of the disease symptoms, and an attempt is then made to purify the toxic substance. Purification of a toxin is no mean task, however, and the success of this operation is based, in part, on the choice of a suitable bioassay to monitor biological activity of the toxic preparation as it is fractionated during a series of separatory procedures. It is desirable for the bioassay to be based on part of the natural symptom complex and usually involves the use of plants or cuttings. Ideally, a pure toxic substance is eventually obtained that has been structurally characterized and causes part or all of the symptoms of the disease when applied to the host plant. The second stage is to determine whether or not the compound is produced in the diseased tissue. This is often difficult but may be approached in the following ways:

(1) Extract and purify the toxic compound from the natural lesion.

(2) Detect the toxin in the lesion, for example, through identification of characteristic physiological changes in the diseased tissue that are known to be caused specifically by the toxin, or if the toxin is antigenic, antiserum raised against it can be used to identify the toxin in extracts of the lesion (Scheffer & Elgersma, 1981)

(3) Obtain non-toxigenic mutants of the pathogen and use these in experiments based on the principle that, if the toxin were produced in the disease caused by the wild-type pathogen, it would give rise to characteristic symptoms that would be absent from infections caused by the non-toxigenic mutants. Similarly, toxin-resistant mutants of the host can be used.

Although mutants provide powerful tools for the study of toxins, caution must be exercised in interpreting results of investigations that have not included critically important evidence that the mutant differs from the wild type only in the character being investigated.

This two-stage approach to the identification of a toxin has the advantage that it is often technically simpler to extract and purify a toxic compound from an axenic culture of the pathogen before attempting to purify the unknown compound directly from the lesion which may also contain other toxic substances of plant origin. A serious disadvantage of this approach is the possibility that a toxin may only be produced by the pathogen growing in host tissues. *Erwinia amylovora*, causal agent of the fireblight disease of apple and pear, for example, produces the toxic polysaccharide amylovorin when grown in association with its host, but not in axenic culture on a variety of media (Goodman *et al.*, 1973).

Mechanisms of action

It may at first seem surprising that the many toxins believed or known to be involved in plant disease are able to cause only three major visible symptoms, wilting, chlorosis or necrosis, singly or in combination. The explanations for this are that plants possess a limited capacity for the visual expression of damage and quite different forms of primary damage caused by various toxins lead to a common secondary damage which in turn gives rise to the symptom. A less convincing alternative explanation is that structurally different toxins responsible for a common symptom type have an identical primary effect on the plant. This issue is partially clarified in the following sections on the mechanisms by which toxins cause symptoms.

Chlorosis

Alternaria tenuis causes an irreversible, variegated chlorosis of seedlings of many hosts including cotton and citrus (Templeton, 1972). A toxin, tentoxin, has been purified from culture filtrates of the fungus and identified as a cyclic tetrapeptide comprising the amino acids glycine, N-methyl dehydrophenylalanine, leucine and N-methylalanine. This structure has been confirmed by chemical synthesis (Rich *et al.*, 1978). Clues to the mechanism of action of tentoxin came from observations that it caused chlorosis only if applied to seedlings within the first 32 h after germination. Later studies revealed that the development of proplastids into chloroplasts was blocked by tentoxin; chlorosis occurred because the affected tissues failed to become green. Steele *el al.* (1976) used radioactive tentoxin to investigate its primary action and found that tentoxin bound to purified chloroplast coupling factor (CF_1), which catalyses the synthesis of ATP from ADP and phosphate in a reaction driven by, or coupled to, the electron flow produced during the light reactions of photosynthesis. Tentoxin was found to inhibit coupled ATP synthesis in intact chloroplasts as efficiently as it bound to purified CF_1: the affinity constant of both reactions for tentoxin was $1.3 \times 10^7 M^{-1}$. Further evidence that the primary action of tentoxin was to inhibit ATP synthesis was reported by Bennett (1976) who observed that it inhibited RNA and protein synthesis in isolated chloroplasts and that these effects could be reversed by adding ATP to the chloroplast suspension. Because many of the chloroplast structural proteins and enzymes are synthesized

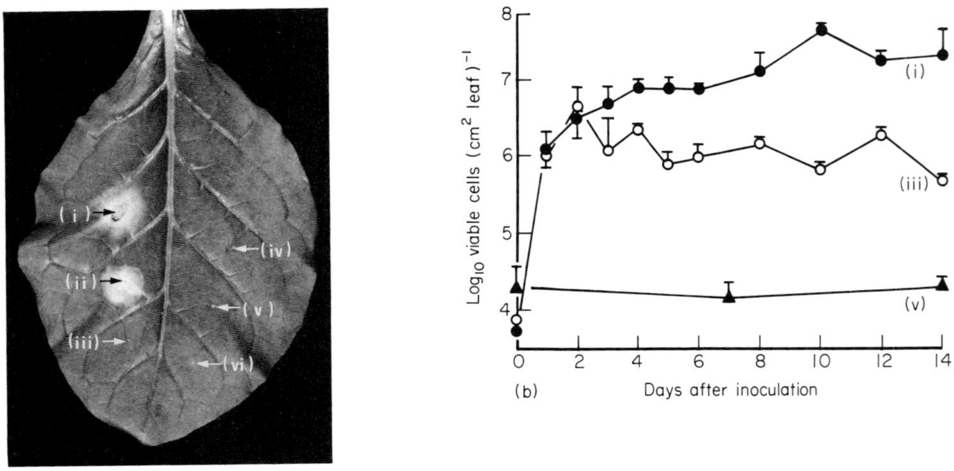

Figure 1.1. Effect of non-toxigenic mutants of *Pseudomonas syringae* pv. *tabaci* on tobacco leaves. Attached leaves were treated by placing a drop (5µl) containing 10^5 bacteria or 350 pmoles tabtoxin on a pinprick wound in the upper epidermis. Treatments were (i) *P. syringae* pv. *tabaci* (toxigenic; causal agent of wildfire disease) (ii) tabtoxin; (iii) *P. syringae* pv. *tabaci*, non-toxigenic mutant I (iv) *P. syringae*, pathogen of bean (v) *P. putida*, saprophyte and (vi) *P. syringae* pv. *tabaci*, non-toxigenic mutant II.
(a) Appearance of leaf after 5 days;
(b) viable counts of bacteria in inoculated tissue (mean of 3 replicates; bars represent standard deviation).

within this organelle it follows that the primary effect of tentoxin, the inhibition of ATP synthesis, could cause seedling chlorosis by preventing the synthesis of essential chloroplast proteins. However, this interpretation is based on results of studies using mature chloroplasts and has been challenged by Duke *et al.* (1982) who suggest that a primary effect of tentoxin occurs before the chloroplast has developed to the stage when it is photosynthetically active.

Pseudomonas syringae pv. *tabaci* is a more invasive pathogen than *A. tenuis*. It causes wildfire disease of tobacco which is characterized by a striking chlorosis that develops around infections in mature leaves (Figure 1a), although these symptoms are typically more severe in seedlings. The chlorosis is caused by tabtoxin, a toxic dipeptide produced by the pathogen. Although tabtoxin has not been isolated directly from infected plants there is indirect evidence for its involvement in the disease. For example, an identical pattern of physiological changes was detected in chlorotic tissue in natural infections and in leaves treated with purified tabtoxin (Turner, 1981). Also, non-toxigenic mutants did not produce tabtoxin nor cause chlorosis although they grew in infected tissue (Figure 1.1). The primary action of tabtoxin is probably the inhibition of glutamine synthetase which occurred 2–3 h after the toxin was injected into tobacco leaves (Turner, 1981). Surprisingly, pure tabtoxin did not inhibit glutamine synthetase *in vitro*. However, this enzyme was inhibited *in vitro* by the amino acid tabtoxinine ß-lactam which was released when tabtoxin was hydrolysed by peptidases of either plant or bacterial origin (Uchytil & Durbin, 1980). This suggests that tabtoxin is also hydrolysed *in vivo*. Glutamine synthetase is a key enzyme in nitrogen metabolism because it is solely responsible for primary N-assimilation into amino acids in plants and also functions in the photorespiratory nitrogen cycle where it recaptures ammonia released by the mitochondrial enzyme, glycine decarboxylase (Keys *et al.*, 1978). Consistent with this, ammonia accumulated in tabtoxin-treated tissue and was derived from the photorespiratory nitrogen cycle, as evidenced by findings that conditions which prevented photorespiration also prevented the accumulation of ammonia (Frantz *et al.*, 1982; Turner & Debbage, 1982) and symptom development (Turner & Debbage, 1982). What caused the leaves to turn yellow was not clear. Possibly, chlorosis depended on the apparent diversion of protein-N into ammonia-N (Turner & Debbage, 1982). However, ammonia uncouples photophosphorylation and this could also have contributed to the pathological changes leading to symptom development.

Another bacterium which produces a chlorosis-inducing toxin is *P. syringae* pv. *phaseolicola* responsible for halo blight disease of beans (*Phaseolus* spp.). This toxin is called phaseolotoxin and has been structurally characterized as the tripeptide (N^{δ}-phosphosulphamyl) ornithylalanylhomoarginine (Mitchell, 1976). Evidence that phaseolotoxin is produced in diseased tissues came largely from the observation that non-toxigenic mutants did not cause chlorosis although they grew in host tissue (Patil *et al.*, 1974). Phaseolotoxin is an inhibitor of the plant enzyme ornithine transcarbamylase which catalyses the synthesis of citrulline from ornithine and carbamyl phosphate. Inhibition of this enzyme appears to be the primary action of phaseolotoxin because the simultaneous addition of phaseolotoxin and citrulline but not ornithine prevented the development of symptoms. It is not known how inhibition of the synthesis of citrulline could give rise to chlorosis although the findings of Smith & Rubery (1982) suggest that chlorosis caused by phaseolotoxin arises from the inhibition of chlorophyll synthesis rather than from the acceleration of its breakdown.

Wilting

Some toxins cause wilting, a symptom which occurs when succulent tissues lose their turgidity. This symptom may result when either (a) the rate of transpiration exceeds the rate of water supply to the transpiring tissue or (b) the plasma membranes of turgid cells become freely permeable to solutes which leak from the cell sap. This causes net efflux of water from the cell. Many toxins do cause host-cell membranes to become freely permeable and as a result the tissues become flaccid and wilt. However, in these cases wilting follows extensive cellular disruption caused by the toxin which is therefore normally regarded as necrosis-inducing and is considered below.

Fusicoccin is a wilt-inducing toxin and is perhaps the best understood of the toxins believed or known to be involved in plant disease (Marrè, 1979). It is a diterpene glucoside produced by *Fusicoccum amygdali*, causal agent of wilt diseases of peach and almond trees. Fusicoccin has been isolated from diseased tissue and causes stomata in leaves to open irreversibly (Turner & Graniti, 1969). Under conditions which permit transpiration, uncontrolled water loss occurs through open stomata and toxin-treated leaves wilt and eventually die. The primary action of fusicoccin is to bind to and stimulate a plasma membrane ATPase responsible for electrogenic H^+-extrusion from the cell. Active H^+-extrusion is coupled to K^+ uptake across the plasma membrane of guard cells and as K^+ accumulates inside these cells, there is a net influx of water by osmosis. As a result, the turgor pressure of the guard cells increases and the stomata open. Malate synthesis is also stimulated following the primary action of fusicoccin and this contributes to the increase in osmotic potential of the guard cell. Thus the initial effect of this wilt-inducing toxin is to promote water uptake into the guard cells.

Other wilt toxins have different primary effects. For example *Erwinia amylovora* produces a toxic polysaccharide, amylovorin, that reproduces the wilt symptoms of the fireblight disease of most species of *Rosaceae* including apple and pear. Amylovorin was shown to block bulk water flow through stem sections of the host and the reduced water supply to the transpiring tissue resulted in the wilt symptoms (Sjulin & Beer, 1978). The biological activity of amylovorin is probably due to the fact that it is large enough (mol. wt. 1.65×10^5) to block the veins that conduct water into the leaf (Van Alfen & Allard-Turner, 1979). Similarly, a glycopeptide (mol. wt. 5×10^6) produced by *Corynebacterium insidiosum*, which causes a wilt disease of alfalfa, reduced water flow through alfalfa stem segments. Wilting developed in shoots that had taken up as little as 2μg of the glycopeptide. Further evidence that the glycopeptide blocked water flow and did not affect membrane permeability came from observations that shoots recovered when the base of the toxin-treated stem was cut off (Van Alfen & Turner, 1975).

Necrosis

Unlike chlorosis and wilting, necrosis may occur in any part of the plant, depending on the disease. A number of toxins that cause necrosis have been identified and of these, T-toxin has probably received most attention. It is produced by *Helminthosporium maydis* race T, which causes southern corn leaf blight. This disease occurs only in maize cultivars carrying the maternally inherited character known as Texas cytoplasmic male sterility factor (T cms) which was introduced to simplify production of hybrid seed. T-toxin causes necrosis only in T cms maize and is therefore a host-specific toxin. The primary action of T-toxin is not clear. Payne *et al.* (1980) found that between 5 and 50 ng T-toxin/ml (a) inhibited coleoptile elongation (b) induced ion leakage from roots (c) inhibited dark CO_2 fixation in leaf discs and (d) increased oxygen uptake by mitochondria. All these responses occurred with material from T cms maize whereas normal maize (N, lacking T cms) was insensitive to concentrations 1000 times higher. The various effects of T-toxin on T cms maize may each be expressions of a common primary action of the toxin, perhaps on mitochondrial oxidation. Alternatively, there may be multiple targets for the toxin. The interpretation of data is further complicated by the finding that T-toxin is a mixture of possibly 10 linear polyketols.

Another host-specific toxin that causes necrosis is victorin which is produced by *Helminthosporium victoriae*, causal agent of a seedling blight of oats. Leakage of potassium from susceptible tissue was detected 4 h after treatment with victorin (Black & Wheeler, 1966) and many subsequent studies suggested a primary action of victorin on the plasma membrane (Gardner & Scheffer, 1973). The hypothesis for a primary action of victorin at the plasma membrane could be tested experimentally if labelled toxin were available to enable measurement of its binding to membranes from oat cultivars. However, these critical experiments must await the development of a satisfactory procedure for the purification of victorin.

Pseudomonas syringae causes disease in species of more than 30 genera and it produces a toxin, syringomycin, that appears to be involved in disease development. Syringomycin is a small peptide, but its structure is not known. It has surfactant properties and will, for example, lyse red blood cells. The primary action of syringomycin is probably the detergent-like lysis of host membranes (Backman & DeVay, 1971).

The foregoing survey illustrates the structural diversity of toxins and the wide range in their primary actions that give rise to the syptoms that characterize the disease.

Toxins as determinants of pathogenicity

The role of a toxin in infection and colonization and the related problem of its contribution to pathogenicity of the pathogen is of critical importance. Many of the definitive experiments on these problems have been based on comparisons of non-toxigenic mutants with the toxin-producing parent strain. Also, studies with toxin-resistant mutants of the host have been valuable.

H. maydis race T not only causes southern corn leaf blight of T cms maize but is also responsible for a mild disease of maize cultivars carrying N-cytoplasm. Non-toxigenic cultures derived from race T did not cause the typical and severe symptoms of southern corn leaf blight on T cms maize (Yoder & Gracen, 1975). However, these cultures retained the ability to cause the milder disease on N as well as on T cms maize cultivars and were therefore very similar to *H. maydis* race 0 which does not produce T-toxin and is also responsible for a mild disease of T cms and N maize. Evidently, T-toxin had modified the pathogenicity of *H. maydis* in a manner that is easily reconciled with the selective effect of this toxin on T cms maize. Leonard (1973) suggested that race T had appeared only recently as a toxigenic mutant of race 0. Because race T and the non-toxigenic cultures of *H. maydis* caused two distinct diseases of T cms maize that differed in kind rather than in degree, it follows that the pathogenicity of these two isolates was also qualitatively different. Therefore, the ability to produce T-toxin is a partial determinant of the pathogenicity of *H. maydis* race T. This interpretation differs from that of Yoder (1980) who suggests that T-toxin is a determinant of virulence. Virulence, however, usually refers to quantitative aspects of the ability to cause disease and T-toxin could only be considered as a determinant of virulence if there was evidence that the severity of southern corn leaf blight infections was causally related to differences in the amount of T-toxin produced in the plant by different isolates of the *H. maydis* race T. It can be argued similarly that phaseolotoxin is a partial determinant of the pathogenicity of *P. syringae* pv. *phaseolicola* because non-toxigenic mutants of this bacterium did grow in host tissues and caused necrosis but not chlorosis (Patil *et al.*, 1974). By contrast, disease caused by the toxigenic wild type was characterized by both necrosis and chlorosis.

It should be evident from the foregoing that toxigenicity is, by definition, at least a partial determinant of pathogenicity. In some cases toxigenicity is more decisive in the expression of pathogenicity. For example, a series of non-toxigenic mutants of *P. syringae* pv. *tabaci* were found to grow in their host but these caused no chlorosis or necrosis (Figure 1.1). Clearly, these non-toxigenic mutants were still parasitic on their host but did not cause recognizable disease. This strongly suggests that tabtoxin is a sole determinant of the pathogenicity, although what determines the parasitic ability of this bacterium remains unclear. By contrast, some non-toxigenic mutants derived from other cultures of the pathogen did give rise to necrotic symptoms (Braun, 1937). Victorin provides a more striking example of a toxin that is the sole determinant of pathogenicity. For example, non-toxigenic isolates of *H. victoriae* were unable even to infect oat cultivars that were susceptible to the parent culture. Similarly, toxigenic cultures did not infect oat lines selected for resistance to victorin. Victorin therefore determines not only the host range of *H. victoriae* but it also defines the pathogenic and parasitic ability of this pathogen (Scheffer, 1976).

Toxins are clearly involved in causing the damage associated with a variety of different plant diseases. Although toxins may not be involved in all plant diseases a great deal more remains to be learned about the mechanism of plant disease from the study of toxins.

References

Backman P.A. & DeVay J.E. (1971) Studies on the mode of action and biogenesis of the phytotoxin syringomycin. *Physiological Plant Pathology* **1**, 215–34.

Bennett J. (1976) Inhibition of chloroplast development by tentoxin. *Phytochemistry* **15**, 263–5.

Black H.S. & Wheeler H. (1966) Biochemical effects of victorin on oat tissues and mitochondria. *American Journal of Botany* **53**, 1108–12.

Braun A. (1937) A comparative study of *Bacterium tabacum* Wolf and Foster and *Bacterium angulatum* Fromme and Murray. *Phytopathology* **27**, 283–304.

Dimond A.E. & Waggoner P.E. (1953) On the nature and role of vivotoxins in plant disease. *Phytopathology* **43**, 229–35.

Duke S.O., Wickliff J.L., Vaughn K.C. & Paul R.M. (1982) Tentoxin does not cause chlorosis in greening mung bean leaves by inhibiting photophosphorylation. *Physiologia Plantarum* **56**, 387–98.

Durbin R.D. (1972) Bacterial phytotoxins: a survey of occurrence, mode of action and composition. In: *Phytotoxins in Plant Diseases* (Ed. by R.K.S. Wood, A. Ballio & A. Graniti), pp. 19–32. Academic Press, New York.

Durbin R.D. (1981) (Ed.) *Toxins in Plant Disease*. Academic Press, New York.

Frantz T.A., Peterson D.M. & Durbin R.D. (1982) Sources of ammonium in oat leaves treated with tabtoxin or methionine sulphoximine. *Plant Physiology* **69**, 345–8.

Gardner J.M. & Scheffer R.P. (1973) Effects of cycloheximide and sulfhydryl-binding compounds on sensitivity of oat tissues to *Helminthosporium victoriae* toxin. *Physiological Plant Pathology* **3**, 147–57.

Goodman R.N., Huang J.S. & Huang P.-Y. (1973) Host-specific phytotoxic polysaccharide from apple tissue infected by *Erwinia amylovora*. *Science* **183**, 1081–2.

Keys A.J., Bird I.F., Cornelius M.J., Lea P.J., Wallsgrove R.M. & Miflin B.J. (1978) Photorespiratory nitrogen cycle. *Nature* (London) **199**, 299–300.

Leonard K.J. (1973) Association of mating type and virulence in *Helminthosporium maydis*, and observations on the origin of the Race T population in the United States. *Phytopathology* **63**, 112–5.

Marrè E. (1979) Fusicoccin: a tool in plant physiology. *Annual Review of Plant Physiology* **30**, 273–88.

Mitchell R.E. (1976) Isolation and structure of a chlorosis-inducing toxin of *Pseudomonas phaseolicola*. *Phytochemistry* **15**, 1941–7.

Patil S.S., Hayward A.C. & Eammons R. (1974) An ultraviolet induced non-toxigenic mutant of *Pseudomonas phaseolicola* of altered pathogenicity. *Phytopathology* **64**, 590–5.

Payne G., Knoche H.W., Kono Y. & Daly J.M. (1980) Biological activity of purified host specific pathotoxin produced by *Bipolaris (Helminthosporium) maydis* race T. *Physiological Plant Pathology* **16**, 227–39.

Rich D.H., Bhatnagar P., Mathiaparanam P., Grant J.A. & Tam J.P. (1978) Synthesis of tentoxin and related dehydro cyclic peptides. *Journal of Organic Chemistry* **43**, 296–302.

Rudolph K. (1976) Non-specific toxins. In: *Physiological Plant Pathology* (Ed. by R. Heitefuss & P.H. Williams), pp. 270–315. Springer-Verlag, Berlin.

Scheffer R.P. (1976) Host-specific toxins in relation to pathogenesis and disease resistance. In: *Physiological Plant Pathology* (Ed. by R. Heitefuss and P.H. Williams), pp. 247–69. Springer-Verlag, Berlin.

Scheffer R.P. & Briggs S.P. (1981) Introduction: a perspective of toxin studies in plant pathology. In: *Toxins in Plant Disease* (Ed. by R.D. Durbin), pp. 1–17. Academic Press, New York.

Scheffer R.P. & Elgersma D.M. (1981) Detection of a phytotoxic glycopeptide produced by *Ophiostoma ulmi* in elm by enzyme-linked immunospecific assay (ELISA). *Physiological Plant Pathology* **18**, 27–32.

Scheffer R.P. & Pringle R.B. (1967) Pathogen-produced determinants of disease and their effects on host plants. In: *The Dynamic Role of Molecular Constituents in Plant Parasite Interaction* (Ed. by C.J. Mirocha & I. Uritani), pp. 217–36. Bruce Publishing Company, St. Paul, Minnesota.

Sjulin T.M. & Beer S.V. (1978) Mechanism of wilt induction by amylovorin in *Cotoneaster* shoots and its relation to wilting of shoots infected by *Erwinia amylovora*. *Phytopathology* **68**, 89–94.

Smith A.G. & Rubery P.M. (1982) Investigation of the mechanism of action of a chlorosis-inducing toxin produced by *Pseudomonas phaseolicola*. *Plant Physiology* **70**, 932–8.

Steele J.A., Uchytil T.F., Durbin R.D., Bhatnager P. & Rich D.H. (1976) Chloroplast coupling factor 1: a species specific receptor for tentoxin. *Proceedings of the National Academy of Sciences U.S.A.* **73**, 2245–8.

Sykes J.B. (1976) *The Concise Oxford Dictionary of Current English, New Edition*. Oxford University Press, Oxford.

Templeton G.E. (1972) Alternaria toxins related to pathogenesis in plants. In: *Microbial Toxins*, Vol. VIII (Ed. by S. Kadis, A. Ciegler & S.J. Ajl), pp. 169–92. Academic Press, New York.

Turner J.G. (1981) Tabtoxin, produced by *Pseudomonas tabaci*, decreases *Nicotiana tabacum* glutamine synthetase and causes accumulation of ammonia. *Physiological Plant Pathology* **19**, 57–67.

Turner J.G. & Debbage J.M. (1982) Tabtoxin-induced symptoms are associated with the accumulation of ammonia formed during photorespiration. *Physiological Plant Pathology* **20**, 223–33.

Turner N.C. & Graniti A. (1969) Fusicoccin. A fungal toxin that opens stomata. *Nature* (London) **223**, 1070–1.

Uchytil T.F. & Durbin R.D. (1980) Hydrolysis of tabtoxins by plant and bacterial enzymes. *Experientia* **36**, 301–2.

Van Alfen N.K. & Allard-Turner V. (1979) Susceptibility of plants to vascular disruption by macromolecules. *Plant Physiology* **63**, 1072–5.

Van Alfen N.K. & Turner N.C. (1975) Changes in alfalfa stem conductance induced by *Corynebacterium insidiosum* toxin. *Plant Physiology* **55**, 559–61.

Wheeler H.E. & Luke H.H. (1963) Microbial toxins in plant disease. *Annual Review of Microbiology* **17**, 223–42.

Yoder O.C. (1980) Toxins in pathogenesis. *Annual Review of Phytopathology* **18**, 103–29.

Yoder O.C. & Gracen V.E. (1975) Segregation of pathogenicity types and host-specific toxin production in progenies of crosses between races T and O of *Helminthosporium maydis* (*Cochliobolus heterostrophus*). *Phytopathology* **65**, 273–6.

2 The role of cell wall-degrading enzymes in infection and damage

R.M. COOPER
School of Biological Sciences, University of Bath,
Bath BA2 7AY, UK

Introduction

Plant pathogenic micro-organisms must repeatedly encounter and penetrate cell walls of their hosts. Most but not all parasites can produce extracellular, cell wall-degrading enzymes (CWDE) capable of cleaving the major glycosidic linkages in plant cell walls, but this ability is not unique to plant pathogens and may often reflect an alternative saprophytic existence. The contrasting strategies of most necrotrophs and biotrophs is particularly reflected in the way that they deal with host walls (Figure 2.1). Degradation of primary walls is often rapid and extensive during infection by facultative parasites which cause necrosis, and most evidence implicates endo-polygalacturonidases (pectic enzymes) as major factors in pathogenicity. As substantial breakdown of primary walls usually entails disruption of contiguous protoplasts (see below), it follows that infection by ecologically obligate parasites must involve minimal changes to host walls.

This account will outline the involvement of CWDE in infection and damage caused by diverse microbial parasites. The possible role of host cell wall–microbial enzyme interactions in resistance and specificity are discussed elsewhere (Cooper, 1977, 1983).

Structure of plant cell walls and types of wall-degrading enzymes

Recent significant advances in our understanding of the structure of cell wall polymers are a result of improved approaches and techniques such as cultivated plant cells, controlled solubilization of wall polymers with pure CWDE, methylation analysis and gas chromatography/mass spectroscopy (Albersheim, 1976; Darvill *et al.*, 1980).

Primary walls mainly consist of polysaccharides and are laid down by undifferentiated, growing cells. Secondary walls may be added by deposition of other polysaccharides and in certain cells lignin (e.g. xylem) or suberin (e.g. endodermis) may be incorporated.

Cell walls are a two-phase system of organized cellulose microfibrils in a hydrated, amorphous matrix. They are not inert structures because walls undergo

Wood R.K.S. & Jellis G.J. (1984) *Plant Diseases: infection, damage and loss.* Blackwell Scientific Publications, Oxford.

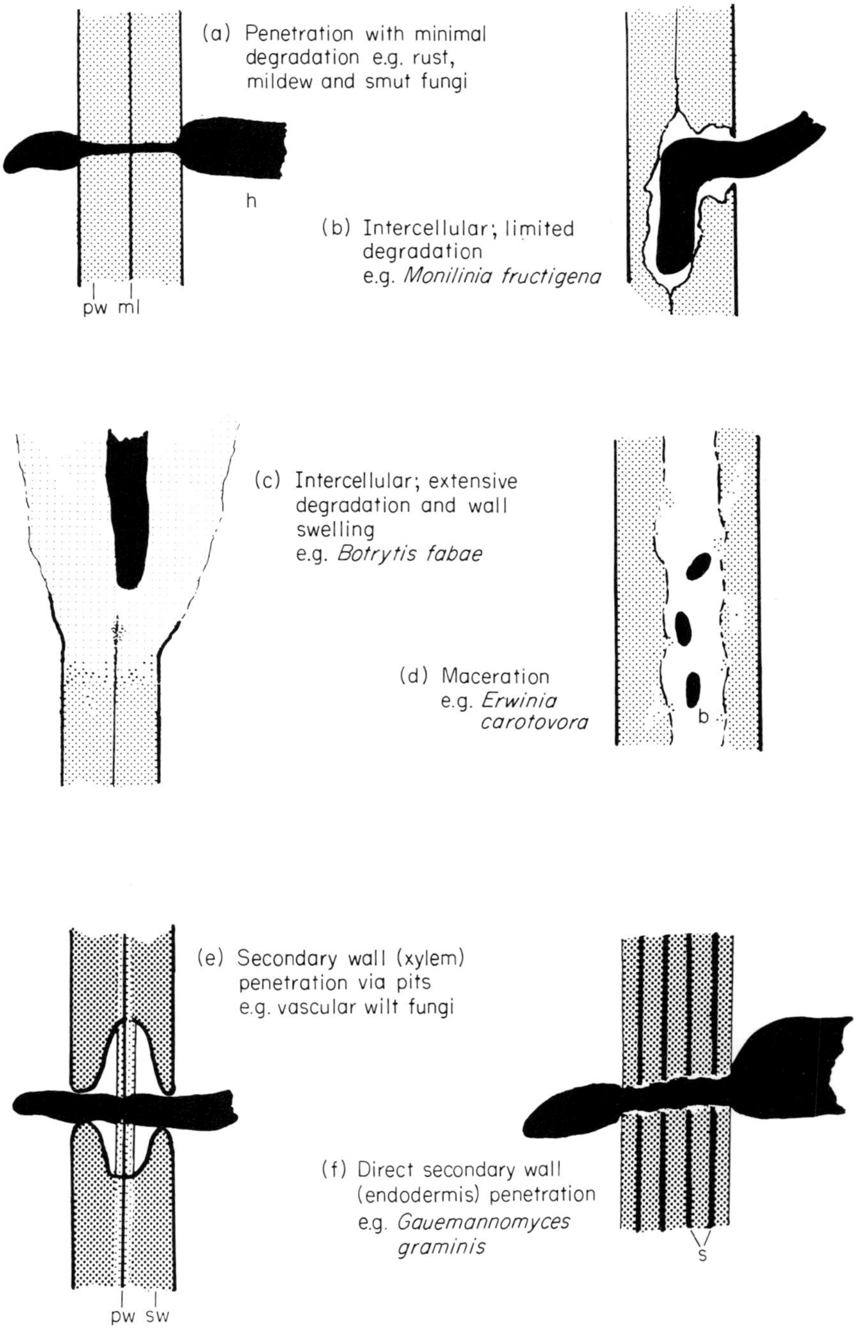

Figure 2.1. Diagrammatic representation of host cell wall penetration by various plant pathogens. pw, primary wall; sw, secondary wall; ml, middle lamella; s, suberin lamellae; h, hypha; b, bacteria. Diagrams are based on original micrographs (Cooper, 1981, 1983).

substantial physiochemical changes during growth and contain bound molecules which have various potential physiological effects once released.

Pectic polymers

These are predominant in, but not exclusive to, the intercellular region or middle lamella — a zone of particular importance in parasitism as its degradation has profound effects on host cells and is exploited by many pathogens which grow between cells. The key polymer, a rhamnogalacturonan, comprises chains of α-1,4-linked galacturonic acid residues interspersed with 1,2-linked rhamnose (Figure 2.2). The carboxyls of galacturonic acid may be methylated (Figure 2.3) or cross-linked by calcium, which can considerably affect wall properties and amenability to polygalacturonidases. To this backbone are linked araban (α-1,3- and α-1,5-, highly branched) and galactan (β-1,4-, linear) side chains but many other linkages can exist between the monomers (Darvill *et al*., 1980). Pectic polymers constitute a substantial proportion (*c*. 35%) of primary walls of dicots but monocot walls generally contain much less (8–9%).

Rhamnogalacturonan is degraded by hydrolytic polygalacturonases (PG), or by lyases (PL) which result in an unsaturated bond between carbons 4 and 5 at the non-reducing end of the broken chain (Figure 2.3). Both may attack the chain internally at random (*endo-*) or terminally (*exo-*) but some combine the two modes of action and attack chains at random, then release mono- or oligomers (*multiple* attack) (Cooper *et al*., 1978). Specificity is shown for regions either with methylated or free carboxyls. PGs usually have acidic pH optima (4–5) whereas lyases are most active at high pH (8–10) and have a requirement for Ca^{2+}. The methoxy groups on pectinic acids are removed by pectin methylesterase (PME) which can therefore facilitate the action of depolymerases specific for free carboxyls. Different activities are often difficult to separate. Thus, PME and PL of *Clostridium multifermentans* are asso-

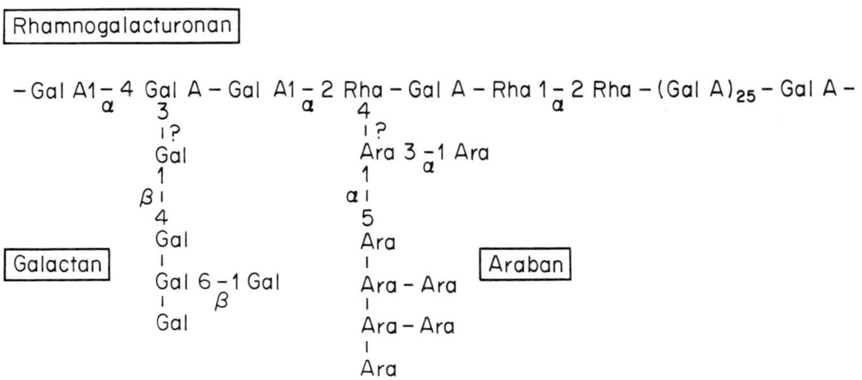

Figure 2.2. Possible structure of 'pectic' polymers — rhamnogalacturonan and covalently linked neutral polysaccharides. GalA, galacturonic acid; Rha, rhamnose; Gal, galactose; Ara, arabinose. (Albersheim, 1976; Darvill *et al*. 1980).

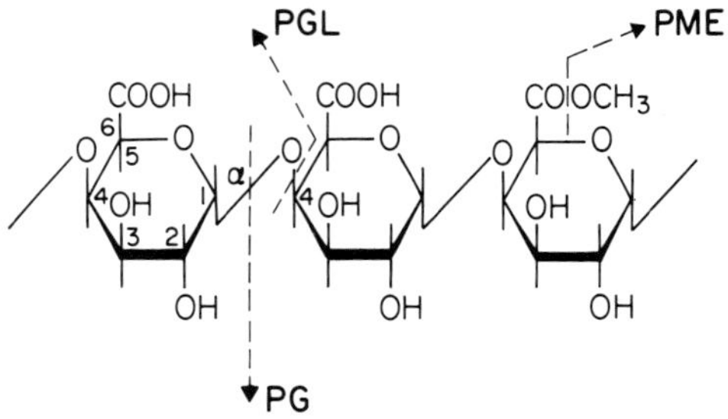

Figure 2.3. Cleavage of an α-1,4-linked polygalacturonide chain by *endo*-polygalacturonase (PG) and *endo*-polygalacturonide lyase (PGL), and demethoxylation by pectin methylesterase (PME). Note the numbering of carbons on the galacturonosyl units, the α configuration of the glycosyl bond, and the methylated or free carboxyls.

ciated as a complex and act synergistically in this bound state (Sheiman *et al.*, 1976).

The neutral polysaccharides linked to rhamnogalacturonan, and other matrix polymers described below, may be cleaved by combinations of *endo-* and *exo-* glycanases and glycosidases; however, breakdown of araban is almost invariably by *exo-* enzymes which release monomeric arabinose (Cooper *et al.*, 1978).

Hemicelluloses

In primary walls of dicots the main hemicellulose is xyloglucan, a polymer with a cellulose-like backbone of β-1,4-glucose with terminal branches of 1,6-xylose and fucose, and capable of hydrogen bonding to cellulose microfibrils. This property may unite the amorphous and fibrillar phases of the wall. Degradation of xyloglucan by endo-β-1,4-glucanases produces small fragments which can no longer bind to cellulose (Bauer *et al.*, 1973) and may therefore have implications in loss of wall structure. Small amounts of xyloglucan may also occur in monocots together with non-cellulosic 1,4- and 1,3-linked glucans (Labavitch & Ray, 1978).

The predominant component of monocot primary walls is β-1,4-xylan with side chains of α-1,3-arabinose and α-1,2-glucuronic acid. In contrast, xylans are mainly found in secondary walls of dicots.

Cellulose

Cellulose comprises extremely long (8000–12 000 units) linear chains of β-1,4-glucose stabilized by inter- and intra-molecular hydrogen bonding into fibrils with highly ordered crystalline regions or micelles (Preston, 1974). Compared to the parallel lamellae laid down in secondary walls, fibrils are randomly orientated and more dispersed in primary walls and can be virtually absent from the middle lamella.

Enzymic degradation of native cellulose is restricted by its highly ordered structure which can physically exclude cellulases (Reese, 1977), and in mature tissue by its association with lignin. Conversion to soluble products requires a complex of enzymes, originally predicted by Reese *et al.* in 1950 and named C_1, C_x and cellobiase. The mechanism remains controversial but generally C_x enzymes are endoglucanases and C_1 are exoglucanases which must act in synergism to degrade native cellulose. It is now thought that C_x initiates the attack and the function of C_1 might be to prevent reformation of broken glucosidic linkages (Wood & McCrae, 1979).

Wall interconnections

A recent picture of the primary cell wall of dicots (Darvill *et al.*, 1980) stresses the intimate association between different polymers but questions the level of dependence on covalent linkages as proposed in an earlier model (Keegstra *et al.*, 1973) in which it was envisaged that polymers in different regions of the wall are interlinked as one macromolecule. Thus, in the original model xyloglucan was conceived as a monolayer, hydrogen bonded to cellulose fibrils *and* covalently linked to pectic polysaccharides. Also it was apparent that a hydroxyproline-rich glycoprotein (which constitutes up to 10% of dicot primary walls) is covalently linked via arabinogalactan to rhamnogalacturonan.

Evidence for covalent links between neutral polymers and rhamnogalacturonan is convincing from structural analyses, the release of galactose and arabinose by pure pectic enzymes, and of galacturonic acid and arabinose by pure endogalactanase (Talmadge *et al.*, 1973; Jarvis *et al.*, 1981). However some of the pectic material in walls is not linked to other polymers but is bound by calcium, as revealed by the release of a substantial proportion of uronic acid (31–69%) by a non-destructive chelating agent (Jarvis, 1982). Calcium confers rigidity to cell walls, probably because Ca^{2+} ions could fit between and chelate to oxygen atoms of four galacturonosyl residues distributed between two polygalacturonide chains, somewhat analogous to eggs in a box as in the model proposed by Rees' group (Grant *et al.*, 1973). Cell walls of monocots and dicots may be arranged on similar lines despite some chemical dissimilarity. However, there may be a greater reliance on the binding of xylan to cellulose and even to itself (Darvill *et al.*, 1980).

Involvement of cell wall-degrading enzymes in pathogenicity

Attempts to implicate CWDE in infection are often based on circumstantial evidence and speculation. Many conditions need to be satisfied but only for very few diseases have critical studies been made. Some criteria that should be considered are given in Table 2.1 (Cooper, 1983).

Polygalacturonide degradation in pathogenesis

Rapid and extensive degradation of primary walls is found in diseases such as damping off, soft and dry rots, and in some stem and leaf lesions. However, it should

Table 2.1.

Criteria	Comments
1 Ability to produce CWDE *in vitro*	No proof of involvement in disease but gives potential range of activities, regulation of synthesis, and properties (Cooper & Wood, 1975).
2 Ability of enzyme to degrade insoluble host cell walls or tissue	Not all polysaccharidases can do this! (Garibaldi & Bateman, 1971; Jarvis *et al.*, 1981).
3 Detection of CWDE in infected tissue	CWDE should be found before or co-incident with symptoms; often not detected because of inactivation by oxidized host phenolics or by specific inhibitors; need to distinguish from CWDE of host or secondary invaders.
4 Depletion of host wall polysaccharides from infected tissue	Loss of specific sugars implies previous action by a CWDE (Bateman & Basham, 1976). However, some single CWDE can also cause reduction in sugars covalently linked to main chain (Basham & Bateman, 1975; Jarvis *et al.*, 1981).
5 Correlation of CWDE with pathogenicity	Induction of mutants may be preferable to screening natural, highly variable populations; plasmid transfer amenable for some bacteria (Chatterjee & Starr, 1977).
6 Microscopic changes in walls of infected tissue	Ultrastructure can reveal nature of wall degradation (Cooper, 1981) e.g. location of degraded substrate; whether CWDE are extracellular or bound; loss of affinity for specific stains, e.g. ferric hydroxylamine and polygalacturonide; loss of birefringence of degraded cellulose.
7 Microscopic localization of CWDE activity *in vivo*	Ultrastructural localization with labelled antibodies to pure CWDE (Kolattukudy, 1980); cytochemical visualization of glycosidases but likely interference by host enzymes.
8 Reproduction of wall changes or disease symptoms with pure CWDE	Molecular form of Koch's postulates, but problem that this approach cannot simulate conditions during infection (Cooper & Wood, 1980).
9 Inhibition or repression of CWDE during pathogenesis	Probably only feasible with antisera specific to CWDE during penetration of host surfaces (Kolattukudy, 1980). Most other approaches are too non-specific to interpret.

be stressed that necrosis can occur without wall degradation, as in fireblight in which the causal organism, *Erwinia amylovora*, appears to lack endo-polysaccharidases (Cooper & R.N. Goodman, unpublished).

The only clear involvement of CWDE in pathogenesis by facultative parasites concerns those enzymes which randomly degrade rhamnogalacturonan. The key role of this polymer in plant cell wall structure is revealed by the drastic effects caused by endopolygalacturonidases which not only solubilize galacturonides but result in a marked depletion of the covalently linked neutral sugars, arabinose, galactose, and rhamnose (Basham & Bateman, 1975; Talmadge *et al.*, 1973).

They are the first CWDE produced *in vitro* when pathogens such as

Verticillium albo-atrum and *Colletotrichum lindemuthianum* are grown on insoluble, extracted host cell walls which act as highly effective inducers of synthesis when used as sole carbon source (Cooper, 1977). Under these conditions CWDE production is sequential in the order of pectic enzymes, hemicellulases, cellulase, and therefore seems to reflect the physico-chemical susceptibilities of corresponding wall polymers. This sequence may relate to the claim that polygalacturonide degradation is a prerequisite to action of other CWDE (Keegstra *et al.*, 1973), although most CWDE can act independently, at least on extracted host walls (Cooper *et al.*, 1978).

Analogously, in some diseases pectic enzymes are the first to appear and cellulases may not be detectable until the host is moribund, as in leaf spots of pea (e.g. *Mycosphaerella pinodes*), take-all of wheat (*Gaeumannomyces graminis*), *Pyrenochaeta terrestris* rot of onions, and Verticillium wilt of lucerne (*Verticillium albo-atrum*) (Cooper, 1977). Rapid PG or PL production during infection may also be evident from the concomitant depletion of galacturonic acid from walls of infected tissue e.g. $\leqslant$ 90% loss from young lesions on bean hypocotyls caused by *Sclerotium rolfsii* and *Rhizoctonia solani* (Bateman & Basham, 1976).

Changes in infected tissue may favour the action of polygalacturonidases, as in *Sclerotinia sclerotiorum* lesions on sunflower stems in which wall pectin was almost completely ($\leqslant$ 93%) demethoxylated and pH decreased from 6.2 to 4.5, which should facilitate the 82% loss of galacturonic acid caused by its PG (Hancock, 1966). Secretion of oxalic acid by this organism and *S. rolfsii* not only lowers pH but chelates calcium from the wall and renders the cross-linked polygalacturonide more amenable to degradation (Bateman & Beer, 1965).

The presence of endo-PG in ungerminated conidia of pathogens such as *Botrytis cinerea* and *Geotrichum candidum* (Barash 1968; Verhoeff & Liem, 1978) might suggest an involvement in the earliest stages of infection, and ultrastructural studies have revealed profound wall changes in epidermal cells even during the first few hours of penetration by *B. cinerea* and *B. fabae* (McKeen, 1974; Mansfield & Richardson, 1981).

Wall damage frequently occurs well in advance of pathogens which produce high levels of endo-polygalacturonidases, e.g. *Erwinia carotovora* and *R. solani*, and often appears as a selective dissolution of the polygalacturonide-rich middle lamella (Cooper, 1981). This type of degradation is invariably accompanied by disruption of host cytoplasm and appears to be a common strategy employed by necrotrophs, presumably to prevent host defences and obtain further nutrition, e.g. *B. fabae* grows almost entirely within walls of its host *Vicia faba* and causes them to swell markedly in advance of invasion with derangement of contiguous protoplasts. Also, for 52 induced mutants of *B. fabae* virulence was correlated with their ability to kill host cells following wall degradation (Mansfield & Richardson, 1981). This action apparently suppresses production of the broad bean phytoalexins and seems partly to explain the success of *B. fabae* compared with that of *B. cinerea* which produces much lower levels (*c.* 1/20) of endo-PG (unpublished results) and remains confined to inoculation sites which contain inhibitory levels of wyerone derivatives. Similarly, rapid synthesis of high levels of a lyase by *E. carotovora* in inoculated potato tubers (rotting begins after only 2–3 h) allows colonization before deposition in walls adjacent to inoculation wounds of suberin which prevents ingress of saprophytes (Zucker & Hankin, 1970). The role of PL but not PG in maceration caused by the related *E. chrysanthemi* has been established by conjugational transfer of a plasmid coding for lyase synthesis (Chatterjee & Starr, 1977).

Studies with purified *endo*-PGs and PLs from many plant pathogenic bacteria and fungi have clearly demonstrated their unique ability to cause maceration (cell separation) and simultaneous damage to protoplasts (Bateman & Basham, 1976). Exo-acting polygalacturonidases and other CWDE are ineffective. Various evidence suggests that cell killing results from damage to the plasma membrane as the weakened wall can no longer contain the turgid protoplast (Basham & Bateman, 1975). However, this may be an over-simplification as leakage of ions from enzyme-treated cells occurs almost immediately, and the only critical ultrastructural studies attempted have not provided evidence for protoplasts rupturing through damaged walls. In fact, walls often appeared not to be substantially altered when adjacent cellular injury was evident (Hislop *et al.*, 1979; Keon, 1982). This implies that insufficient weakening has occurred at this stage to allow even limited expansion of the contained protoplast.

Cell walls have been considered as part of the cell's lysosome system (Hislop *et al.*, 1979), and it remains possible that components are released from degraded walls to damage adjacent protoplasts, but so far attempts to establish toxicity *per se* of any enzymically solubilized factors have been unsuccessful (Bateman & Basham, 1976), although cell wall fragments released by acid hydrolysis may be phytotoxic (Yamazaki *et al.*, 1983). However, the experimental approach of adding factors released from degraded walls to 'healthy' plant cells is problematical, as the potential 'toxin' must pass through entire cell walls before encountering the plasma membrane, and if unstable or labile may well be detoxified within the wall, e.g. detoxification of superoxide-free radicals by superoxide dismutase (McCord, 1974) or peroxides by wall-bound peroxidase (Mussell & Strand, 1977). Extending this argument, the plasma membrane may only be damaged when in direct contact with its own damaged wall. Thus it is of interest that in at least some cell walls the polygalacturonide substrate (as revealed by iron hydroxylamine staining) for endo-pectic enzymes can be found in high levels adjacent to the plasma membrane (Keon, 1982), as well as concentrated in the more usual location of the middle lamella. The need to further investigate this membrane–wall interface is evident.

Whatever the mechanism of cell damage, in view of the frequency of wall degradation in plant diseases, and the ability of only low levels of polygalacturonidases to induce in walls and in host cells ultrastructural changes typical of those found during invasion by necrotrophs (Baker *et al.*, 1980; Hislop *et al.*, 1979), it follows that polygalacturonidase-induced injury may be one of the most common means of cell damage during necrotrophic parasitism.

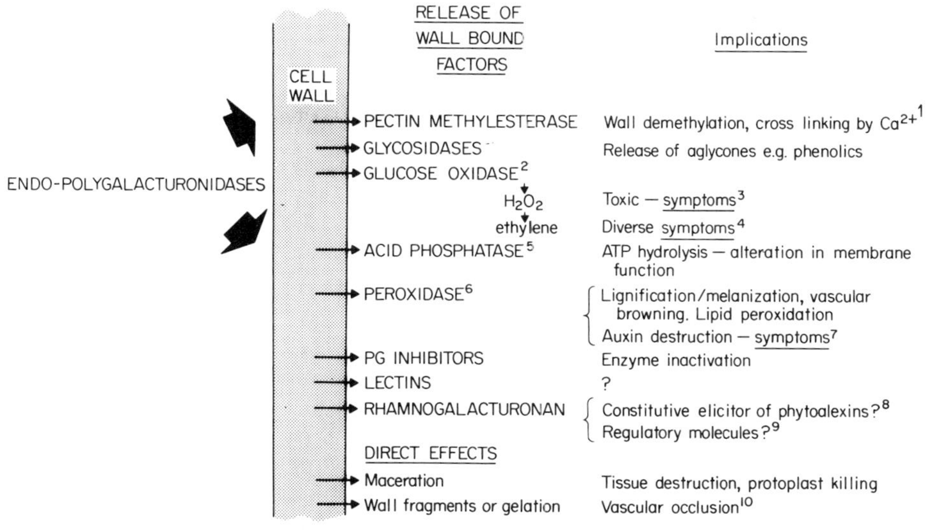

Figure 2.4. Possible diverse effects resulting from enzymic degradation of host wall polygalacturonide. 1, Bateman (1964); 2, Lund & Mapson (1970); 3,6,7, Mussell & Strand (1977); 4, Pegg, this volume; 5, Hislop *et al.* (1979); 8, Nothnagel *et al.* (1983); 9, Ryan *et al.* (1981); 10, Cooper & Wood (1980).

In addition, numerous side-effects may result from degradation of host rhamnogalacturonan and give rise to diverse symptoms and damage. Many molecules with potential physiological or damaging effects are bound to plant cell walls and may be exposed or released following wall breakdown. Some possibilities are summarized in Figure 2.4 but space does not allow detailed consideration. Attention should perhaps be drawn to rhamnogalacturonan, fragments of which can have regulatory properties at a distance (wound hormone?), e.g. induction of synthesis of a proteinase inhibitor in tomato plants (Ryan *et al.*, 1981), and may function as the constitutive (endogenous) elicitor of phytoalexins (Nothnagel *et al.*, 1983). If the latter is the case, there must be a fine balance between suppression of host defences by rapid cell killing and triggering of host resistance because endo-polygalacturonidases can also 'elicit' phytoalexin production (West, 1981; Davis *et al.*, 1983).

Also to be considered in this context, are vascular wilt diseases in which symptoms are various and the mechanisms are apparently complex. Water stress is probably the main cause of yield loss and derives largely from occluded xylem vessels (Cooper & Wood, 1980; Street & Cooper, 1984; Talboys, this volume). Endo-polygalacturonidases have been implicated because xylem vessels in infected plants are often occluded by pectic gels which arise from the primary wall at pit membranes. Other symptoms which may originate indirectly from degradation of host walls, are abscission and epinasty (from enhanced ethylene), chlorosis (resulting from generation of H_2O_2, claimed as the damaging agent in wilt of cotton caused by *Verticillium dahliae* (Mussell & Strand, 1977)), and vascular browning (from solubilization of peroxidase). For verticillium wilts of cotton, lucerne, and tomato PG or PL have been detected in the xylem before onset of symptoms, and when purified and reintroduced into vessels have reproduced symptoms such as chlorosis and wilting of leaves (Cooper & Wood, 1980), reduced vascular flow, and solubilized wall-bound peroxidase (Street & Cooper, unpublished).

Pathogens may produce PG, PL or both. *V. albo-atrum* from tomato produces both types, but only the lyase appears to be involved in pathogenesis because it was the only fungal CWDE found in infected vessels, and an induced mutant deficient for PG but with normal capacity for PL synthesis retained full pathogenicity (unpublished results).

Degradation of neutral matrix polymers and cellulose

The significance in pathogenesis of CWDE which degrade polymers other than rhamnogalacturonan is little understood. Many pathogens readily produce enzymes such as arabanases, xylanases and galactanases *in vitro*, and corresponding wall sugars may be depleted in infected tissue which contains the appropriate CWDE(s) (Bateman & Basham, 1976). However these enzymes have generally been relegated to a nutritional role as they do not cause maceration or, so far as is known, any significant loss of wall strength. However, overall loss of wall integrity bears little relation to the highly localized penetration effected by most fungal pathogens. Various evidence suggests that the contribution of these CWDE may be underestimated.

Thus, degradation of neutral polysaccharides may be significant in infection of tissues especially rich in these polymers. The wall matrix of potato tubers contains exceptionally high levels (45%) of galactan which is rapidly depleted during invasion by *Phytophthora infestans*, presumably as a result of its high endogalactanase activity (Jarvis *et al.*, 1981).

In spite of its *exo*-mode of action, α-arabinosidase production by *Monilinia fructigena* has been correlated with pathogenicity and the enzyme and the free breakdown product, arabinose, can be detected in infected apple fruits (Howell, 1975).

Most studies on CWDE have been on pathogens of dicots, but the predominant matrix polymer in primary walls of monocots is (glucuronoarabino) xylan. It is therefore intriguing that xylanase appears to be a key CWDE in infection by cereal pathogens such as *Pseudocercosporella herpotrichoides*, *Rhizoctonia cerealis*, *Fusarium culmorum*, (W. Carlile, pers. comm.), *Gaeumannomyces graminis* (J. Friend, pers. comm.), *Ustilago maydis* (unpublished results) and possibly *Helminthosporium maydis* as xylanase production is constitutive (Bateman *et al.*, 1973) rather than specifically induced as are the CWDE of most pathogens (Cooper & Wood, 1975).

Direct penetration of secondary walls is infrequent and cellulases usually appear only in later stages of infection, but their activity may be important in chronic diseases, e.g. cellulase has been detected in eyespot of wheat (*P. herpotrichoides*), and presumably contributes to the considerable yield losses resulting from lodging.

acid oligosaccharide from plant cell walls elicits phytoalexins. *Plant Physiology* **71**, 916–26.

Preston R.D. (1974) *The Physical Biology of Plant Cell Walls.* Chapman and Hall, London.

Reese E.T. (1977) Degradation of polymeric carbohydrates by microbial enzymes. In: *The Structure, Biosynthesis and Degradation of Wood* (Ed. by F.A. Loewus & V.C. Runeckles). *Recent Advances in Phytochemistry* No. **11**, pp. 311–67. Plenum Press, New York.

Ride J.P. (1980) The effect of induced lignification on the resistance of wheat cell walls to fungal degradation. *Physiological Plant Pathology* **16**, 187–96.

Ryan C.A., Bishop P., Pearce G., Darvill A.G., McNeil M. & Albersheim P. (1981) A sycamore cell wall polysaccharide and a chemically related tomato leaf oligosaccharide possess similar proteinase inhibitor inducing activities. *Plant Physiology* **68**, 616–18.

Sheiman M.I., Macmillan J.D., Miler L. & Chase T. (1976) Coordinated action of pectinesterase and polygalacturonate lyase complex of *Clostridium multifermentans. European Journal of Biochemistry* **64**, 565–72.

Street P.F.S. & Cooper R.M. (1984) Quantitative measurement of vascular flow in petioles of healthy and *Verticillium albo-atrum*-infected tomato. *Plant Pathology* (in press).

Talmadge K.W., Keegstra K., Bauer W.D. & Albersheim P. (1973) The structure of plant cell walls. I. The macromolecular components of the walls of suspension-cultured sycamore cells with a detailed analysis of the pectic polysaccharides. *Plant Physiology* **51**, 158–73.

Verhoeff K. & Liem J.I. (1978) Presence of endo-polygalacturonase in conidia of *Botrytis cinerea* before and during germination. *Phytopathologische Zeitschrift* **91**, 110–5.

West C.A. (1981) Fungal elicitors of the phytoalexin response in higher plants. *Naturwissenshaften* **68**, 447–57.

Wood T.M. & McCrae S.I. (1979) Synergism between enzymes involved in the solubilization of native cellulose. In: *Advances in Chemistry,* Vol. 181. *Hydrolysis of Cellulose: Mechanisms of Enzymatic and Acid Catalysis* (Ed. by R.D. Brown & L. Jurasek), pp. 181–209. American Chemical Society.

Yamazaki N., Fry S.C., Darvill A.G. & Albersheim P. (1983) Host–pathogen interactions. XXIV. Fragments isolated from suspension-cultured sycamore cell walls inhibit the ability of cells to incorporate ^{14}C-leucine into proteins. *Plant Physiology* **72**, 864–9.

Zeroni M. & Hall M.A. (1980) Molecular effects of hormone treatment on tissue. In: *Encyclopedia of Plant Physiology,* Vol. 9 (Ed. by J. MacMillan), pp. 511–86. Springer-Verlag, New York.

Zucker M. & Hankin L. (1970) Regulation of pectate lyase synthesis in *Pseudomonas fluorescens* and *Erwinia carotovora. Journal of Bacteriology* **104**, 13–8.

3 The role of growth regulators in plant disease

G.F. PEGG,

Department of Horticulture, University of Reading, Earley Gate, Reading RG6 2AU, UK

Introduction

Our knowledge of the existence of growth-regulating substances goes back almost exactly 100 years to the work of Darwin and Sachs. It was not until 1934, however, with the almost simultaneous discovery that indol-3-yl-acetic acid (IAA) could be synthesized by higher plants and fungi, that micro-organisms were associated with growth substance activity. This work by Kögl *et al.* (1934) and Thimann (1935) in which the chemical structure of IAA was determined, was a landmark in plant physiology, paving the way for all further work on plant hormones.

The recognition that chlorosis, necrosis and foliar abscission reminiscent of a disease could be induced by a growth-promoting air contaminant came from the pioneering work of Fahnstock in 1858; this preceded work on the hormonal control of healthy plants. Neljubov (1900) showed that the active principle was ethylene, which very much later was recognized as an endogenous growth promoter of higher plants (Osborne, 1968).

Toxins, with rare and notable exceptions, are all metabolites of pathogens. The same is not strictly true for enzymes, but in practice enzymes in enzyme-induced diseases can be attributed almost exclusively to pathogens. With growth-regulating substances on the other hand, the same type of molecule can be shown to be a constitutive compound of both host and pathogen, making difficult an interpretation of the role of each.

Examples of recognized classes of growth regulators and toxins with growth-regulating properties are shown in Table 3.1.

Most plant growth regulators were first discovered as toxic metabolites of pathogens e.g. gibberellic acid, ethylene, cytokinin and more recently abscisic acid. A number of hormone-like molecules are known only as pathogenic toxins, viz. helminthosporol, fusicoccin, hadicin from *Penicillium frequentans*, the diaporthins and the ophiobolins.

Much work has been carried out on the effect of exogenously-applied synthetic growth regulators on healthy and diseased plants. This present review is confined to studies on changes in naturally-occurring substances in relation to disease or disease resistance. There are several detailed reviews on growth regulators

Wood R.K.S. & Jellis G.J. (1984) *Plant Diseases: infection, damage and loss.* Blackwell Scientific Publications, Oxford.

Table 3.1. Naturally occurring plant growth-regulating substances and phytotoxins with growth regulator properties.

Class of Compound	Chemical Name	Producing organism
Cytokinin	6-Δ^2 isopentylamino purine (2 i P)	*Corynebacterium fascians* (Klambt *et al.*, 1966; Helgeson & Leonard, 1966)
	6-(4 hydroxy-3-methyl-*trans* 2 butenylamino purine) (zeatin)	*Rhizopogon roseolus* (Miller, 1967; Laloue & Hall 1973)
Gibbane cyclic diterpenes	Gibberellic acid (GA_3)	*Gibberella fujikuroi* (Brian *et al.*, 1954)
	Gibberellin A_4	*Sphaceloma manihoticola* (Zeigler *et al.*, 1980)
Auxin	Indol-3-yl-acetic acid (IAA)	Numerous fungi (Gruen, 1959)
		Pseudomonas savastanoi (Smidt & Kosuge, 1978)
Unsaturated hydrocarbon gases	Ethylene (ethene)	*Fusarium oxysporum* f.sp *tulipae* (Beijersbergen, 1973)
		Verticillium dahliae (Weise & DeVay, 1970)
Abscisic acid	3 methyl-5-(1′hydroxy-4′-oxo-2′, 6′,6′-tri methyl-2′ cyclohexen--1′-y′) *cis* trans-2,4-pentadienoic acid (*cis* ABA)	*Cercospora rosicola* (Assante *et al.*, 1977; Norman *et al.*, 1983)
Sesquiterpenoid alcohol	Helminthosporol (gibberellin-like)	*Drechslera sorokiniana* (Briggs, 1966; Hashimoto & Tamura, 1967)
Carbotricyclic diterpene glucoside	Fusicoccin	*Fusicoccum amygdali* (Ballio *et al.*, 1964)

and plant disease by Sequeira (1963, 1973), Pegg (1976a), Daly & Knoche (1976), Dekhuijzen (1976) and Pegg (1981a). Many of the studies on host–parasite interactions reported in earlier papers have not been re-examined using modern techniques and the various claims and interpretations remain unchallenged and are here accepted at face value. This chapter deals with selected aspects of the topic and for the most part with work which has emerged since the review by Pegg (1981a). Studies of virus-infected plants are regarded as of especial importance in that the ambiguity of whether host, pathogen or both contribute to the overall growth regulator activity of the plant does not arise.

Auxins

Auxin is a generic term for growth-promoting substances based on the indole

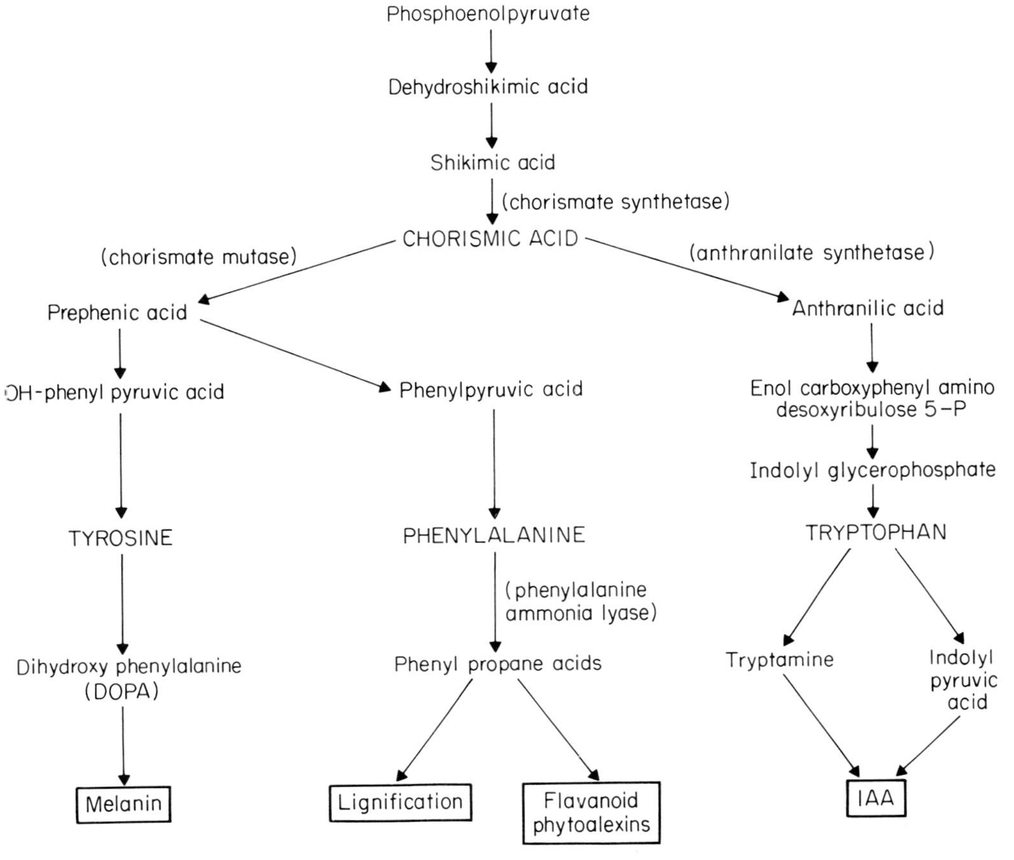

Figure 3.1. Pathways in the biosynthesis of IAA and phenolic compounds.

nucleus and characterized by indol-3-yl-acetic acid (IAA). Several types of indolyl compounds are found in plants or are produced *in vitro* by a wide spectrum of micro-organisms including pathogens; these are all thought to be precursors of IAA, which is probably the only functional indolyl hormone (Pegg, 1976a).

The synthesis of IAA in higher plants and micro-organisms is via the primary deamination or decarboxylation of tryptophan (Figure 3.1).

Phenolic compounds in plants and indolyl auxins share a common ancestry in shikimic acid. The full relationships between the effects of auxin and phenols in plant disease have yet to be investigated. The common origins of melanization, lignification, isoflavanoid phytoalexins and IAA synthesis, however, are indicated in Figure 3.1. The key enzymes determining the metabolic products of chorismic acid are chorismate mutase and anthranilate synthetase. Both of these enzymes have isozymes which are regulated by a complex feedback mechanism. Another mechanism regulating auxin production is the stimulation or inhibition of IAA-oxidase-peroxidase by different cinnamic acid derivatives (Pegg, 1976a).

The classic examples of hyperauxiny are diseases associated with tissue overgrowths such as galls, neoplasms and epinasty, involving hypertrophic and hyperplastic growth. IAA is a principle growth substance of healthy plants and where IAA is found in excess in diseased tissue the presumption has been made that this is the cause of hypertrophic growth. There are several reasons, however, for interpreting IAA-gall relations with caution. Thus:

(a) Many diseases, e.g. cereal rusts, exhibit hyperauxiny in the absence of overgrowth symptoms.

(b) Levels of auxins found in diseased tissue do not induce galls in healthy tissue.

(c) Growth substances are interactive and interdependent *in vivo*, e.g. ethylene, cytokinins and abscisic acid, all of which have been shown to affect IAA synthesis at different concentrations. IAA is also involved in ethylene biosynthesis.

(d) A diseased plant showing tissue proliferation may show simultaneous increases in IAA, gibberellins, cytokinins and ethylene.

Increased levels of IAA in plant tissue may result from one or more of the following:

(a) Enhanced host synthesis — loss of a cellular control mechanism.

(b) *In vivo* hormone secretion by the pathogen.

(c) Destruction of a host IAA oxidizing enzyme.

Conversely, reduced levels of auxin may result from the production of a pathogen extra-cellular IAA oxidase (Sequeira & Steeves, 1954) or decreased host synthesis. Saprophytic bacteria including those present as contaminants in tissue brei have a considerable propensity for the metabolism of free tryptophan to IAA. The work of Libbert *et al.* (1968), Libbert & Manteuffel (1970) and Libbert & Silhengst (1970) provides a caution to prospective workers in this research field. There is a voluminous literature on auxins in the healthy as well as the diseased plant, but IAA has been conclusively identified in only 17 studies (Bearder, 1980).

The best evidence to date for a positive and exclusive role for IAA in pathogenesis, is work by Smidt & Kosuge (1978) on oleander galls induced by *Pseudomonas savastanoi*. In the UK *P. savastanoi* causes a debilitating canker of ash (*Fraxinus excelsior*) very reminiscent of an hormonal imbalance. Beltra (1961) reported that galls on olive contained increased levels of a growth-promoting substance. Magie & Wilson (1962) confirmed the presence of IAA and envisaged a role for it in gall formation. The pathway for IAA synthesis in *P. savastanoi* is via oxidative decarboxylation of tryptophan to indolylacetamide (IAM) which is subsequently hydrolysed to IAA (Kosuge *et al.*, 1966; Magie *et al.*, 1963). IAA synthesis is regulated by feedback inhibition of tryptophan decarboxylase by IAA and IAM. Smidt & Kosuge (1978) derived mutants of *P. savastanoi* by growing the wild type on a medium containing α-methyl tryptophan. Mutants resistant to this tryptophan analogue showed an altered anthranilate synthetase with reduced sensitivity in some mutants to tryptophan feedback inhibition. Thus some mutants were IAA-less while others produced more IAA than the wild type. Similar effects had been obtained earlier in *Escherichia coli* by Somerville & Yanopsky (1965) using 5-methyl tryptophan. Smidt & Kosuge (1978) showed that the ability to induce galling in oleander was correlated precisely with the ability to produce IAA. Gall size was proportional to the amount of IAA produced *in vitro* by the mutant. IAA-less mutants did not induce galls. Both mutants and wild type isolates grew similarly in infected tissue.

By far the majority of auxin studies have been concerned with levels of free IAA readily extractable from tissues over a short period in organic solvents.

There is a growing recognition, however, of the importance of bound forms of IAA, with particular implications for plant disease. IAA occurs as conjugates in the following forms:

(a) Indoleacetamino acid form (Andreae & Good, 1955).
(b) Glycosidic form (Butcher *et al.*, 1974).
(c) Glucan polysaccharide form (Piskornik & Bandurski, 1972).

Bound forms of auxin have been known since 1955 when Andreae & Good (1955) reported indolylacetylaspartic acid in peas. Following the removal of free auxin from tissue, additional auxin may be extracted by prolonged incubation of the tissue or by mild (ester) or strong alkali (amide) hydrolysis. Strong alkali hydrolysis is required for the release of glucan polysaccharide bound auxin (Piskornik & Bandurski, 1972). Hangarter & Good (1981) have shown that IAA-glycine or IAA-alanine conjugates provide a source of slow-release free auxins under the control of an enzyme which splits acylamino acid bonds. A similar release can be envisaged from glycosidic bound forms involving 1,4-β-glucosidase and 1,3-β-glucanase. In Hangarter & Good's work (1981) ethylene was released from tissues fed with bound IAA in proportion to the release of free IAA which was then decarboxylated. An important element in the availability of free auxin for involvement in cellular proliferation is the role of host or pathogen IAA decarboxylation or oxidase-peroxidase systems (Sequeira, 1973; Pegg, 1976a). In the bound form IAA is protected from peroxidase catalysed oxidation but the physiological significance of this with bound IAA in tissue with high IAA oxidative capacity, as measured in cell macerates on a free IAA substrate, is not clear. (Hangarter & Good, 1981; Cohen & Bandurski, 1978).

A fundamental element in the control of growth in healthy and in diseased tissue must be the compartmentation of hormones in the cell in relation to the availability of auxin to membrane binding sites, and similarly the availability to IAA of enzymes both hydrolysing bound forms and destroying free IAA. Little information is available on these topics.

Club root of brassicas caused by *Plasmodiophora brassicae* is a classic example of a root-hypocotyl gall indicative of abnormal hormone physiology (Dixon, this volume). Brassicas accumulate IAA in the form of 3-indolylmethylglucosinolate which is under the enzymic control of myrosinase (thioglucoside glucohydrolase) which liberates the IAA precursor indolylacetonitrile (IAN). Tamura *et al.* (1970) showed the presence of high levels of IAN in clubbed tissue. The control of free and bound levels of IAA in the cell is not clear. Grob & Matile (1979) consider that enzyme and substrate are separated in cell compartments. Searle *et al.* (1982) showed that ^{14}C IAN was found in *P. brassicae*-induced clubbed swede tissue fed with ^{14}C indole-methylglucosinolate in a reaction enhanced by ferrous ions. It was postulated that *P. brassicae* breaks down membranes which compartment myrosinase. The availability of IAA to support the growth of clubbed tissue is under the control of nitrilase which is present in higher levels in clubbed tissue than in controls (Rausch *et al.*, 1981). In this work and in the many studies on crown gall, in the galls of which auxin levels are also higher than in non-galled tissue, there is no indication that IAA *initiates* cell hypertrophy but instead merely sustains the growth substance requirement in a system which has lost a regulatory control mechanism. An alternative hypothesis comes from studies on tumorous and non-tumorous species of *Nicotiana* by Liu *et al.*, (1978) and others. The synthesis of IAA conjugates was higher in tumorous hybrids and amphidiploid tumour plants than in non-tumorous mutants. Genetic tumours appear in *Nicotiana* hybrids at centres which become meristematic following a signal which appears to be a reduction in endogenous IAA (Ames, 1974; Ames & Mistretta, 1975), and the subsequent activation of an IAA synthetase system which may be demonstrated readily in tumorous hybrid plants but is almost non-existent in parental species. The initiation of tumours according to Liu *et al.*, (1978) is due to the reduced level of free IAA following its sequestering into a bound form.

The intimate interrelationship between ethylene and IAA in the physiology of plants referred to elsewhere in this chapter is also illustrated in auxin conjugation. Epstein (1982) showed that leaves and callus of olive (*Olea europea*) exposed to ethylene changed little in free IAA content but the level of amide-bound IAA increased 3-fold compared to the non-gassed controls. This finding may be of significance in ethylene-auxin studies of diseased plants.

In most of the foregoing examples the pathogens were able to synthesize IAA and related compounds. With virus infection, auxins, or more specifically, IAA have been associated with stunting or with lesion development. Tomato spotted wilt virus of tomato (Grieve, 1943), potato leaf-roll virus (Jahnel, 1939), tobacco mosaic virus (TMV) in tobacco (Pavillard, 1952), beet curly top virus-infected tomato plants (Smith *et al.*, 1968) have all been associated with reduced auxin content of plants. Rajagopal (1977) showed 95% decreases in IAA and phenylacetic acid (PAA) derived from phenylalanine, within 12 h of infection of tobacco by TMV and corresponding increases in tryptophan and phenylalanine. This suggested the transfer of a messenger molecule in the plant faster than virus replication. Increased IAA oxidase activity was also found in cowpea mosaic virus-infected plants (Lockhart & Semancik, 1970); similarly Lobenstein & Linsey (1963) and Van Loon & Geelen (1971) showed increased IAA peroxidase activity.

Conversely barley yellow dwarf infection caused no change in auxin content although plants showed considerable stunting. Paradoxically auxin levels were higher in pea mosaic virus-infected lupins (Van Steveninck, 1959) and in TMV-induced local lesions (Van Loon & Berbee, 1978). In Samsun NN tobacco reacting hypersensitively to TMV, Van Loon & Berbee (1978) found that 10^{-4}M IAA applied within 12 h of inoculation reduced lesion size by 30%. In the systemically-infected isoline Samsun, IAA inhibited virus multiplication. This lesion limiting effect of IAA is not clear; whereas IAA stimulated peroxidase (PO) levels in healthy controls, in TMV-infected plants PO levels were reduced, but this may result indirectly from decreased necrosis. Curiously 2,4-D, which is used ubiquitously as an auxin analogue, reduced lesion size at 10^{-1}M but increased it at 10^{-3}M.

It will be apparent from the foregoing that no simple explanation of *the* role or *a* role for IAA in virus diseases can be advanced even for a single well-studied system such as TMV. At the present state of our knowledge it is not possible to separate the contributory effects of ethylene and IAA even before a consideration of other growth regulators such as cytokinin, gibberellin and abscisic acid (ABA) which are also changed following infection and which interact (Trewas, 1981).

Ethylene

Ethylene was the first growth regulator to be implicated in the induction of disease-like symptoms in plants. It is a gas produced ubiquitously by higher plants and micro-organisms and is extremely reactive in physiological and biochemical processes (Abeles, 1973). Previous reviews (Sequeira, 1973; Pegg, 1976a, 1981a, 1981b) have considered a multiple role for ethylene in the diseased plant. Indeed, it is precisely because it is involved in so many cellular processes and with other growth hormones that it is impossible to present a simplified role for it in pathogenesis (Lieberman, 1979; Pegg, 1981a, 1981b).

Several symptoms of plant disease, some resembling premature senescence, can be attributed to ethylene, namely foliar chlorosis, petiolar epinasty, foliar abscission, and tissue distortion and necrosis. In the induction of these effects ethylene produced either by pathogen or host is regarded as a phytotoxin. Paradoxically, by acting on other metabolic pathways, ethylene can induce resistance to different pathogens, e.g. alleviation of symptoms caused by *Verticillium albo-atrum* in tomato (Pegg, 1976b), reduced bitter rot in apples (Lockhart *et al.*, 1968); increased pisatin synthesis in pea (Chalutz & Stahmann, 1969). Recently there has been a great interest in the involvement of ethylene in virus-infected plants in which its production, in contrast to other diseases, must be the result of disordered host physiology.

In general, local lesion and necrosis-inducing virus infection leads to increased ethylene production. This is not so with systemic symptomless infection. Ross & Williamson (1951) showed that a physiologically active emanation produced during local lesion development mimicked the triple response in etiolated pea seedlings, namely stunting, thickening and diageotropy. This was confirmed as ethylene by Balázs *et al.* (1969). Ethylene production is stimulated 3–13 h before the appearance of local lesions caused by TMV and tobacco necrosis virus (TNV) in tobacco (Gaborjanyi *et al.*, 1971; Pritchard & Ross, 1975), but the largest increases occurred

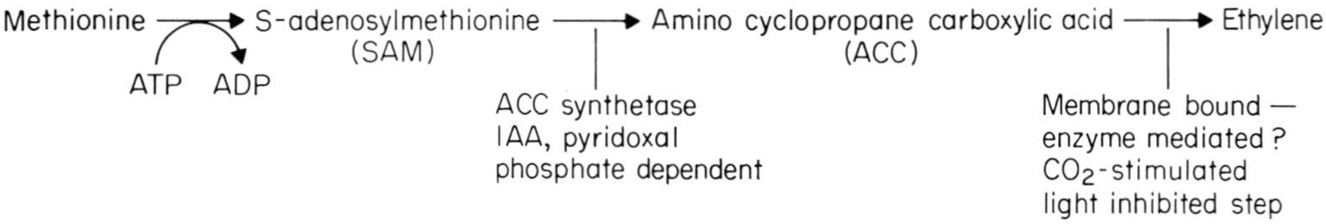

Figure 3.2. Stages in ethylene biosynthesis.

subsequently. Many other plant virus infections have been associated with increased ethylene production, e.g. cowpea mosaic virus and cucumber mosaic virus (Lockhart & Semancik, 1970; Kato, 1976).

The evidence for the primary role of ethylene in promoting necrotic lesions is conflicting. It has long been known that the sulphur-containing amino acid methionine is involved in ethylene biosynthesis. Adams & Yang (1979) showed that the immediate precursor of ethylene is amino cyclopropane carboxylic acid (ACC). This is derived from S-adenosylated methionine (SAM) (Figure 3.2) under the control of ACC synthetase.

In tobacco responding hypersensitively to TMV, Pritchard & Ross (1975) found that an agent blocking ethylene synthesis, succinic acid 2,2′-dimethyl hydrazide, successfully inhibited ethylene production but did not limit lesion size or number. Conversely Bailiss *et al.* (1977) found that normally chlorotic lesions induced in *Tetragonia expansa* by bean yellow mosaic virus became necrotic in the presence of ethylene. When CO_2 (a competitive inhibitor of ethylene) was introduced, both chlorosis and necrosis were prevented. De Laat & Van Loon (1982, 1983) showed that healthy and TMV infected Samsun NN tobacco leaves produce ethylene via the ACC pathway 36 h after inoculation; large increases in ACC occurred 12 h before lesion appearance. Plants fed with UL-^{14}C-methionine showed a peak of labelled ACC coinciding with lesion appearance. Thereafter the capacity for ACC to convert to ethylene increased sharply, peaking 4 h after lesion appearance. Methionine and SAM, unlike ACC, were limiting factors in ethylene production. As ethylene increased, so the peak of endogenous ACC declined.

The interrelations between IAA and ethylene are well known. IAA applied to leaf laminae will prevent ethylene-induced petiolar abscission. Auxin is known to affect ethylene synthesis. Recently Jones & Kende (1979) showed that etiolated pea tissue fed with ^{14}C methionine produced ^{14}C ethylene and unlabelled ethylene only in the presence of 0.1 M IAA. IAA stimulated ethylene production by increasing the activity of ACC synthetase in the conversion of SAM to ACC. It was not known whether the increased activity represented *de novo* synthesis or activation of the enzyme. These results supported the work of Yu *et al.* (1979) who showed that the conversion of ACC to ethylene in mung bean hypocotyls was not dependent on IAA. The rhizobitoxin analogue aminoethoxyvinyl glycine (AVG) in the presence of IAA depressed the synthesis of ACC and ethylene. In a subsequent study of TMV and TNV infection in the tobacco isolines Samsun and Samsun NN (containing the hypersensitive gene N), De Laat & Van Loon (1983) showed that virus-induced ACC and ethylene were not dependent on the N gene or on the type of infecting virus. Inhibitors of RNA and protein synthesis inhibited ACC and ethylene but to different extents. The synthesis or increased activity of ACC synthetase depended on events occurring during virus synthesis and the development of the visible symptoms. The authors conclude that the necrotic reaction is not the cause of ethylene production and that increased ethylene does not lead to tissue necrosis. Challenge inoculation led to an immediate burst of ethylene which may be responsible for limited lesion enlargement in leaves exhibiting systemic acquired resistance. Similar results on limiting lesion size were obtained by Van Loon (1977) using ethephon treatments. Ethephon (chloroethylphosphonic acid) liberates ethylene under alkaline conditions. Peroxidase isozyme activity induced by ethephon and TMV were shown to be similar. Caution should be observed in interpreting results obtained with this compound since under certain conditions PO_4^{3-} anions are released which interact toxically with ethylene (for this and other work on ethylene involvement in resistance and pathogenesis, Pegg, 1981b).

Abscisic acid

ABA has many physiological effects in plants. The major ones are: inhibition of growth and RNA synthesis, accelerated senescence, abscission, dormancy, enzyme stimulation, cold hardiness and seed germination inhibition.

Relatively few studies have been carried out on diseased plants, possibly because prior to the work of Assante *et al.* (1977) ABA was the only example of the known growth regulator classes not shown to be produced by a plant pathogen.

A principal role for ABA in both healthy and diseased plants appears to be as a regulator of stomatal activity under conditions of drought stress. Under these conditions *cis* ABA is released rapidly from bound glycosidic forms causing stomata to close (Wright & Hiron, 1969; Jones & Mansfield, 1970; Milborrow, 1978). This effect is opposite to that caused by fusicoccin. Although ABA can be shown to be an inhibitor of growth in sensitive assays and also of seed germination, there is little or no general evidence to show that it is a cause of stunting in diseased plants. Where wilting is a feature of disease, enhanced ABA will automatically be produced but it may not accumulate to levels that inhibit growth.

One of the problems in an interpretation of the role of ABA and indeed of other growth regulators, is its interaction with other hormones in the plant. Assante *et al.* (1977) showed that *Cercospora rosicola* produced ABA in liquid media as a secondary metabolite. This system was used by Norman *et al.* (1983) to study the effects of other hormones on ABA biosynthesis in the absence of the pool of endogenous substances. A range of cytokinins with effects opposite to those of ABA (e.g. growth promotion, ribonucleic acid (RNA) synthesis, stomatal opening and retardation of senescence) were found to inhibit biosynthesis of ABA from ^{14}C farnesyl pyrophosphate. Benzylaminopurine (BAP) was the most active compound, inhibiting synthesis by 74% and 98% at 100 and 500 μm respectively. Of the naturally

Table 3.2. Effect of abscisic acid (ABA) on disease reaction and phytoalexin accumulation in potato slices.

Treatment	Reaction	Rishitin	Lubimin
5% MeOH — Water	0	1	0
5% MeOH — Elicitor from $Race_4$ *P. infestans*	Hypersensitive	283	23
ABA — Elicitor from $Race_4$ *P. infestans*	0	3	8
5% MeOH — $Race_4$ *P. infestans*	Hypersensitive	350	108
ABA — $Race_4$ *P. infestans*	Compatible	96	14
5% MeOH — $Race_0$ *P. infestans*	Hypersensitive	278	210
ABA — $Race_0$	Compatible	92	16
5% MeOH — $Race_{1,2,3,4}$ *P. infestans*	Compatible	3	4
ABA — $Race_{1,2,3,4}$ *P. infestans*	Compatible	9	0

Within 1 h of being cut, the upper surface of each slice was treated with 0.1 ml 5% methanol containing 100 μg abscisic acid or with 0.1 ml 5% methanol. Eight hours later, an elicitor preparation of $Race_4$ *P. infestans* or 5×10^4 zoospores of compatible [Race 1,2,3,4] or incompatible [$Race_4$] [$Race_0$] *P. infestans* in 5% MeOH were applied to the same surface. Slices were evaluated visually 72 h after inoculation. 0 = no visible change in the appearance of tissue. Hypersensitive = browned tissue without apparent fungal growth. Compatible = abundant fungal growth and sporulation.
Based on data from Henfling *et al.* (1979).

occurring compounds, 2-isopentenyl adenine (2 i P) and its riboside caused up to 89% inhibition. These results suggest that a cytokinin/ABA control mechanism could exist in healthy and diseased plants.

Recent work has suggested a potentially important role for ABA and cytokinins in the control of phytoalexin synthesis in plant parts or in cell suspension cultures. Dixon & Fuller (1978) have shown that the isoflavanoid phaseollin is produced by non-induced cell suspension cultures of *Phaseolus vulgaris* and in increased amounts following treatment with sterile culture filtrates from *Botrytis cinerea*. Dichlorophenoxyacetic acid (2,4-D) and napthaleneacetic acid (NAA) as auxin analogues both stimulated phytoalexin production in non-induced cultures, but were inhibitory to induced cultures at 2×10^{-6} M and 2×10^{-4} M respectively. Gibberellic acid (GA_3) and ABA both stimulated non-induced phaseollin but were not studied in *B. cinerea*-induced cultures. Kinetin enhanced phaseollin in induced culture filtrates at low (5×10^{-7} M) concentrations in a range of 2,4-D or NAA concentrations. Higher concentrations of kinetin were only stimulatory in the presence of low (2×10^{-6} and 2×10^{-7} M) synthetic auxins. Goosens & Vendrig (1982) demonstrated that 10^{-4} M ABA stimulated phaseollin accumulation in excised cotyledons in light and dark whereas BAP increased these levels only in light. With kievitone, however, ABA and BAP were effective only in dark-incubated cotyledons. ABA for the most part depressed phytoalexin accumulation. The subtlety of the regulatory role of ABA is seen in the work of Henfling *et al.* (1979) (Table 3.2).

Pre-treatment of 30 mm diam. potato discs with 100 μg ABA in 5% MeOH prior to challenging with incompatible races of *Phytophthora infestans*, greatly reduced the accumulation of rishitin and lubimin while spectacularly changing the reaction type from hypersensitive to compatible. ABA applied to potato tuber discs was shown similarly to change to compatible the hypersensitive resistance induced by an elicitor preparation from *P. infestans* Race 4. When *Cladosporium cucumerinum*, a non-pathogen of potato was used, ABA treatment led to an increase in terpenoid synthesis but, paradoxically, to a compatible reaction.

A similar result was obtained when ABA applied to cucumber leaves, 2900-fold in excess of endogenous levels, failed to reduce growth to an extent comparable to that resulting from CMV infection. However, 25 ppm ABA applied to virus-infected leaves had a growth inhibitory effect additive to that of the virus. (Aharoni *et al.*, 1977).

Pegg (1981a) found no significant difference in overall levels of *cis* ABA in resistant and susceptible lines of tomato infected with *Verticillium albo-atrum* and showing reduced leaf area and stem height. Apical leaves and shoot apices of susceptible infected plants had a higher ABA content (0.05–0.25 μg/g fresh wt). Treatment of leaves with an abnormally high level (1000 μg) of exogenous ABA caused slight inhibition only in the target and sub-tending leaves. The endogenous ABA level of the target leaf, however, was 2 μg/g fresh wt, some 10-fold higher than infected stunted leaves.

Whenham & Fraser (1981) in similar experiments on TMV infected tobacco where the endogenous concentration matched the exogenous level applied, claimed that the reduced growth of virus-infected plants could be attributed to increased inhibition of cell division by ABA. Neither study considered bound ABA or its enzymic release during pathogenesis. Weiler (1980) and Whenham & Fraser (1981) considered that levels of bound ABA (glucosyl esters and glycosides, Milborrow, 1978; Sembdener *et al.*, 1980) may exceed free ABA in the leaf; the importance of sequestering of ABA by chloroplasts, and or other spatial or chemical binding may thus obscure a true interpretation of its role in infected plants.

Cytokinins

The role of cytokinins in plant disease has been reviewed in detail by Dekhuijzen (1976). The naturally occurring cytokinins are substituted adenine derivatives with an isoprenoid side chain. By definition these substances are capable of inducing cell division in callus cultures and it is assumed that they play a major role in hyperplastic growth in galls and other types of tissue proliferation. Synthesized for the most part in root tissues, cytokinins stimulate the growth of lateral buds and leaf area in healthy plants and are thus opposite to auxins and ABA in their effects.

The pathological systems most studied are fasciation and leafy galls typified by *Corynebacterium fascians*, from which 2 i P has been identified. In contrast to certain other growth regulators, applications of 2 i P to healthy plants in the concentrations found in diseased tissue, induce most of the symptoms of the disease (Thimann & Sachs, 1966; Sachs & Thimann, 1967). Crown gall caused by *Agrobacterium tumefaciens*, club root caused by *Plasmodiophora brassicae* and maize smut tumour caused by *Ustilago maydis*, are other examples of diseases associated with cytokinins (Pegg, 1981a).

Transfer RNA species for the initiation of serine and tyrosine have been shown to contain 2 i P (Hall, 1968). This suggests a subtle mode of action involving very low levels of compound; indeed 'free' cytokinins from plant tissue are present in the order of less than 10 μg kg fresh weight. Most of the published work on cytokinins in diseased tissues has been based on non-critical extraction methods and a callus bioassay (usually soybean callus) with results expressed in kinetin or benzyladenine (BA) equivalents. Horgan (1978), using nine cytokinins, has shown the inefficiency of solvent partition as a purification stage and the bioassay methods most commonly used have been critically evaluated by Dekhuijzen (1976). Only physicochemical methods, GC-MS, HPLC-MC and multiple-ion monitoring techniques are likely to give unequivocal results (Horgan, 1978), but few studies to date have used these methods such that critical identification and accurate quantification are mostly lacking.

An important aspect of cytokinins in cellular metabolism is their role in stimulating ribosomal protein synthesis, reversing one of the effects of ethylene. This is best seen in diseased tissue, particularly with biotrophic pathogens, in the 'green island' effect. Green islands represent interveinal green areas surrounded by chlorotic or necrotic tissue. Since a number of pathogens are capable of synthesizing cytokinins *in vivo* it is sometimes assumed that extracellular cytokinins from the pathogen are responsible for the green island, but see Dekhuijzen *et al.* (1967). These green islands act as nutrient sinks, redirecting photosynthate to the infection pustule. The evidence for this comes from autoradiography of $^{14}CO_2$-fed rust and mildew infected leaves, from positive cytokinin responses, in several bioassays, of extracts from rust and mildew infected leaves and from the simulation of green islands using droplets of cytokinin on healthy leaves approaching senescence. (Mothes & Engelbrecht, 1961; Livne & Daly, 1966; Bushnell, 1967; Dekhuijzen & Staples, 1968.)

The relationship between resistance to virus infection and cytokinins is an area of especial interest in view of the sequestering of host RNA and protein synthesis by the virus.

Selman (1964) showed that kinetin reduced multiplication of tomato spotted wilt virus in tomato and local lesion size and number in *Petunia*. Kuriger & Agrios (1977) similarly showed that lesions caused by tobacco ring spot virus (TRSV) in cowpea (*Cicer arietinum*) were reduced by 75% in plants treated 9 days before inoculation. This effect could not be reproduced by kinetin applied three days before inoculation (Tavantzis *et al.*, 1979). They showed that endogenous cytokinin activity thought to be caused by zeatin or zeatin riboside in TRSV-infected *Nicotiana glutinosa* or cowpea was reduced. Since kinetin is known to bind reversibly to plant ribosomes (Berridge & Ralph, 1969), depending on concentration, it is postulated

Fahnstock G.W. (1858) Memoranda of the effects of carburetted hydrogen gas upon a collection of exotic plants. *Proceedings of the Academy of Natural Sciences, Philadelphia*, 118–34.

Fraser R.S. (1972) Effects of two strains of tobacco mosaic virus on growth and RNA content of tobacco leaves. *Virology* **47**, 261–9.

Gaborjanyi R., Balázs E. & Király Z. (1971) Ethylene production, tissue senescence and local virus infection. *Acta Phytopathologica Academiae Scientiarum Hungaricae* **6**, 51–5.

Ghosh S.K. (1982) Growth promotion in plants by rice necrosis mosaic virus. *Planta* **155**, 193–8.

Goossens J.F.V. & Vendrig J.C. (1982) Effects of abscisic acid, cytokinins and light on isoflavanoid phytoalexin accumulation in *Phaseolus vulgaris* L. *Planta* **154**, 441–6.

Graniti A. (1964) The role of toxins in the pathogenesis of infections by *Fusicoccum amygdali* Del. on almond and peach. In: *Host-Parasite Relations in Plant Pathology* (Ed. by Z. Király & G. Ubriziy) pp. 211–7. Research Institute for Plant Protection, Budapest.

Grieve B.J. (1943) Studies in the physiology of host-parasite relations. 4. Some effects of tomato spotted wilt virus on growth. *Australian Journal of Experimental Biology and Medical Sciences* **21**, 89–101.

Grob K. & Matile P.H. (1979) Vacuolar location of glucosinolates in horse-radish root cells. *Plant Science Letters* **14**, 327–35.

Gruen, H (1959) Auxin and fungi. *Annual Review of Plant Physiology* **10**, 405–40.

Hall R.H. (1968) Cytokinins in the transfer RNA: their significance to the structure of RNA. In: *Biochemistry and Physiology of Plant Growth Substances* (Ed. by F. Wightman & G. Setterfield), pp. 47–56. Runge Press, Ottawa.

Hangarter R.P. & Good N.E. (1981) Evidence that IAA conjugates are slow-release sources of free IAA in plant tissue. *Plant Physiology* **68**, 1424–7.

Hashimoto T. & Tamura S. (1967) Physiological activities of helminthosporol and helminthosporic acid. II. Effects of excised plant parts. *Plant Cell Physiology* **8**, 35–45.

Helgeson J.P. & Leonard N.J. (1966) Cytokinins: identification of compounds isolated from *Corynebacterium fascians. Proceedings of the National Academy of Sciences USA* **56**, 60–3.

Henfling J., Bostock R. & Kuć, J. (1979) Effect of abscisic acid on rishitin and lubimin accumulation and resistance to *Phytophthora infestans* and *Cladosporium cucumerinum* in potato tissue slices. *Phytopathology* **70**, 1074–8.

Horgan R. (1978) Analytical procedures for cytokinins. In: *Isolation of Plant Growth Substances* (Ed. by J.R. Hillman), pp. 97–114. Cambridge University Press, Cambridge.

Jahnel H. (1939) Wuchsstoffuntersuchungen an abbaukranken Kartoffeln. *Phytopathologische Zeitschrift* **12**, 312–7.

Jones J.F. & Kende H. (1979) Auxin-induced ethylene biosynthesis in subapical stem sections of etiolated seedlings of *Pisum sativum* L. *Planta* **146**, 649–56.

Jones R.J. & Mansfield T.A. (1970) Suppression of stomatal opening in leaves treated with abscisic acid. *Journal of Experimental Botany* **21**, 714.

Kato Y. (1976) Ethylene production during lipid peroxidation in cowpea leaves infected with CMV. *Annals of the Phytopathological Society of Japan*, **43**, 587–9.

Klambt D., Thies G. & Skoog F. (1966) Isolation of cytokinins from *Corynebacterium fascians. Proceedings of the National Academy of Sciences USA* **56**, 52–9.

Kögl F., Haagen-Smit A.J. & Erxelben H. (1934) Über ein neues auxin (Heteroauxin) aus Harn xi, *Zeitschrift für Physiologica Chemie* **228**, 104–12.

Kosuge T., Heskett M.G. & Wilson E.E. (1966) Microbial synthesis and degradation of indole-3-acetamide by an enzyme system from *Pseudomonas savastanoi. Journal of Biological Chemistry* **241**, 3738–44.

Kuriger W.E. & Agrios G.N. (1977) Cytokinin levels and kinetin-virus interactions in tobacco ringspot virus-infected cowpea plants. *Phytopathology* **67**, 604–9.

Lado P., Rasi-Caldogno R. & Colombo R. (1974) Promoting effect of fusicoccin on seed germination. *Physiologia Plantarum* **31**, 149–52.

Laloue M. & Hall R.H. (1973) Cytokinins in *Rhizopogon roseolus*-secretion of N-[9-(β-D-ribofuranosyl-9H) purin-6-yl-carbomyl] threonine. *Plant Physiology* **51**, 559–62.

Letham D.S. (1978) Cytokinins. In: *Phytohormones and Related Compounds: a Comprehensive Treatise*, Vol. 1 (Ed. by D.S. Letham, P.B. Goodwin and T.J.V. Higgins), pp. 205–64. Elsevier/North Holland

Biomedical Press, Amsterdam.

Libbert E., Wichner S., Duerst E., Kaiser W., Kunert R., Manicki A., Manteuffel R., Riecke E. & Schroder R. (1968) Auxin content and auxin synthesis in sterile and non-sterile plants, with special regard to the influence of epiphytic bacteria. In: *Biochemistry and Physiology of Plant Growth Substances* (Ed. by F. Wightman & G. Setterfield), pp. 213–30. Runge, Ottawa.

Libbert E. & Manteuffel R. (1970) Interactions between plants and epiphytic bacteria regarding their auxin metabolism, VII. *Physiologia Plantarum* **23**, 93–8.

Libbert E. & Silhengst P. (1970) Interactions between plants and epiphytic bacteria regarding their auxin metabolism, VIII. *Physiologia Plantarum* **23**, 480–7.

Lieberman M. (1979) Biosynthesis and action of ethylene. *Annual Review of Plant Physiology*, **30**, 533–91.

Liu S-T., Gruenert D. & Knight C.A. (1978) Bound form indole-3-acetic acid synthesis in tumorous and non-tumorous species of *Nicotiana*. *Plant Physiology*, **61**, 50–3.

Livne A. & Daly J.M. (1966) Translocation in healthy and rust-affected beans. *Phytopathology* **56**, 170–5.

Lobenstein G. & Linsey N. (1963) Effect of virus infection on peroxidase activity and C_6/C_1 ratios. *Phytopathology* **53**, 350.

Lockhart C.L., Forsyth F.R., & Eaves C.A. (1968) Effect of ethylene on development of *Gloeosporium album* in apple and on growth of the fungus culture. *Canadian Journal of Plant Science* **48**, 557–9.

Lockhart B.E. & Semancik J.S. (1970) Growth inhibition, peroxidase and 3-indolacetic acid oxidase activity, and ethylene production in cowpea seedlings. *Phytopathology*, **60**, 553–6.

Magie A.R. & Wilson E.E. (1962) Expressions of virulence among isolates of *Pseudomonas savastanoi* from olive and oleander. *Proceedings of the American Phytopathological Society, Phytopathology*, **52**, 741.

Magie A.R., Wilson E.E. & Kosuge T. (1963) Indoleacetamide as an intermediate in the synthesis of indoleacetic acid in *Pseudomonas savastanoi*. *Science*, **141**, 1281–2.

Marré E. (1979) Fusicoccin: A tool in plant physiology. *Annual Review of Plant Physiology*, **30**, 273–88.

Milborrow B.V. (1978) Abscisic acid. In: *Phytohormones and Related Compounds: A Comprehensive Treatise*, Vol. 1 (Ed. by D.S. Letham, P.B. Goodwin and T.J.V. Higgins), pp. 295–347. Elsevier/North Holland Biomedical Press, Amsterdam.

Miller C.O. (1967) Zeatin and zeatin riboside from a mycorrhizal fungus. *Science* **157**, 1055–6.

Mothes K. & Englebrecht L. (1961) Kinetin-induced directed transport of substances in excised leaves in the dark. *Phytochemistry* **1**, 58–62.

Neljubov D. (1900) Über die horizontale mutation der stengel von *Pisum sativum* und einiger anderer pflanzen. *Botanisches Zentralblatt* **10**, 128–39.

Norman S.M., Bennett R.D., Maier V.P. & Poling S.M. (1983) Cytokinins inhibit abscisic acid biosynthesis in *Cercospora rosicola*. *Plant Science Letters* **28**, 255–63.

Osborne D.J. (1968) Ethylene as plant hormone. *Society of Chemical Industry* (London), *Monograph* **31**, 236–49.

Pavillard J. (1952) Researches sur la croissance des plantes viroses, virus et auxines. *Comptes Rendues des Séances de L'Academie des Sciences*, **235**, 87–8.

Pegg G.F. (1976a) Endogenous auxins in healthy and diseased plants. The involvement of ethylene in plant pathogenesis. Endogenous gibberellins in healthy and diseased plants. Endogenous inhibitors in healthy and diseased plants. In: *Encyclopedia of Plant Physiology*, New Series, Vol. 4 (Ed. by R. Heitfuss & P.M. Williams), pp. 560–616. Springer-Verlag, Berlin.

Pegg G.F. (1976b) The response of ethylene-treated tomato plants to infection by *Verticillium albo-atrum*. *Physiological Plant Pathology* **9**, 215–26.

Pegg G.F. (1981a) The involvement of growth regulators in the diseased plant. In: *Effects of Disease on the Physiology of the Growing Plant* (Ed. by P.G. Ayres), pp. 149–177. Cambridge University Press.

Pegg G.F. (1981b) *Fungal Wilt Diseases of Plants* (Ed. by M.E. Mace, A.A. Bell and C.H. Beckman), pp. 193–253. Academic Press, London.

Pilet P.E. (1976) Fusicoccin and auxin effects on root growth. *Plant Science Letters* **7**, 81–4.

Piskornik Z. & Bandurski R.S. (1972) Purification and partial characterisation of a glucan containing indole-3-acetic acid. *Plant Physiology* **50**, 176–82.

Pritchard D.W. & Ross A.F. (1975) The relationship of ethylene to formation of tobacco mosaic virus lesions in hypersensitive responding tobacco leaves with and without induced resistance. *Virology* **64**, 295–307.

Radley M. (1956) The distribution of substances similar to gibberellic acid in higher plants. *Nature* (London) **178**, 1070–1.

Rajagopal R. (1977) Effect of tobacco mosaic virus infection on the endogenous levels of indoleacetic, phenylacetic and abscisic acids of tobacco leaves in various stages of development. *Zeitschrift Pflanzenphysiologie* **83**, 403–9.

Rausch T., Butcher D.N. & Hilgenberg W. (1981) Nitrilase activity in club root diseased plants. *Physiologia Plantarum* **52**, 467–70.

Ross A.F. & Williamson C.E. (1951) Physiologically active emanations from virus-infected plants. *Phytopathology* **41**, 431.

Sachs T. & Thimann K.V. (1967) The role of auxins and cytokinins in the release of buds from dominance. *American Journal of Botany* **54**, 136–44.

Searle L.M., Chamberlain K. Rausch T. & Butcher D.N. (1982) Conversion of 3-indolemethylglucosinolate to 3-indolylacetonitrile by myrosinase, and its relevance to the clubroot disease of the Cruciferae. *Journal of Experimental Botany* **33**, 935–42.

Selman I.W. (1964) The effect of kinetin in infection of petunia and tomato leaves with tomato spotted wilt virus. *Annals of Applied Biology* **53**, 67–76.

Sembdner G., Datter W., Kefeli V.I. & Kutacek M. (1980) Abscisic and other naturally occurring plant growth inhibitors. In: *Plant Growth Substances* (Ed. by F. Skoog), pp. 254–61. Springer-Verlag, Berlin.

Sequeira L. (1963) Synthesis of indoleacetic acid by *Pseudomonas solanacearum*. *Phytopathology* **54**, 1240–6.

Sequeira L. (1973) Hormone metabolism in diseased plants. *Annual Review of Plant Physiology* **24**, 353–80.

Sequeira L. & Steeves T.A. (1954) Auxin inactivation and its relation to leaf drop caused by the fungus *Omphalia flavida*. *Plant Physiology* **29**, 11–6.

Smidt M. & Kosuge T. (1978) The role of indole-3-acetic acid accumulation by alpha methyl tryptophan-resistant mutants of *Pseudomonas savastanoi* in gall formation on oleanders. *Physiological Plant Pathology* **13**, 203–14.

Smith S.H., McCall S.R. & Harris J.H. (1968) Auxin transport in curly top virus-infected tomato. *Phytopathology* **58**, 1669–70.

Somerville R.L. & Yanopsky C. (1965) Studies on the regulation of tryptophan biosynthesis in *Escherichia coli*. *Journal of Molecular Biology* **11**, 747–59.

Squire G.R. & Mansfield T.A. (1972) Studies on the mechanism of action of fusicoccin the fungal toxin that induces wilting and its interaction with abscisic acid. *Planta* **105**, 71–8.

Sziraki I. & Balázs E. (1975) The effect of infection by TMV on cytokinin level of tobacco plants and cytokinins in TMV RNA. In: *Current Topics in Plant Pathology* (Ed. by Z. Király), pp. 345–52. Akadémiai Kiadó, Budapest.

Sziraki I. & Balázs E. (1979) Cytokinin activity in the RNA of tobacco mosaic virus. *Virology* **92**, 578–82.

Sziraki I., Balázs E. & Király Z. (1980) Role of different stresses in inducing systemic acquired resistance to MTV and increasing cytokinin levels in tobacco. *Physiologia Plantarum* **16**, 277–84.

Sziraki I. & Gaborjanyi R. (1974) Effect of systemic TMV infection on cytokinin level of tobacco leaves and stems. *Acta Phytopathologica Academiae Scientiarum Hungaricae* **9**, 195–9.

Tamura S. (1971) Gibberellins and other biologically active substances produced by microorganisms. In: *Biochemical & Industrial Aspects of Fermentation* (Ed. by K. Sakaguchi, T. Uemura & S. Kinoshita), pp. 155–73 Kodansha Scientific Books, Tokyo.

Tamura S., Namoto M. & Nagao M. (1970) Isolation and characterisation of indole derivatives in club roots of Chinese cabbage. In: *Plant Growth Substances* (Ed. by D.J. Carr), pp. 127–32. Springer-Verlag, Berlin.

Tavantzis S.M., Smith S.H. & Witham F.H. (1979) The influence of kinetin on tobacco ringspot virus infectivity and the effect of virus infection on the cytokinin activity in intact leaves of *Nicotiana glutinosa* L. *Physiological Plant Pathology* **14**, 227–33.

Thimann K.V. (1935) On the plant hormone produced by *Rhizopus suinus*. *Journal of Biological Chemistry* **109**, 279–91.

Thimann K.V. & Sachs T. (1966) The role of cytokinins in the 'fasciation' disease caused by *Corynebacterium fascians*. *American Journal of Botany*, **53**, 731–9.

Trewas, A. (1981) How do plant growth substances work? *Plant Cell & Environment* **4**, 203–28.

Van Loon L.C. (1977) Induction by 2-chloroethylphosphonic acid of viral-like lesions, associated proteins and systemic resistance in tobacco. *Virology* **80**, 417–20.

Van Loon L.C. & Berbee A.T. (1978) Endogenous levels of indolacetic acid in leaves of tobacco reacting hypersensitively to tobacco mosaic virus. *Zeitschrift Pflanzenphysiologie* **89**, 373–5.

Van Loon L.C. & Geelen J.L.M.C. (1971) The relation of polyphenoloxidase and peroxidase to symptom expression in tobacco var. 'Samsun NN' after infection with tobacco mosaic virus. *Acta Phytopathologica Academiae Scientiarum Hungaricae* **6**, 9–20.

Van Steveninck R.F.M. (1959) Factors affecting the abscission of reproductive organs of yellow lupins (*L. luteus* L.) III. Endogenous growth substances in virus-infected and healthy plants and their effects on abscission. *Journal of Experimental Botany* **10**, 367–76.

Weiler E.W. (1980) Radioimmunoassays for the differentiation and direct analysis of free and conjugated abscisic acid in plant extracts. *Planta* **148**, 262–72.

Weise M.V. & Devay J.E. (1970) Growth regulator changes in cotton associated with defoliation caused by *Verticillium albo-atrum. Plant Physiology* **45**, 304–9.

West C.A. & Phinney B.O. (1956) Properties of gibberellin-like factors from extracts of higher plants. *Plant Physiology* **31**, Supplement XX.

Whenham R.J. & Fraser R.S.S. (1981) Effect of systemic and local-lesion-forming strains of tobacco mosaic virus on abscisic acid concentration in tobacco leaves: consequences for the control of leaf growth. *Physiological Plant Pathology* **18**, 267–78.

Wright S.T.C. & Hiron R.P.W. (1969) (+)-abscisic acid, the growth inhibitor induced in detached wheat leaves following a period of wilting. *Nature* (London) **224**, 719.

Yu Y., Adams D.O. & Yang S.F. (1979) Regulation of auxin-induced ethylene production in mung bean hypocotyls. Role of 1, amino cyclopropane-1-carboxylic acid. *Plant Physiology* **63**, 589–90.

Zeigler R.S., Powell L.E. & Thurston H.D. (1980) Gibberellin A_4 production by *Sphaceloma manihoticola. Phytopathology* **70**, 589–93.

4 Mechanisms of biotrophy in fungal pathogens

JOHN L. GAY

Department of Pure and Applied Biology, Imperial College, London SW7 2BB, UK

Introduction

Biotrophy is a mode of nutrition whereby parasites obtain nutrients from living host cells (Thrower, 1966). It contrasts with necrotrophic parasitism where host cells are killed and saprotrophy where only non-living sources are utilized. Biotrophy may not extend throughout the whole period of parasitism (hemibiotrophy) but this distinction need not concern us. Many economically important plant pathogens are biotrophic.

If biotrophy is considered by examining the nutritional requirements of parasities, no significant differences between many biotrophs and saprotrophs are found. Thus *Claviceps spp.* and *Ustilago spp.* may be grown on normal culture media and even *Puccinia graminis*, which has races with a high degree of host specificity, will grow in axenic culture over long periods with media which are no more elaborate than those for a saprotroph such as *Saprolegnia*. *Puccinia graminis* requires a carbon source such as a sugar or sugar alcohol, ammonium ion as a bulk nitrogen source, a sulphur-containing amino acid and the normal range of mineral salts (Maclean, 1982). Vitamins are not required. These comparisons indicate that, for vegetative growth, mechanisms by which nutrients are derived from the host are more important in determining biotrophy than the availability of particular compounds.

In order that a biotroph may obtain nutrients, it should be close to the source. Therefore it must enter the host and grow between the cells, usually those of a particular organ, or perhaps attack individual cells directly. Thus the developmental sequences in which germ hyphae locate stomata, form appressoria and ultimately produce penetrating structures, all contribute to biotrophy but in this chapter only mechanisms which are directly concerned with nutrient exchange will be examined. Since much of the data on nutrient transport from host to pathogen has been reviewed recently (Gay & Manners, 1981; Hancock & Huisman, 1981; Whipps & Lewis, 1981; Manners & Gay, 1983), a general approach, attempting to seek principles, and so provide a rationale for future research, will be adopted.

From a nutritional viewpoint, it would seem most advantageous to a parasite if its cells fuse with those of its host so that direct cytoplasmic continuity exists between them. This condition is only found in a few fungal parasites of fungi which

Wood R.K.S. & Jellis G.J. (1984) *Plant Diseases: infection, damage and loss.* Blackwell Scientific Publications, Oxford.

form plasmodesma-like connections or open pores to their hosts (Hoch, 1977; 1978). In most parasitic associations, plasma membranes separate the cells of the two organisms. Some chytrids enter the host cytoplasm so that their plasma membrane is the only barrier to nutrient exchange. Infections by *Olpidium brassicae* have this type of organization (Bracker & Littlefield, 1973). In most biotrophic plant pathogens the plasma membranes of both organisms have to be traversed and this applies to intracellular structures such as haustoria and the plasmodia of *Plasmodiophora brassicae* in addition to intercellular hyphae.

It is instructive to consider the hypothetical relationship of two adjacent cells of different organisms. If one cell is able to utilize a particular nutrient or carry out photosynthesis, it will have an immediate competitive advantage but because its internal solute concentrations and surface areas are then increased, it will tend to lose nutrients at a greater rate and these will become available to the other organism. When this simple model is applied to invaders of intercellular spaces in higher plant tissues, it is found that some organisms are able to grow well in unspecialized associations. For example, *Pseudomonas pisi* initially multiplies with a generation time of just under 2 h in the intercellular spaces of tobacco leaves (Sigee & Al-Issa, 1982). Furthermore, Hancock and Huisman (1981) have estimated that the concentration of nutrients in intercellular spaces is sufficient to sustain fungal growth. However, it is noteworthy that in *Taphrina deformans*, which is confined to intercellular spaces of leaves, the hyphae are most abundant around epidermal cells and in the upper mesophyll (Syrop, 1975). Similarly, smut fungi also develop abundantly at specific locations but form few hyphae at the systemic phase. These examples suggest that nutrients may be especially concentrated at some intercellular locations and that at some sites, although nutrients may have been

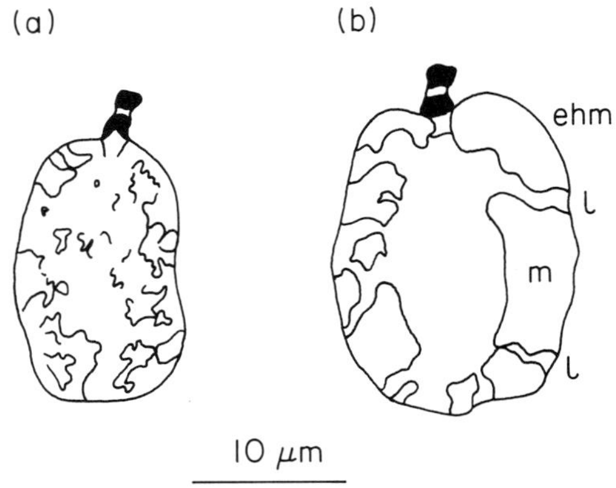

Figure 4.1. Diagrams of freshly isolated haustorial complexes from pea infected with *Erysiphe pisi* (a) in isolation medium containing 100-mM sucrose the extrahaustorial membrane follows the contour of the haustorium (b) mounted in water showing the expanded extrahaustorial matrix (m) and distended extrahaustorial membrane (ehm) attached to the neck and the tips of the haustorial lobes (l). The significance of the lobe attachment is unclear. The bar represents 10 μm. Based on Gil & Gay (1977). Reproduced by permission of Academic Press Inc. (London) Ltd.

initially sufficient, their replenishment, determined by the rate of diffusion out of host cells, may be a factor limiting continued growth. It follows that a pathogen may subsequently succeed if it is able to maintain or increase the efflux of nutrients from host cells. However, if the host cells are to remain alive, thus maintaining the biotrophic relationship, a precise mechanism of control is essential. This probably represents a substantial problem to an intercellular pathogen which necessarily operates in an open system (i.e. one in which the volume is not defined) as fungal products (toxins) which modify plasma membranes cause irreversible damage (see Turner, this volume). It is probably significant in this context, that production of diffusible toxins has not been associated with biotrophy. Biotrophy thus demands that host cells lose metabolites in quantities sufficient for the parasite's needs while retaining metabolities to maintain themselves. These two cellular functions, which at first seem mutually exclusive, are achieved in two systems in plants. In both instances the host cells are structurally polarized; in one, the cells are infected by haustoria and in the other, the pathogen occupies a site where nutrients are in the process of being transferred from one population of host cells to another.

Haustorial infections

Haustoria are produced by a variety of biotrophic fungi but until recently their function has been obscure. Powdery mildew infections offer highly advantageous systems for investigations of their physiology. Almost all of the fungus resides over the plant surface and thus host and most of the parasite can be separated easily. Feeding with radioactive substrates followed by fractionation and analysis have provided evidence of assimilation, transformation and accumulation of specific compounds and estimates of the rates of transport through the haustoria situated between the source in the leaf and the sink in the mycelium. Additionally, in *Erysiphe pisi* and some other powdery mildews, haustoria can be isolated in quantities sufficient for chemical analysis (Gil & Gay, 1977; Manners & Gay, 1977). Thus the interface and the immediate conduit of assimilates can also be studied, and in conjunction with light and electron microscopy, the solute pathway is open to detailed investigation.

Haustoria of *E. pisi*, in common with those of other fungi, are contained in an invagination of the plasma membrane of the infected host cell and this structure remains around them when they are isolated. The composite structure is the haustorial complex (Gil & Gay, 1977). The semipermeability of the invaginated region has been demonstrated by suspending isolated haustorial complexes successively in water and in sucrose solutions. The invagination distends and retracts as the extrahaustorial matrix which it encloses swells and shrinks in response (Figure 4.1). This also indicates that the invaginated membrane is tightly bound to the haustorial neck by a structure called a neckband. Further evidence of the membrane's semipermeability and of the seal around the neck, is provided by experiments using diffusion markers. When isolated haustorial complexes were

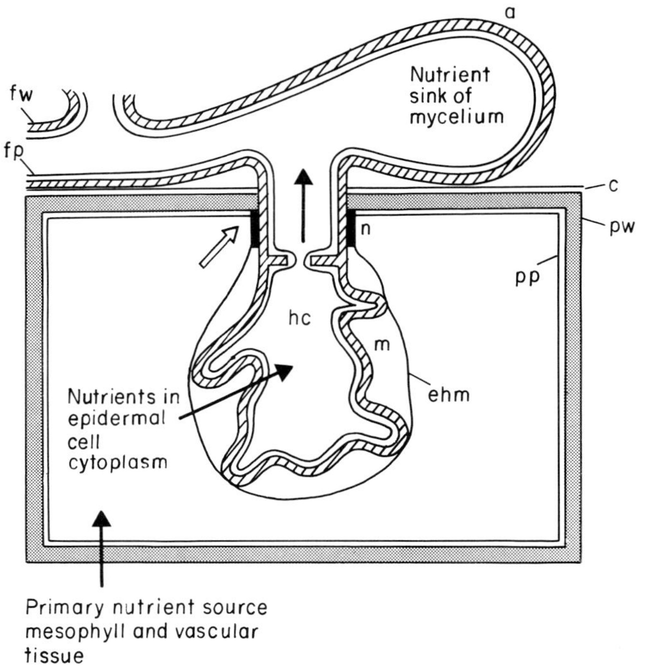

Figure 4.2. Diagram to show the interfaces, and to illustrate the pathways and barriers to nutrient flux, between *Erysiphe pisi* and *Pisum sativum*. The haustorium occupies an invagination (extrahaustorial membrane, ehm) of the epidermal plasma membrane (pp). When the mycelium is removed from the surface of the leaf, or the haustorium is released from the cell during isolation, the neck (n) is broken at the host cell wall (open arrow) and the extrahaustorial membrane remains attached at the neck-bands (shown here as one) so that it and the enclosed matrix (m) are retained with the haustorium as a structural unit (haustorial complex). The haustorial cytoplasm (hc) is retained by closure of the pore in the neck septum. The solid arrows indicate the postulated nutrient pathway. a, appressorium; c, cuticle; fp, fungal plasmalemma; fw, fungal cell wall: pw, plant cell wall. From Manners & Gay (1978). Reproduced by permission of Academic Press Inc. (London) Ltd.

suspended in uranyl acetate and then the uranyl ion was precipitated by addition of phosphate, electron microscopy demonstated that uranyl phosphate crystals were never present in the matrix or in the haustorial cytoplasm unless the structures were obviously damaged. Similarly, the extrahaustorial membrane did not stain with fluorescent plasma membrane-binding reagents applied to infected epidermal cells unless the latter were damaged.

These experiments demonstrate that the zone (extrahaustorial matrix) between each haustorium and the portion of host plasma membrane around it, is not in direct diffusion continuity with the host apoplast (intercellular spaces and cell walls) (Figure 4.2). The same conclusion has been reached in studies of tissues infected with a rust fungus (Heath, 1976). Furthermore, many studies of a variety of haustorial infections have shown that the structure of the invaginated region of the host plasma membrane is distinct from the uninvaginated, wall-lining region (Table 4.1). The observation which bears most directly on the function of the membrane concerns the distribution of plasma membrane ATPase; activity in the wall-lining

Table 4.1. Examples of comparisons of properties of the extrahaustorial and wall-lining regions of host plasma membranes.

Fungus	Host	Test	Special properties of extrahaustorial region	Reference
Erysiphe graminis	*Hordeum vulgare*	OsO_4 fixation + EM	~ 1.5x thicker	Bracker, 1968
		Abnormal conc's $KMnO_4$ + EM	Undamaged	"
Erysiphe pisi	*Pisum sativum*	Osmotic stress + LM	Stronger	Gil & Gay, 1977
		Aldehyde-OsO_4 fixation + EM	~ 2.3x thicker	"
		Enzymic digestion + EM	More carbohydrate	"
		Calcofluor white M2R fluorescence + LM	Abundant β-linked carbohydrate	Gil, 1976
		Freeze fracture + EM	No intra-membrane particles	"
		Enzymology + EM	No ATPase	Spencer-Phillips & Gay, 1981
Melampsora lini	*Linum usitatissimum*	Plasma membrane stain + EM	± no reaction	Littlefield & Bracker, 1972
		Freeze-fracture + EM	No intra-membrane particles	"
Uromyces appendiculatus	*Phaseolus vulgaris*	Enzymology + EM	No ATPase	Spencer-Phillips & Gay, 1981
Peronospora pisi	*Pisum sativum*	Plasma membrane stain + EM	± no reaction	Hickey & Coffey, 1978
Albugo candida	*Cardamine hirsuta*	Plasma membrane stain + EM	± no reaction but extensions into cytoplasm normal	Woods & Gay, 1983
		Carbohydrate stain + EM	No difference (weak reaction over neck)	"
		Enzymology + EM	No ATPase	"

Key: LM = light microscopy
EM = electron microscopy

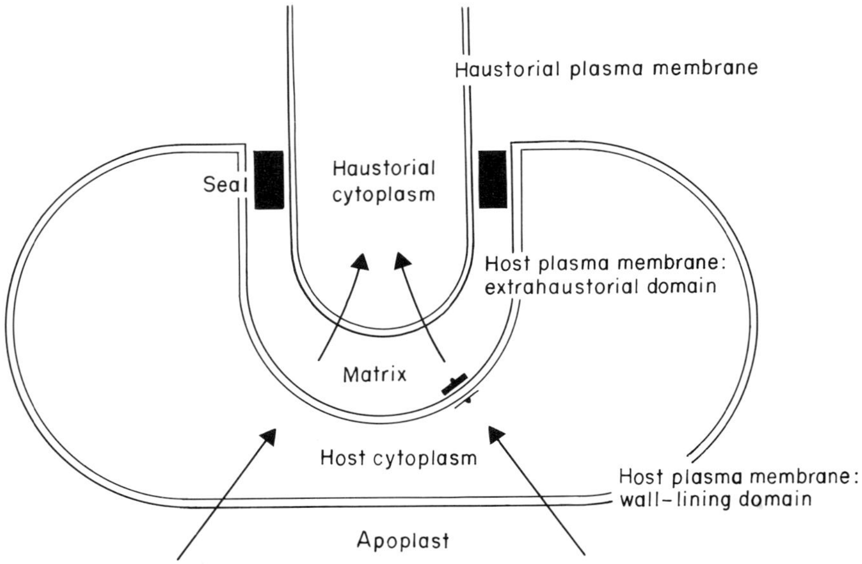

Figure 4.3. Diagram to show the proposed mechanisms of solute transport through the plasma membranes of pea epidermal cells and haustoria of *Erysiphe pisi*. The arrows indicate ATPase-activated influxes through the wall lining domain of the host and the haustorial plasma membranes. It is proposed that, because the matrix is maintained as a closed compartment by the seal between host and haustorial plasma membranes, efflux through the invaginated domain of the host plasma membrane is effected by either a high concentration difference of the transported species (left) or interaction of complementary ionic products of electrogenic pumps at the ATPase active membranes (right). The seal also maintains the segregation of the two domains of the host plasma membrane. From Spencer-Phillips & Gay (1981). Reproduced by permission of the trustees of the *New Phytologist*).

region is unaffected by infection but is absent from the invagination. The transition is abrupt and in every species examined, occurred at the edge of a neckband (Spencer-Phillips & Gay, 1981; Woods & Gay, 1983). The distribution of ATPase in *E. pisi* is shown diagrammatically in Figure 4.3. This indicates that the host plasma membrane has a diminished control of solute transport at its haustorial aspect while still functioning normally in the wall-lining region.

These results may be interpreted in conjunction with data from analytical and kinetic studies (Manners & Gay, 1982). Sucrose is the major soluble component of haustorial complexes extracted immediately after exposure of infected leaves for 30 min to $^{14}CO_2$. *E. pisi* mycelium infecting one epidermis of a pea leaf, collects up to 30% of the leaf's photosynthate and this represents a rate of solute efflux from the extrahaustorial invagination several orders of magnitude higher than rates at which plant cells normally lose solutes. The efflux approaches the rates at which plasma membranes of *Neurospora crassa* take in glucose (Jennings, 1976) and transfer cells of beet load the phloem with sucrose (Sovonick *et al.*, 1974). Thus it is seen that by polarizing individual infected cells, haustoria do not impair the normal loading activity of those cells, or others in the tissues, and are even able to utilize the pumps in the host plasma membranes which present assimilates to them.

The distribution of ATPase activity of plasma membranes of cells infected with a rust (*Uromyces appendiculatus*) (Spencer-Phillips & Gay, 1981) or an oomycete (*Albugo candida*) (Woods & Gay, 1983) are essentially the same as in pea infected with powdery mildew. However, whereas the plasmalemmas of *E. pisi* haustoria possess ATPase, no activity was found in haustorial plasma membranes of the other fungi. Thus, in *E. pisi* the haustorium may be expected actively to take up nutrients presented to its interface with the extrahaustorial matrix, but in the others the only factors which can be envisaged as assisting haustorial loading are the pumps in plasma membranes of the infected cells, production or accumulation of solutes by the host, and polymerization of solutes by the pathogen. Thus in addition to being dependent on the hosts as a source of nutrients, these haustoria most probably also depend to a greater or lesser degree, on host activity for solute transport. A similar mechanism may operate in *Plasmodiophora brassicae* infections. Plasmodia are enveloped in a membrane whose structure is distinct from that of a plasma membrane (Aist, 1974) and is presumed to be equivalent to the invagination around haustoria. Thus a portion of apoplast is isolated without the necessity of a neckband seal and recent investigations indicate absence of ATPase activity at the interfacial membranes (Gay & Martin, unpublished).

In most powdery mildews, only the haustoria have direct contact with host cells and thus are likely to provide the only avenue for nutrient uptake. Ellingboe and his colleagues (Mount & Ellingboe, 1969; Slesinski & Ellingboe, 1971) showed that radio-isotopes presented to host tissues did not enter the superficial structures until haustoria had developed. In rust and oomycete infections little is known of the relative contributions to uptake of haustoria and intercellular hyphae. Autoradiography has shown that some solutes supplied to host tissues are incorporated into the fungi before haustoria are formed whereas others are not incorporated until afterwards (Onoe *et al.*, 1973; Andrews, 1975). Both sets of results suggest specificity in uptake by the two components of the fungi. It is not surprising that in *E. pisi* the plasma membranes of superficial hyphae, which are in an aerial environment, lack ATPase activity (Spencer-Phillips & Gay, 1981). However, ATPase is also absent from plasma membranes of intercellular hyphae of *Uromyces appendiculatus* (Spencer-Phillips & Gay, 1981) and in *Albugo candida* only slight activity was recorded (Woods & Gay, 1983). The result for the rust suggests that intercellular hyphae are unable to take up nutrients at concentrations lower than those internally. This observation may be significant in light of reports that mycelial axenic cultures of *Puccinia graminis* released into the culture medium larger quantities of amino acids, including essential ones containing sulphur, than they incorporated into protein (Maclean, 1982). It is not known whether the excreted compounds are also continuously taken up and reutilized. However, results of studies on uptake of 3-o-methyl-D-glucose by cultured mycelia of *P. graminis* showed that the compound, which is not metabolized, accumulated so that its internal concentration exceeded the external (Manners *et al.*, cited by Maclean, 1982). In contrast to the result for the amino acids, this indicates an active process so that there may be specificity in uptake.

Transfer-intercept infections

The hypothesis that solute transport by haustoria depends on the functional polarization of individual cells leads us to consider whether the concept may be applied also to tissues. The model of the interaction of a pair of unelaborated cells proposed above (page 50) is clearly much simpler than obtains in plant tissues where many cells are interconnected by plasmodesmata. These structures conduct solutes at rates over one thousand times greater than the same area of a plasma membrane (Gunning & Hughes, 1976) and therefore solutes lost by one cell may be quickly replenished by others directly or indirectly connected to it. Furthermore, it is important to note that the replenishment does not involve transport through the extra-cytoplasmic compartment (apoplast) occupied by the pathogen. It may be that pathogens do damage the plasma membranes of cells immediately adjacent to them but death is avoided because solute losses caused by the damage are made good by plasmodesma transport. Additionally, higher plant tissues include cells which normally lose solutes at high rates and replenish the losses *via* plasmodesmata. An hypothesis is proposed, therefore, that some pathogens intercept nutrients which normally would be *en route* through the apoplast between two groups of host cells which are in close proximity and individually interconnected by plasmodesmata. Transfer cells are the most obvious candidates for this rôle but interception between other types of cell should not be excluded from the concept.

Transfer cells are parenchymatous, frequently associated with vascular elements and the internal surface of their walls has elaborate reticular thickenings. Their plasma membranes are thus extensive (20-fold amplification has been recorded) and these are much richer in ATPase than they are in most other plant cells. The cells have numerous plasmodesmata connecting them with at least some of their neighbours and their cytoplasmic organelles indicate high metabolic activity. Transfer cells occur frequently at nodes, at the bases of embryo sacs, in embryos, in glands and are associated with minor vascular tissues in leaves. The function of transfer cells has been largely inferred from their structure and situation (Gunning & Pate, 1974; Bentwood & Cronshaw, 1978). It seems likely that they control the transport of ions and assimilates between their cytoplasms and the surrounding apoplast. Their extensive plasma membranes and high ATPase activity, which activates transmembrane pumps (Baker, 1978), presumably promote interchange. Furthermore, their plasmodesma connections to neighbouring cells, which in turn are connected to other cells, enable them to control the solutes in a substantial volume of cytoplasm (symplast). This is especially so when they are connected to a sieve element, which has symplastic continuity throughout the vascular tissue.

Ergot infections provide a good example of the way in which this type of biotrophic system probably operates. The uninfected embryo has no direct (symplastic) connection with the plant's vascular tissues or with the associated transfer cells (Gunning & Pate, 1974). However, at its surface it has transfer cells which are close to the vascular-associated transfer cells and, as in phloem loading,

(Geiger, 1975; Giaquinta, 1976) these cells take solutes from the apoplast. The infecting ergot mycelium competes with the developing embryo for apoplastic nutrients and replaces it by growing faster (Luttrell, 1980; Shaw & Mantle, 1980). Although solute fluxes into developing sclerotia have not been measured they are probably substantial. The fungal structures achieve greater weights than the grain which they replace and additionally they produce conidia and excrete honeydew (mainly sucrose) during their development. Loose smuts probably operate in a similar way at the stage immediately preceeding sporulation. However, it seems likely that the systemic vegetative mycelium is sparse because then it is not associated with a transfer cell system.

Ergot infections are accompanied by the production of a glucanase which almost certainly stimulates the flow of nutrients from the host by preventing deposition of callose (β1−3 glucan) in adjacent host cells (Dickerson *et al.*, 1978). Similarly, invertase formed in ergot (Dickerson *et al.*, 1976) and *Ustilago maydis* (Callow *et al.*, 1980) infections may be regarded as assisting nutrient availability. However, it is suggested now that the primary mechanism resembles haustorial infections in utilizing a polarized system but whereas the system is induced in haustorial infections, transfer intercept infections exploit the normal function of a plant tissue.

Conclusions

It is clear that the exact details of cellular associations are of paramount importance in biotrophy. Host cells are extremely diverse in their function and the particular type and location involved can greatly influence the course of an infection.

Haustorial infections merit special attention because the biotrophy in which they participate can be directly related to the differentiation induced in infected cells. Development of both a new region of plasma membrane with unique transport properties, and haustorial neckbands occur in infections by members of several distinct groups of fungi. In powdery mildews, rust fungi and probably in oomycetes a neckband isolates the apoplast around each haustorium so that the transport systems of hosts and pathogen are tightly coupled in series. This may even eliminate the necessity for active transport by the pathogen. Furthermore, it seems inevitable that the extrahaustorial domain of the plasma membrane of infected cells differentiates under fungal control. Isolation of the extrahaustorial apoplast is also of paramount importance in this context because it could restrict the movement of a diffusable initiator of differentiation (toxin) produced by the fungus and prevent it from reacting with the whole plasma membrane of plant cells, thereby avoiding the undesirable consequences mentioned at the beginning of this chapter. In retrospect, it is not surprising that toxins have seldom been found in haustorial infections. Searches should be concentrated on the extrahaustorial apoplast where results indicate that they are most likely to occur. Furthermore, the transfer-intercept infecting fungi have an alternative strategy to toxin formation and utilize polarized tissues normally present in plants.

It is perhaps wise to conclude with a cautionary note on studies of metabolite leakage from tissues. Comparative measurements with floating disks cut from uninfected and infected leaves may be exceedingly difficult to interpret in light of the points made above. Certain regions are isolated so that they are not represented in the samples obtained and others are normally specialized for apoplastic transport. Additionally, effluents represent the resultant of efflux from certain cells and influx by others and thus, although tissues may be losing abnormal quantities of solutes, this activity may not be recorded where the infecting organism is a powerful scavenger. Biotrophy demands highly structured systems and experimentation to investigate it should be designed accordingly.

References

Aist J.R. (1974) A freeeze-etch study of membranes of *Plasmodiophora* — infected and non-infected cabbage root hairs. *Canadian Journal of Botany* **52**, 1441–9.

Andrews J.H. (1975) Distribution of label from ^{3}H-glucose and ^{3}H-leucine in lettuce cotyledons during the early stages of infection with *Bremia lactucae. Canadian Journal of Botany* **53**, 1103–15.

Baker D.A. (1978) Proton co-transport of organic solutes by plant cells. *New Phytologist* **81**, 485–97.

Bentwood B.J. & Cronshaw J. (1978) Cytochemical localisation of adenosine triphosphatase in the phloem of *Pisum sativum* and its relation to the function of transfer cells. *Planta* **140**, 111–20.

Bracker C.E. (1968) Ultrastructure of the haustorial apparatus of *Erysiphe graminis* and its relationship to the epidermal cell of barley. *Phytopathology* **58**, 12–30.

Bracker C.E. & Littlefield L.J. (1973) Structural concepts of host-pathogen interfaces. In: *Third Long Ashton Symposium* 1971, *Fungal Pathogenicity and the Plant's Response* (Ed. by R.J.W. Byrde & C.V. Cutting), pp. 159–318. Academic Press, London.

Callow J.A., Long D.E. & Lithgow E.D. (1980) Multiple molecular forms of invertase in maize smut infections. *Physiological Plant Pathology* **16**, 93–107.

Dickerson A.G., Mantle P.G. & Nisbet L.J. (1976) Carbon assimilation by *Claviceps purpurea* growing as a parasite. *Journal of General Microbiology* **97**, 267–76.

Dickerson A.G., Mantle P.G., Nisbet L.J. & Shaw B.I. (1978) A role for β-glucanases in the parasitism of cereals by *Claviceps purpurea. Physiological Plant Pathology* **12**, 55–62.

Gay J.L. & Manners J.M. (1981) Transport of host assimilates to the pathogen. In: *Effects of Disease on the Physiology of the Growing Plant* (Ed. by P.G. Ayres), pp. 85–100. Cambridge University Press, Cambridge.

Geiger D.R. (1975) Phloem loading. In: *Encyclopedia of Plant Physiology*, New Series. Vol. 1. *Transport in Plants 1. Phloem Transport* (Ed. by M.H. Zimmerman & J.A. Milburn), pp. 395–431. Springer-Verlag, Berlin.

Giaquinta R.T. (1976) Evidence for phloem loading from the apoplast. *Plant Physiology* **57**, 872–5.

Gil F. (1976) Ultrastructural and physiological properties of haustoria of powdery mildews and their host interfaces. Ph.D. Thesis. University of London.

Gil F. & Gay J.L. (1977) Ultrastructural and physiological properties of the host interfacial components of haustoria of *Erysiphe pisi in vivo* and *in vitro. Physiological Plant Pathology* **10**, 1–12.

Gunning B.E.S., & Hughes J.E. (1976) Quantitative assessment of symplastic transport of pre-nectar into the trichomes of *Abutilon* nectaries. *Australian Journal of Plant Physiology* **3**, 619–37.

Gunning B.E.S. & Pate J.S. (1974) Transfer cells. In: *Dynamic Aspects of Plant Ultrastructure* (Ed. by A.W. Robards), pp. 441–80. McGraw Hill, London.

Hancock J.G. & Huisman O.C. (1981) Nutrient movement in host-pathogen systems. *Annual Review of Phytopathology* **19**, 309–31.

Heath M.C. (1976) Ultrastructural and functional similarity of the haustorial neckband of rust fungi and the Casparian strip of vascular plants. *Canadian Journal of Botany* **54**, 2484–9.

Hickey E.L. & Coffey M.D. (1978) A cytochemical investigation of the host-parasite interface in *Pisum sativum infected by the downy mildew fungus Peronospora pisi. Protoplasma* **97**, 201–20.

Hoch H.C. (1977) Mycoparasitic relationships. III. Parasitism of *Physalospora obtusa* by *Calcarisporium parasiticum*. *Canadian Journal of Botany* **55**, 198–207.

Hoch H.C. (1978) Mycoparasitic relationships. IV. *Stephanoma phaeospora* parasitic on a species of *Fusarium*. *Mycologia* **70**, 370–9.

Jennings D.H. (1976) Transport and translocation in filamentous fungi. In: *The Filamentous Fungi* Vol. 2 (Ed. by J.E. Smith & D.R. Berry), pp. 32–64. Edward Arnold, London.

Littlefield L.J. & Bracker C.E. (1972) Ultrastructural specialisation of the host-pathogen interface in rust-infected flax. *Protoplasma* **74**, 271–305.

Luttrell E.S. (1980) Host-parasite relationships and development in the ergot sclerotium in *Claviceps purpurea*. *Canadian Journal of Botany* **58**, 942–58.

Maclean D.J. (1982) Axenic culture and metabolism of rust fungi. In: *The Rust Fungi* (Ed. by K.J. Scott & A.K. Chakravorty), pp. 37–120. Academic Press, London.

Manners J.M. & Gay J.L. (1977) The morphology of haustorial complexes isolated from apple, barley, beet and vine infected with powdery mildews. *Physiological Plant Pathology* **11**, 261–6.

Manners J.M. & Gay J.L. (1978) Uptake of ^{14}C photosynthates from *Pisum sativum* by haustoria of *Erysiphe pisi*. *Physiological Plant Pathology* **12**, 199–209.

Manners J.M. & Gay J.L. (1982) Transport, translocation and metabolism of ^{14}C-photosynthates at the host-parasite interface of *Pisum sativum* and *Erysiphe pisi*. *New Phytologist* **91**, 221–44.

Manners J.M. & Gay J.L. (1983) The host-parasite interface and nutrient transfer in biotrophic parasitism. In: Biochemical Plant Pathology (Ed. by J.A. Callow) pp. 163–95. Wiley, London.

Mount M.S. & Ellingboe A.H. (1969) ^{32}P and ^{35}S transfer from susceptible wheat to *Erysiphe graminis* f.sp. *tritici* during primary infection. *Phytopathology* **59**, 235.

Onoe T., Tani T. & Haito N. (1973) The uptake of labelled nucleosides by *Puccinia coronata* grown in susceptible oat leaves. *Report of the Tottori Mycological Institute* (Japan) **10**, 303–12.

Shaw B.I. & Mantle P.G. (1980) Host infection by *Claviceps purpurea*. *Transactions of the British Mycological Society* **75**, 77–90.

Sigee D.C. & Al-Issa A.N. (1982) The hypersensitive reaction in tobacco leaf tissue infiltrated with *Pseudomonas pisi*. 2. Changes in the population of viable, actively metabolic and total bacteria. *Phytopathologische Zeitshrift* **105**, 71–86.

Slesinski R.S. & Ellingboe A.H. (1971) Transfer of ^{35}S from wheat to the powdery mildew fungus with compatible and incompatible parasite/host genotypes. *Canadian Journal of Botany* **49**, 303–10.

Sovonick S.A., Geiger D.R. & Fellows R.J. (1974) Evidence for active phloem loading in the minor veins of sugar beet. *Plant Physiology* **54**, 886–91.

Spencer-Phillips P.T.N. & Gay J.L. (1981) Plasma membrane ATP-ase domains and transport through infected cells. *New Phytologist* **89**, 393–400.

Syrop M. (1975) Leaf curl disease of almond caused by *Taphrina deformans* (Berk.) Tul. 1. A light microscope study of the host-parasite relationship. *Protoplasma* **85**, 39–56.

Thrower L.B. (1966) Terminology for plant parasites. *Phytopathologische Zeitschrift* **49**, 460–9.

Whipps J.M. & Lewis D.H. (1981) Patterns of translocation, storage and interconversion of carbohydrates. In: *Effects of Disease on the Physiology of the Growing Plant* (Ed. by P.G. Ayres), pp. 47–83. Cambridge University Press, Cambridge.

Woods A.M. & Gay J.L. (1983) Evidence for a neckband delimiting structural and physiological regions of the host plasmalemma associated with haustoria of *Albugo candida*. *Physiological Plant Pathology* **23**, 73–88.

which produces consistent and reproducible results and a single race of pathogen the next step is to decide what measurements to make.

Enhancement of photosynthesis rates during the infection process

Reports exist of transient stimulations of leaf photosynthesis soon after infection of leaves with pathogens (Ayres, 1979) and of apparently high rates of photosynthesis late in the infection process (Last, 1963). Unpublished work at the Edinburgh School of Agriculture suggests that, for a range of cultivars, there is no statistically valid evidence that net photosynthesis of infected leaves increases at any stage of the infection process if the responses of infected and uninfected parts of the leaf are pooled (Figure 5.1, Line P_T). Figure 5.1 Line P_I, shows how this is altered if only the actively photosynthesizing non-infected leaf area is considered. Here there is a statistically significant and compensatory increase in the rate of photosynthesis of uninfected tissue in infected leaves. The rate rises as the area of infected leaf increases until senescence becomes dominant in the final stages of leaf collapse. It has been suggested that the increase in the initial stages is linked with changes in photorespiration (Ayres, 1979). The increase at later stages of infection is probably related to changes in photosynthate transport between the leaf and the rest of the plant and the increasing size of the sink represented by the respiration of infected host cells (Daly, 1976). Photosynthetic activity is in part dependent on the rate of removal of photosynthate from the site of synthesis (Habeshaw, 1973; Lui *et al.*, 1973). The efficiency of photosynthate removal from uninfected tissue should increase as its proportion declines. This trend is shown in Figure Line P_I. So far as the plant is concerned, there is undoubtedly a reduction in overall net assimilation by the leaf. However it appears that the pathogen is able to assume some control over the photosynthetic CO_2 fixation in adjacent but uninfected leaf tissue and increase its efficiency before senescence leads to the collapse or abscission of the leaf.

Physiological evidence for toxin activity — rhynchosporoside

An interesting feature of the host-pathogen relationship in barley infected with *E. graminis* f.sp. *hordei* was the relatively minor effect on the net photosynthetic uptake of leaf tissue not in the immediate vicinity of fungal colonies. Other host–pathogen relationships are less benign and observations of leaf symptoms have led many workers to suspect that host cell photosynthesis in uncolonized areas of the leaf may be affected by pathotoxins of which some are known to act directly on photosynthesis, stomata, water retention and transport in leaves (Daly, 1981). Photosynthetic rates would be affected by changes in carbon dioxide flux caused by alterations in stomatal resistance or by biochemical changes which would be reflected in the mesophyll resistance. *Rhynchosporium secalis* produces a toxin rhynchosporoside (Auriol *et al.*, 1978) and, although there is some uncertainty about

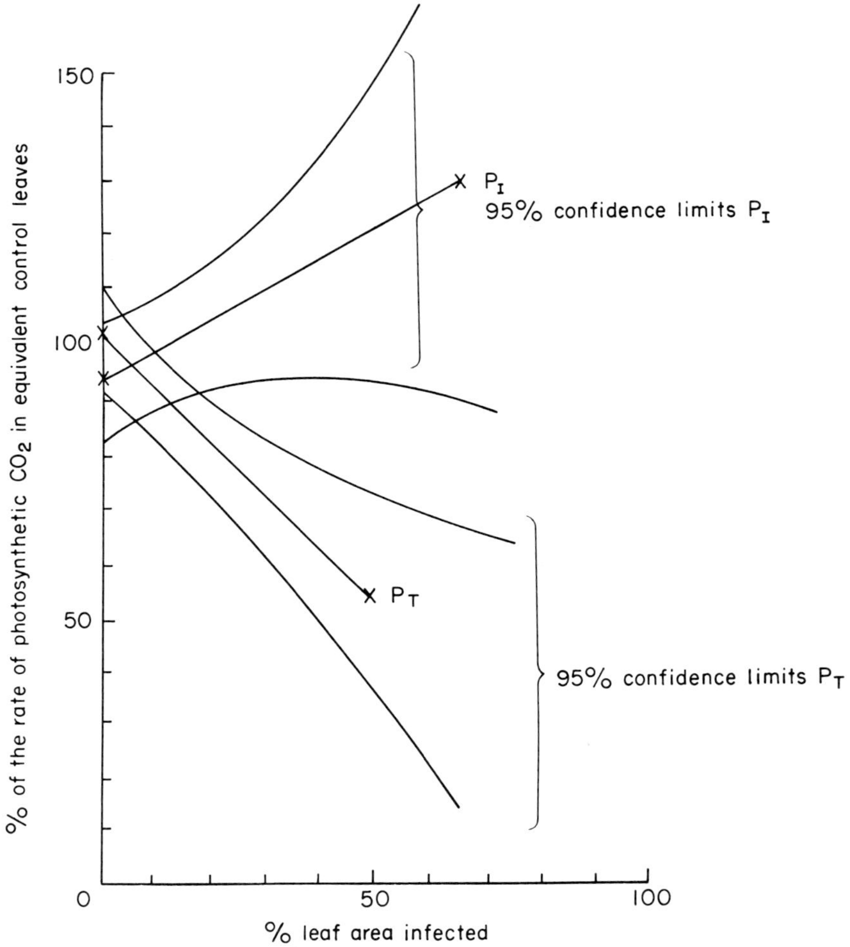

Figure 5.1. Changes in photosynthetic rates with increases in leaf area infected (regression lines using individual leaf values; 'green island' leaves omitted).
P_T = Rate of photosynthesis in infected leaves under standard conditions expressed as a percentage of the rate in the equivalent control leaf.
P_I = Rate of photosynthesis in the uninfected area of infected leaves under standard conditions expressed as a percentage of the rate in the equivalent control leaf.

its chemical nature and mode of action (Yoshiki *et al.*, 1981), a physiological investigation of the photosynthetic uptake of barley leaves infected with *R. secalis* could be expected to provide evidence of pathotoxin activity if photosynthetic carbon dioxide and water fluxes are monitored. Table 5.2 shows the results of such an experiment. The mean photosynthetic rates of samples of infected and uninfected leaves of the same age range and plant position were not significantly different. The loss of water from infected leaves was significantly greater than from the control leaves. The increase in water flux without change in carbon dioxide flux could be explained as a rise in mesophyll resistance. The results of a calculation for mesophyll resistance using the mean carbon dioxide and water fluxes (under standard experimental conditions) are shown in Table 5.2.

Table 5.2. Carbon dioxide and water fluxes and leaf resistances in barley leaves infected with *Rhynchosporium secalis.*

		Uninfected control	Infected leaves	Uninfected portion of infected leaf
$F_{(CO_2)}$	Number of leaves in sample	33	32	32
	Mean value $F_{(CO_2)}$ g/m²/s	0.000268	0.000252	0.000309
	SE of difference between means	0.000032		0.000081 (Mean)
	Homogeneity of variance	0.00122	0.00167	—
	df	61		—
	t	0.0655		—
	t (0.05; 60 df)	2.000		—
$F_{(H_2O)}$	Number of leaves in sample	20	18	
	Mean value $(F_{(H_2O)})$ g/m²/s	0.0209	0.0427	
	SE of difference between means	0.0080		
	Homogeneity of variance	0.1319	1.0140	
	df	22		
	t	2.7030		
	t (0.05; 22 df)	2.080		
Resistances calculated from means s/m				
CO_2	Total r_q	1881	2000	1631
	Boundary r_a	12	12	12
	Stomatal r_s	550	265	265
	Mesophyll r_m	1319	1723	1354
H_2O	Total r_q	337	166	166
	Boundary r_a	7	7	7
	Stomatal r_s	330	159	159

Note that the mesophyll resistance has to be increased by approximately 30% to account for the observed rate of photosynthesis under the lowered stomatal resistance of infected leaves. However, the infected leaves consist of fungal lesions surrounded by healthy tissue. The fungal lesions, while water-soaked and thus able to lose water, were bleached and necrotic and thus could be presumed to do little photosynthesis. A recalculation of the rate of photosynthesis of uninfected tissues in infected leaves allowing for the non-photosynthetically active fungal lesions (Table 5.2) showed that the photosynthesis of the uninfected regions had increased relative to that of control leaves. Using the value for the carbon dioxide flux of the uninfected portions of the infected leaves, calculations of the values of the leaf resistances can be made and are shown in Table 5.2 column 3. The mesophyll resistance is now only about 3% higher than the control value so the reduced amount of photosynthesis in whole leaves can be explained by the fact that only a portion of the infected leaf is photosynthesising, not by an increase in mesophyll resistance. The higher rates in the uninfected portions of infected leaves are a consequence of a reduced stomatal resistance in infected leaves. There is no doubt about the higher water flux rate from infected leaves. Some toxins, notably fusicoccin, are known to open stomata (Daly, 1981) and rhynchosporoside could be acting similarly.

Perhaps a more relevant consideration is how the leaf controls assimilation and transpiration by varying stomatal resistance. Cowan & Farquhar (1977) and Farquhar (1979) have discussed this in terms of the 'gain ratio'. This states that the rates of change in water transpiration ($F_{(H_2O)}$) and net photosynthesis CO_2 flux ($F_{(CO_2)}$) are controlled by the leaf so that the ratio of the two rates is a constant known as the gain ratio. This implies that the control systems of the stomatal aperture of the leaf operate so as to maximize the ratio of carbon fixed to water lost. Thus in an infected leaf not suffering water stress and with a given level of carboxylase enzyme activity, the loss of some leaf photosynthetic area to fungal colonies coupled with continued water loss from these areas will mean that the ratio of carbon dioxide fixed to water lost will fall. Closing the stomata will reduce photosynthesis proportionately more than transpiration in the absence of any change in photosynthetic enzyme concentrations, and the ratio of carbon dioxide fixed to water lost will lessen although the amount of water lost by the leaf will fall. If it is assumed that the system of controlling stomatal aperture always operates to maximize the ratio of carbon fixed to water lost, and if the leaf is not under water stress, then the response of the leaf in an attempt to keep the gain ratio constant will be to open the stomata even further and thus increase photosynthesis by reducing stomatal resistance. It would be expected that the normal response of leaves to *R. secalis* infection would be to decrease the stomatal resistance in the photosynthesizing portions, provided that water loss was not critical. The opening of stomata in infected leaves in this experiment need not be taken as evidence of an effect of toxin but is good evidence that the normal control systems of the leaf are still operating and functioning so as to compensate for the destruction of tissue by the pathogen.

Infection and leaf senescence

However well the undamaged parts of an infected leaf compensate for the local effects of a pathogen, eventually one or more of the leaf physiological control systems fail to function properly or senescence supervenes. In the experiments on the photosynthetic system of the second leaf of young barley plants used to illustrate this discussion, infected leaves showed symptoms of generalized senescence 15 to 20 days after emergence. Data from leaves older than this were not

6 The role of respiration and energy generation in diseased and disease-resistant plants

V. SMEDEGAARD-PETERSEN
Department of Plant Pathology
The Royal Veterinary and Agricultural University,
Thorvaldsensvej 40, DK-1871 Copenhagen V., Denmark

Introduction

It is generally accepted that disease and usually also disease resistance are ultimately a result of abnormal changes in plant metabolism. One of the most striking and well documented examples of such changes is an increased rate of respiration.

Although increased respiration appears to be a general phenomenon in both compatible and incompatible host–pathogen interactions there is good experimental evidence to suggest that in susceptible tissues increased respiration may be part of a disease-promoting system whereas in resistant tissues it may have quite the opposite effect, being part of a defence system. This chapter deals with pathogen-induced respiration in susceptible and resistant plants.

Dark respiration and energy generation

Dark respiration is a main energy-generating system in growing plants. The term respiration refers to the complex series of chemical reactions by which living organisms obtain energy through the oxidation of organic material to simpler compounds.

Degradation of carbohydrates such as starch and sugars to smaller carbon compounds is mediated primarily by the glycolytic (EMP) and the pentose phosphate pathways. The resulting compounds, mainly pyruvic acid which is the end product in glycolysis, are subsequently oxidized via the citric acid (TCA) cycle to carbon dioxide and water.

During the oxidative steps in glycolysis, the pentose phosphate pathway and the citric acid cycle, electrons and H-ions are removed and utilized in the reduction of the pyridine nucleotides NAD (nicotinamide adenine dinucleotide), NADP (nicotinamide adenine dinucleotide phosphate) and FAD (flavin adenine dinucleotide). It is by the reoxidation of these reduced coenzymes in the electron transport

Wood R.K.S. & Jellis G.J. (1984) *Plant Diseases: infection, damage and loss.* Blackwell Scientific Publications, Oxford.

system that most of the energy which becomes available to plant cells is trapped with the formation of high-energy phosphate bonds in ATP (adenosine triphosphate). Figure 6.1 shows some of the main steps in respiration.

Respiration of diseased plants

One of the earliest reports of respiratory changes in diseased plants is that of Allen & Goddard (1938) working with wheat leaves infected by the powdery mildew pathogen, *Erysiphe graminis* f.sp. *tritici*. By mechanical removal of the mycelium and measurements of mycelial respiration on epidermal strips removed from the host, they obtained evidence that the major part of the rise in respiration came from the host tissue, and that the parasite itself only contributed little. These results were confirmed by Millerd & Scott (1956) who found that barley tissue infected with *Erysiphe graminis* f.sp. *hordei* respired at almost the same rate as tissue with the fungus removed.

The significance of increased oxygen uptake by diseased and resistant plants may depend on the mechanisms by which it is caused. Some of the mechanisms which may explain pathological changes in respiration are discussed in this chapter.

Uncoupling of oxidative phosphorylation

Allen & Goddard (1938) suggested that the rise in respiration of wheat infected with mildew was due to diffusible substances which moved into the mesophyll. Fifteen years later Allen (1953) extended this idea by suggesting that enhancement might be caused by a toxin which uncouples respiration from energy-requiring processes by acting on oxidative phosphorylation.

The concept is based on the coupling of oxidative reactions in respiration with phosphorylation (formation of ATP) in the electron transport system. This process can be retarded by certain inhibitors, e.g. DNP (2,4-dinitrophenol), which dissociate or uncouple the electron transfer from phosphorylation so that synthesis of ATP is reduced or prevented, the concentration of ADP increases, and the rate of oxygen uptake increases (Allen, 1953).

The uncoupling hypothesis was supported by Pozsár & Király (1958) who found that the ratio of inorganic to organic phosphate in rust-infected wheat tissue was higher than in normal tissues. Inhibitor studies and marked increases in ascorbic acid oxidation were interpreted as evidence for uncoupling of the oxidative phosphorylation and an increase of the ADP/ATP ratio. The increased ratio of inorganic to organic phosphate was, however, not confirmed by Heitefuss & Fuchs (1963) who instead found a marked increase of organic phosphate in rust-infected wheat leaves.

Uncoupling has also been suggested as an explanation of increased oxygen uptake of oat tissue treated with victorin, the host-specific toxin of *Helminthosporium victoriae* (Krupka, 1959). The evidence for this was obtained in extensive

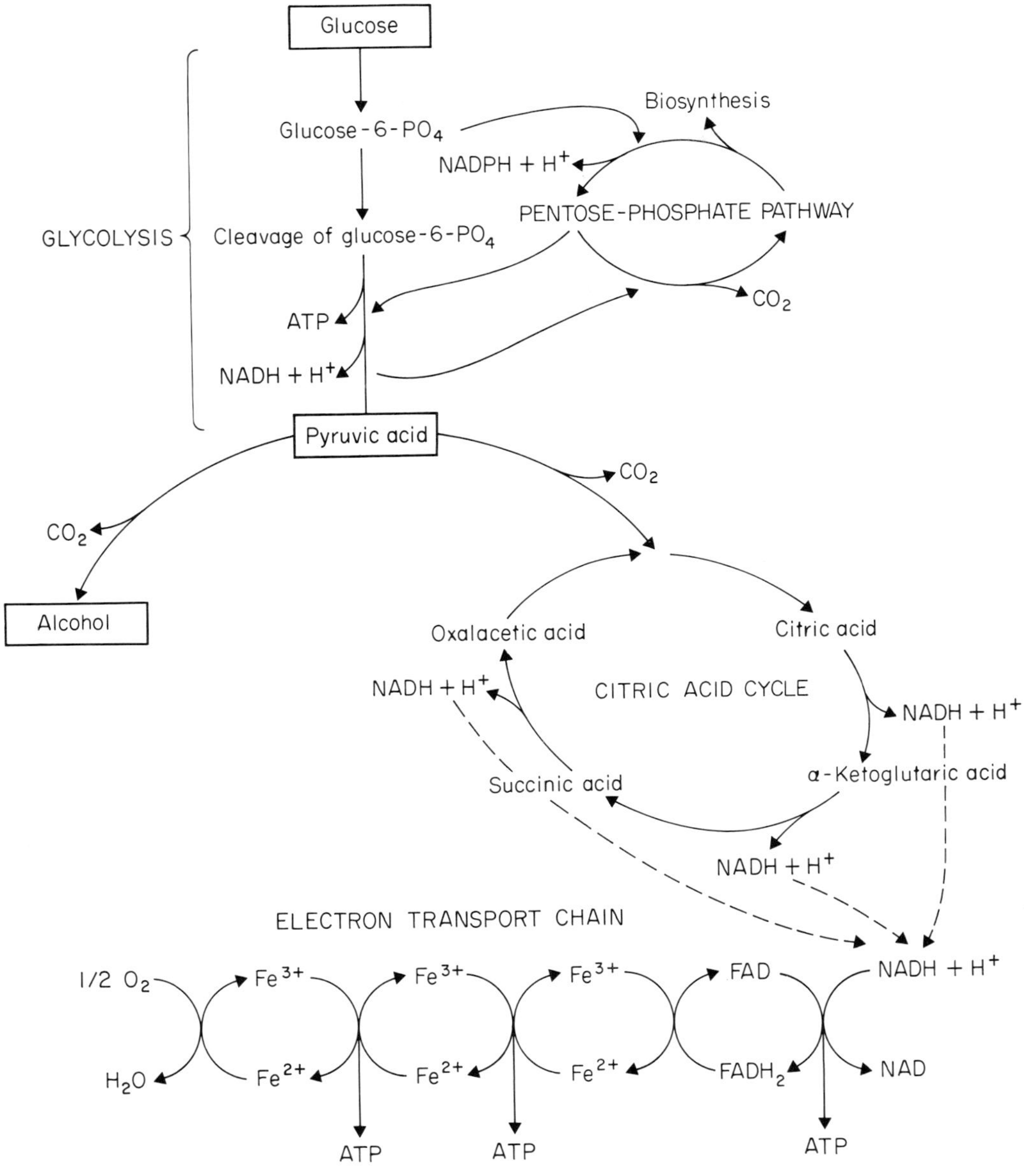

Figure 6.1. Some major steps in the overall process of respiration.

studies of enzyme activity and effects of various respiratory inhibitors in healthy and victorin-treated oat tissues. It was found that 2,4-dinitrophenol had no effect on victorin-treated oat tissue whereas it caused marked increases in oxygen uptake of untreated tissues. Also ascorbic acid oxidase activities were 2–4-fold greater in homogenates of victorin-treated than of untreated leaves and the levels of ascorbic acid in victorin-treated leaves were lower than in untreated leaves. These findings indicate that increased respiration in victorin-treated oat tissue is mediated by an ascorbic acid oxidative system.

In maize there is evidence of a role for uncoupling in the leakage of ions caused by the host-selective toxin of *Helminthosporium maydis* (Miller & Koeppe 1971). Culture filtrates increased respiration and oxidation of malate and caused swelling of mitochondria.

In enzyme-inhibitor studies of respiratory pathways in barley leaves infected with the net blotch pathogen, *Pyrenophora teres*, or exposed to its toxins, Smedegaard-Petersen (1980a) suggested that electron transport in infected leaves is mediated through a non-cytochrome oxidative system resulting in uncoupling of respiration from oxidative phosphorylation.

The uncoupling hypothesis has been challenged by results from experiments showing that victorin and other host-selective toxins caused not only increased respiration of infected tissues but also a rapid loss of ions occurring prior to the respiratory rise (Grimm & Wheeler, 1963; Black & Wheeler, 1966). Similarly Arntzen *et al.* (1973) found that corn tissue treated with the selective toxin of *Helminthosporium maydis* race T produced ion leakage after 10 min, whereas respiratory increases did not occur until after several hours. They suggested that the effect of the toxin on mitochondria is secondary rather than primary.

The fact that not only respiration but also membrane integrity may be affected early in infection complicates interpretation of uncoupling as a primary factor in disease development. However, as stated by Daly (1976), uptake and retention of ions are energy-dependent processes, and any interference with energy generation, membrane-associated or membrane-independent, can account for both leakage and alterations in respiration.

The mechanisms for increased respiration in diseased tissue may be quite different in diseases caused by biotrophic and necrotrophic pathogens (Smedegaard-Petersen, 1980b). In susceptible barley leaves inoculated with *E. graminis* f.sp. *hordei* the increase does not start until 3 days after inoculation and oxygen consumption remains high until conidial production stops and leaf senescence begins (Figure 6.2). In susceptible barley leaves affected by the necrotrophic net blotch pathogen, *Pyrenophora teres*, respiration started to increase within 24 h, reached a maximum after 7 days and then decreased rapidly as necrosis progressed. After 10 days leaves were almost entirely necrotic and there was no respiration (Figure 6.2).

These results support the view that enhanced oxygen uptake in diseases caused by biotrophic fungi is not directly a part of pathogenesis but rather a result of increased biosynthesis by which the host supplies the fungus with nutrients for its growth and sporulation. Also, biotrophic fungi create a metabolic sink which at the expense of host energy draws photoassimilates and other nutrients from surrounding tissues to the sites of infection (Shaw & Samborski, 1956; Pozsár & Király, 1966).

This is in contrast to the situation in diseases caused by necrotrophs such as the net blotch pathogen of barley where it is suggested that increased respiration is a result of a combined toxin-induced uncoupling of oxidative phosphorylation

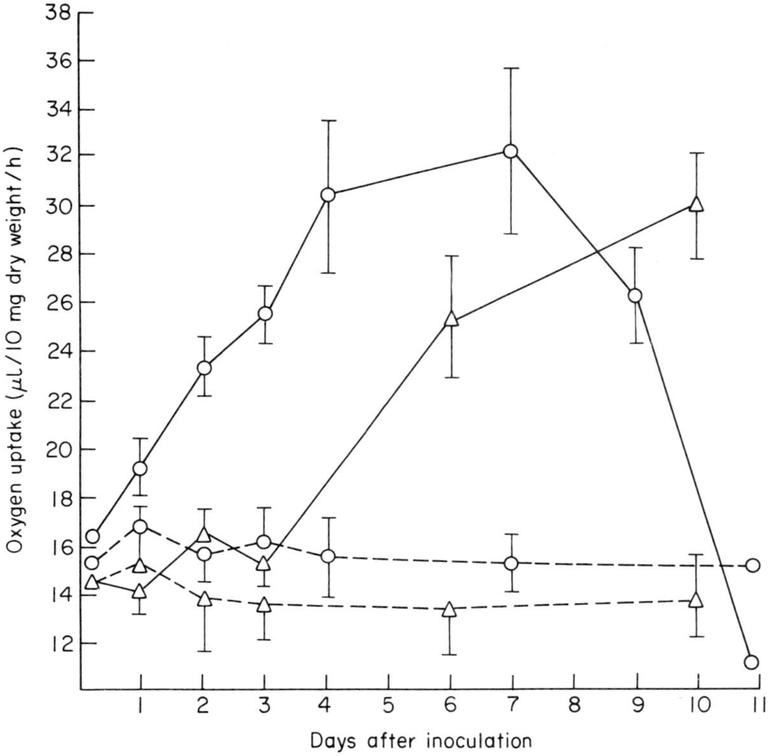

Figure 6.2. Time course of respiration in two susceptible barley cultivars, Wing and Sultan, inoculated respectively with the net blotch pathogen *Pyrenophora teres* (a necrotroph) and the powdery mildew pathogen *Erysiphe graminis* f.sp. *hordei* (a biotroph). Vertical bars represent standard deviations. ○——○ Inoculated with *Pyrenophora teres* ○----○ *Uninoculated control* △——△ *Inoculated with Erysiphe graminis* △----△ *Uninoculated control.*

(Smedegaard-Petersen 1977, 1980a) and an effect on the cell membranes (Smedegaard-Petersen *et al.*, 1984). In such cases increased respiration is not only wasteful expenditure of host energy, but is also directly involved in pathogenesis.

Altered balance of operation pathways and involvement of alternative pathways

The major respiratory pathway in healthy growing plant tissue is through glycolysis, the citric acid cycle and the cytochrome electron-transport chain.

An important feature of pathogen-induced respiration is an altered balance of operating pathways and activation of alternative pathways usually not or much less used in healthy plants.

Several lines of evidence clearly suggest a shift from the glycolytic to the pentose phosphate pathway, especially in diseases caused by biotrophic fungi. Thus Daly *et al.* (1957) found that oxygen uptake in safflower infected with *Puccinia carthami* was not inhibited by sodium fluoride, indicating the operation of a non-glycolytic pathway. In barley leaves infected with *E. graminis* f.sp. *hordei* Scott (1965) found a 2–3-fold increase in activity of enzymes of the pentose pathway but only slight increases in activity of glycolytic enzymes.

A decrease in the C_6/C_1 ratio of CO_2 respired from rust-affected safflower and from cereals infected with the rust and powdery mildew pathogens (Daly *et al.*, 1957; Shaw & Samborski, 1957) also suggests shifts from the glycolytic sequence in favour of the pentose phosphate pathway. As much as 90% of respiration in heavily rust-affected wheat occurred through this pathway (Shaw & Samborski, 1957).

The use of the C_6/C_1 ratio in determining the relative contributions of the pentose phosphate pathway in the degradation of glucose is based on the evolution of $^{14}CO_2$ respired from glucose-1-^{14}C and glucose-6-^{14}C. If glucose is oxidized via the glycolytic pathway then $^{14}CO_2$ should be produced at an equal rate from glucose-1-^{14}C and glucose-6-^{14}C and the C_6/C_1 ratio will be one. However, if glucose is oxidized via the pentose phosphate pathway the C_1 carbon will contribute most of the evolved CO_2 and the C_6/C_1 ratio will fall below one.

The increased participation by the pentose phosphate pathway in some diseases may be important for several reasons. This pathway is also known as the hexose monophosphate shunt or the direct oxidative pathway. During the initial phase of the cycle oxidation of glucose-6-phosphate is followed by decarboxylation of C-1 of the sugar phosphate. These reactions, in which NADP is reduced rather than NAD as in glycolysis and the citric acid cycle, yield ribulose-5-phosphate + CO_2 + 2NADPH. Subsequent reactions produce phosphorylated hexoses and trioses which may enter the usual respiratory pathway via the glycolytic sequence, or enter the pentose pathway.

The NADPH produced in the pentose phosphate pathway is the main source of reducing power in the biosynthetic processes of the plant. Further, the cycle produces a number of intermediates e.g. pentoses and erythrose-4-phosphate, which may be important in the biosynthesis of nucleic acids, phenols and other compounds with a role in pathogenesis as well as in resistance. Hence activation of this cycle in plants infected by biotrophic fungi may result from the much higher biosynthetic activity in infected tissues which supplies the fungus with nutrients.

Changes in the citric acid cycle and the electron transport chain also occur in diseased tissue. Inhibitors of respiratory enzymes are important tools in revealing such changes. Although some of these inhibitors are not specific in their effects they have been widely used with other methods to investigate the function of respiratory pathways. Thus malonic acid, a competitive inhibitor of succinic dehydrogenase, strongly inhibited oxygen uptake in healthy wheat plants but only slightly decreased respiration in rust-affected wheat (Farkas & Király, 1955). Since evidence was provided that malonate acted specifically in these experiments it was suggested that a pathway other than the Krebs cycle is involved in increased respiration in this disease. Similar results were obtained by Scheffer & Pringle (1963) for the effect on respiration of the *Helminthosporium victoriae* toxin. Although pure toxin increased respiration in susceptible oat seedlings it had no effect on the succinic oxidase activity of isolated mitochondria, indicating that the oxygen uptake was not mediated through the Krebs cycle.

Smedegaard-Petersen (1980a) found that increased respiration of barley plants infected with *Pyrenophora teres*, or treated with toxins of this pathogen, were sensitive to sodium fluoride, an inhibitor of enolase in glycolysis, and to malonate, an inhibitor of succinic dehydrogenase. This meant that inhibition was almost equal in diseased and in healthy leaves. Oxygen uptake inhibited by sodium fluoride could be partly restored by addition of pyruvate, and inhibition with malonate was almost entirely restored by addition of malate suggesting a specific effect of sodium fluoride on certain enzymes in glycolysis, presumably enolase, and by malonate on succinic dehydrogenase in the citric acid cycle. The results clearly indicate that the dramatic increase in respiration of barley plants infected with *P. teres* or its toxins (Smedegaard-Petersen, 1977, 1980a, 1980b) occurs through glycolysis and the citric acid cycle. Any substantial participation of the pentose phosphate pathway was excluded since this pathway is highly insensitive to sodium fluoride. This conclusion was supported by the finding that the respiratory quotient (RQ) was about one in healthy, diseased and toxin-treated leaves.

The results obtained with *P. teres* contrast with these for diseases caused by biotrophs which are characterized by a shift in favour of the pentose phosphate pathway (Daly *et al.*, 1957; Shaw & Samborski, 1957; Scott, 1965).

Further studies with inhibitors of terminal respiration showed that thiourea inhibited oxygen uptake in infected but not in healthy tissue (Smedegaard-Petersen, 1980a). Thiourea is considered to be a specific inhibitor of copper-enzymes such as ascorbic acid oxidase and polyphenol oxidase whereas it does not inhibit cytochrome oxidase which contains iron. Therefore thiourea has been used to distinguish the roles of cytochrome and copper oxidases (Hill & Hartree, 1953; James, 1953; Krupka, 1959). The results suggest that electron transport in barley infected with *P. teres* operates through a non-cytochrome oxidative system which results in uncoupling of respiration from oxidative phosphorylation.

Several non-cytochrome oxidative systems which transfer hydrogen (electrons) from reduced pyridine nucleotides to oxygen are known although their importance in normal respiration is still unclear. Thus Mapson & Goddard (1951) provided evidence for a non-cytochrome oxidative system which can transport hydrogen from NADPH to molecular oxygen according to the following scheme:

NADPH→glutathione→dehydroascorbic acid→ascorbic acid→O_2

Activity of non-cytochrome oxidizing enzymes such as ascorbic acid oxidase, polyphenol oxidase and peroxidase in diseased and toxin-affected plant tissue has been suggested by several investigators (Krupka, 1959; Grimm & Wheeler, 1963; Balasubramani *et al.*, 1971). For example, Krupka (1959) found that the marked increase of respiration in oat leaves treated with victorin from *Helminthosporium victoriae*, coincided with increased ascorbic acid oxidase activity. At the same time ascorbic acid decreased in victorin-treated leaves to about one third of the level found in untreated leaves. Similarly Király & Farkas (1957) found that an increased rate of respiration in wheat leaves infected with stem rust paralleled increased ascorbic acid oxidase activity. They suggested that the cytochrome oxidase system was replaced by a pathway using ascorbic oxidase as the terminal oxidase.

Respiration in resistant plants

Many workers have focussed on pathological changes in the respiratory patterns of infected susceptible plants but there has been much less research on infected resistant plants. There is, however, good evidence that incompatibility is associated with intense energy-dependent metabolic activities and increases in the rate of respiration of the plant. To illustrate the potential role of respiration in biosynthesis of compounds associated with incompatibility Daly (1976) calculated, with certain reservations, that if glycolysis is the source of both pyruvate and ATP, a minimum of 4.5 molecules of glucose must be metabolized to synthesize one molecule of rishitin in potato plants.

While workers agree generally about respiratory responses in diseased plants there is much less agreement for resistant plants even among authors working with the same host–pathogen combinations.

For wheat cultivars infected with *Puccinia graminis* f.sp. *tritici*, Samborski & Shaw (1956) found that incompatible combinations showed an earlier and sharper rise in respiration than did compatible combinations. In contrast, Heitefuss (1965) did not find a clear-cut correlation between compatibility and increase in respiration in combinations which included those with the temperature-sensitive Sr6 gene. Antonelli & Daly (1966) also did not observe respiratory differences for near-isogenic wheat lines with Sr6 and Sr6 alleles when inoculated with a specific race of *P. graminis* f.sp. *tritici*. Increases in respiration were parallel until the sixth day after inoculation; later respiration continued to increase only in the compatible combination. Bushnell (1970) found a slower increase in respiration in resistant than in susceptible cultivars.

Millerd & Scott (1956) used barley cultivars with varying degrees of resistance to powdery mildew. One highly resistant cultivar reacted with an earlier rise in respiration than did a susceptible cultivar, whereas in another resistant cultivar there was a decrease in respiration. For semi-resistant and susceptible varieties there were no differences in oxygen uptake during the first 96 h after inoculation.

Scott & Smillie (1966) used leaves of two near-isogenic barley lines infected with *Erysiphe graminis* f.sp. *hordei*. In a highly incompatible combination there was only a slight increase compared to controls. In a compatible combination there were no marked differences between infected and non-infected leaves until 96 h after inoculation. Later respiration was higher in infected leaves.

One reason for the inconsistencies in the results may be that most of the comparisons, except for wheat stem rust and the Sr6 gene (Daly, 1972; Heitefuss,

1965), were made with cultivars of quite different or unknown genetic backgrounds, as pointed out by Johnson (1976) and Daly (1972).

Energy-requiring processes reduce grain yield and quality in barley inoculated with avirulent races of powdery mildew

Interpretation of metabolic changes in infected plants, including changes in respiration, is complicated by the findings that such changes often occur in both susceptible and resistant plants. However, the fact that increases in respiration typically occur earlier in resistant than in susceptible tissues suggests that different mechanisms may be operating. Comparing the respiratory rates of compatible and incompatible host–pathogen interactions, Smedegaard-Petersen (1980b) found that hypersensitivity was associated with a rapid temporary increase in respiration of resistant plants which always occurred earlier in resistant than in susceptible plants. He suggested that disease resistance based on incompatibility is associated with energy-requiring processes — presumably involving defence reactions — which deprive the host of energy and ultimately lead to a yield reduction. Later research with barley powdery mildew has confirmed this hypothesis (Smedegaard-Petersen & Stolen, 1981).

Figure 6.3a shows respiration in the barley cultivar Sultan following a single inoculation with an avirulent and a virulent race of *Erysiphe graminis*. The resistant leaves reacted to inoculation with a very rapid temporary increase starting as early as 12 h after inoculation and returning to about the level of uninoculated controls after 3 days. Figure 6.3b shows a time course study of the same combination in which there were three inoculations at intervals of 2 days (0, 2 and 4 days). Resistant leaves again reacted with a rapid increase which stabilized at a level significantly higher than that of non-inoculated controls. Thus repeated inoculation of barley with an avirulent race caused a permanent increase in respiration.

To establish a role for the respiratory increase in resistance it is important to note that increased oxygen uptake in incompatible tissue can be detected 12 h after inoculation and that it coincides with the time at which incompatibility between barley and *E. graminis* is established (Ellingboe, 1972) and with a series of structural and biochemical events in the host plant, including synthesis of antifungal compounds (Oku *et al.*, 1975; Ouchi *et al.*, 1974), increased synthesis of nucleic acids (Oku *et al.*, 1973), accumulation of new, aromatic substances (Mayama & Shishiyama, 1978), increased peroxidase and ethylene production (Hislop & Stahmann, 1971) and changes in the content of carotenoids (Haspel-Horvatovic & Paulech, 1969).

Although the role of such activities in resistance is debatable, there is good experimental evidence that at least some are involved in defence reactions. Thus Ouchi *et al.* (1974) demonstrated that barley plants inoculated with an avirulent race of *E. graminis* became partially resistant to a race originally virulent, suggesting

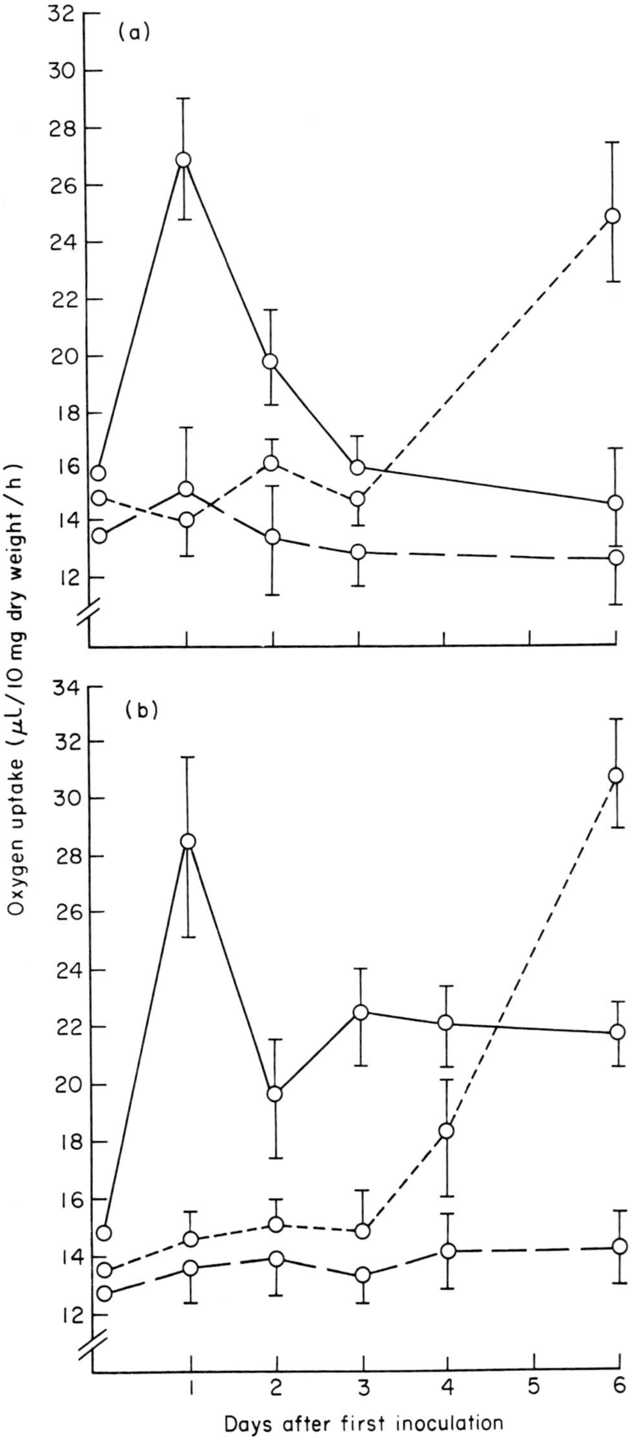

Figure 6.3. Respiration in compatible and incompatible barley/powdery mildew interactions. (a) Plants inoculated once (at zero time) (b) Plants inoculated three times (at 0, 2 and 4 days). Vertical bars represent standard deviations. ○——○ Incompatible ○----○ Compatible ○—·—○ Control.

production of antifungal compounds in response to infection by the avirulent race. Later Oku *et al.* (1975) detected high phytoalexin activity 12–20 h after inoculation in an incompatible reaction.

It is well known that increases in oxygen uptake do not necessarily result in the production of energy. Breakdown of compartmentation in cells may allow enzymes such as peroxidases, polyphenol oxidases, and ascorbic acid oxidases to function and contribute to the oxygen uptake. These kinds of reactions are, however, most prominent during cellular degradation in the later stages of infection and it is unlikely that they could account for marked increases in oxygen uptake during the initial stages of infection.

To investigate whether increased respiration in resistant barley plants inoculated with *E. graminis* consumes host energy to such an extent that growth and yield are influenced, a series of experiments were carried out in growth chambers under highly controlled conditions (Smedegaard-Petersen & Stolen, 1981). During the growing season plants were continually inoculated with either an avirulent or a virulent race. Although plants remained free of symptoms after inoculation with the avirulent race, grain yield was reduced by 7%, kernel weight by 4% and straw length by 5.7 cm. The content of grain protein was reduced from 9.75% to 9.38%, equivalent to 11% reduction in the yield of grain protein. In susceptible, infected plants the grain yield was reduced by 26%, the kernel weight by 11% and straw length by 4.9 cm. The protein content was reduced from 9.75% to 9.25%, equivalent to a reduction in the yield of grain protein by 30%.

In interpreting these results a consideration of inoculum density may be important. Since equal amounts of inoculum of the two races were applied to plants in the growth chambers, a measure of the inoculum density is that 56% of the total leaf area in susceptible plants was covered with mildew at heading.

The fact that highly resistant barley plants did not suffer visible damage by inoculation thus does not mean that the plants were unaffected. The results indicate that resistant plants reacted to inoculation by energy-requiring physiological reactions, probably defence reactions, which caused a drain on the host energy available for other work and ultimately decreased grain yield and quality.

Effect of leaf saprophytes on grain yield of barley

During the growing season the lower, dead leaves in grain crops are colonized by a number of different saprophytic fungi, the spores of which are deposited on the upper, green leaves. Floristic and cytological studies (Tolstrup & Smedegaard-Petersen, 1984) have shown that *Cladosporium* spp. and yeasts are the most common saprophytic fungi on barley leaves in Denmark. Although considered as saprophytes, *Cladosporium* spp. can penetrate epidermal cells of leaves causing stainable haloes around penetration sites and significant increases in rates of respiration, much as do avirulent conidia of powdery mildew. Field trials and extensive floristic studies indicated that the saprophytic leaf fungi caused up to 16% significant reduction in grain yield, although no disease symptoms occurred.

The yield reductions varied from year to year depending on the effects of cultural and environmental conditions for propagation of the saprophytes.

The results summarized above suggest that energy-requiring and yield-reducing physiological reactions are common when resistant plants respond not only to recognized pathogens, but also to saprophytic micro-organisms which normally occur on their surfaces and which until now have been considered as harmless.

References

Allen P.J. (1953) Toxins and tissue respiration. *Phytopathology* **43**, 221–29.

Allen P.J. & Goddard D.R. (1938) A respiratory study of powdery mildew of wheat. *American Journal of Botany* **25**, 613–21.

Antonelli E. & Daly J.M. (1966) Decarboxylation of indoleacetic acid by near-isogenic lines of wheat resistant or susceptible to *Puccinia graminis*. *Phytopathology* **56**, 610–8.

Arntzen C.J., Koeppe D.E., Miller R.J. & Peverly J.H. (1973) The effect of pathotoxin from *Helminthosporium maydis* (race T) on energy-linked processes of corn seedlings. *Physiological Plant Pathology* **3**, 79–89.

Balasubramani K.A., Deverall B.J. & Murphy J.V. (1971) Changes in respiratory rate, polyphenoloxidase and polygalacturonase activity in and around lesions caused by *Botrytis* in leaves of *Vicia faba*. *Physiological Plant Pathology* **1**, 105–13.

Black H.S. & Wheeler, H. (1966) Biochemical effects of victorin on oat tissues and mitochondria. *American Journal of Botany* **53**, 1108–12.

Bushnell W.R. (1970) Patterns in the growth, oxygen uptake, and nitrogen content of single colonies of wheat stem rust on wheat leaves. *Phytopathology* **60**, 92–9.

Daly J.M. (1972) The use of near-isogenic lines in biochemical studies of the resistance of wheat to stem rust. *Phytopathology* **62**, 392–400.

Daly J.M. (1976) The carbon balance of diseased plants: changes in respiration, photosynthesis and translocation. In: *Physiological Plant Pathology* (Ed. by R. Heitefuss & P.H. Williams), pp. 450–79. Springer-Verlag, Berlin.

Daly J.M., Sayre R.M. & Pazur J.H. (1957) The hexose monophosphate shunt as the major respiratory pathway during sporulation of rust of safflower. *Plant Physiology* **32**, 44–8.

Ellingboe A.H. (1972) Genetics and physiology of primary infection by *Erysiphe graminis*. *Phytopathology* **62**, 401–6.

Farkas G.L. & Király Z. (1955) Studies on the respiration of wheat infected with stem rust and powdery mildew. *Physiologia Plantarum* **8**, 877–87.

Grimm R.B. & Wheeler H. (1963) Respiratory and enzymatic changes in Victoria blight of oats. *Phytopathology* **53**, 436–40.

Haspel-Horvatovic E. & Paulech C. (1969) Changes of oxidized and reduced carotenoids, of the respiration and the photosynthesis of barley (*Hordeum sativum* L.) in connection with its resistance against powdery mildew. In: *Progress in Photosynthesis Research* (Ed. by H. Metzner), pp. 396–401. Tübingen.

Heitefuss R. (1965) Untersuchungen zur Physiologie des temperaturgesteuerten Verträglichkeitsgrades von Weizen und *Puccinia graminis tritici*. I. Veränderungen von Sauerstoffaufnahme und Phosphatstoffwechsel. *Phytopathologische Zeitschrift* **54**, 379–400.

Heitefuss R. & Fuchs W.H. (1963) Phosphatstoffwechsel und Sauerstoffaufnahme in Weizenkeimpflanzen nach Infektion mit *Puccinia graminis tritici*. *Phytopathologische Zeitschrift* **46**, 174–98.

Hill R. & Hartree E.F. (1953) Hematin compounds in plants. *Annual Review of Plant Physiology* **4**, 115–50.

Hislop E.C. & Stahmann M.A. (1971) Peroxidase and ethylene production by barley leaves infected with *Erysiphe graminis* f.sp. *hordei*. *Physiological Plant Pathology* **1**, 297–312.

James W.O. (1953) The use of respiratory inhibitors. *Annual Review of Plant Physiology* **4**, 59–90.

Johnson R. (1976) Development and use of some genetically controlled lines for studies of host-parasite interactions. In: *Biochemical Aspects of Plant-Parasite Relationships.* (Ed. by J. Friend & D.R. Threlfall), pp. 25–41. Academic Press. New York.

Király Z. & Farkas G.L. (1957) On the role of ascorbic oxidase in the parasitically increased respiration of wheat. *Archives of Biochemistry and Biophysics* **66**, 474–85.

Krupka L.R. (1959) Metabolism of oats susceptible to *Helminthosporium victoriae* and victorin. *Phytopathology* **49**, 587–94.

Mapson L.W. & Goddard D.R. (1951) The reduction of glutathione by plant tissues. *Biochemical Journal* **49**, 592–601.

Mayama S. & Shishiyama J. (1978) Localized accumulation of fluorescent and u.v.-absorbing compounds at penetration sites in barley leaves infected with *Erysiphe graminis hordei. Physiological Plant Pathology* **13**, 347–54.

Miller R.J. & Koeppe D.E. (1971) Southern corn leaf blight: susceptible and resistant mitochondria. *Science* **173**, 67–9.

Millerd A. & Scott K. (1956) Host pathogen relations in powdery mildew of barley. II. Changes in respiratory pattern. *Australian Journal of Biological Sciences* **9**, 37–44.

Oku H., Ouchi S. & Sato M. (1973) Nucleic acid metabolism in powdery mildew-infected barley leaves as associated with symptom development. *Report Tottori Mycological Institute,* Japan **10**, 511–6.

Oku H., Ouchi S., Shiraishi T., Komoto Y. & Oki K. (1975) Phytoalexin activity in barley powdery mildew. *Annales of the Phytopathological Society of Japan* **41**, 185–91.

Ouchi S., Oku H., Hibino C. & Akiyama I. (1974) Induction of accessibility and resistance in leaves of barley by some races of *Erysiphe graminis. Phytopathologische Zeitschrift* **79**, 24–34.

Pozsár B.I. & Király Z. (1958) Effect of rust infection on oxidative phosphorylation of wheat leaves. *Nature* (London) **182**, 1686–7.

Pozsár B.I. & Király Z. (1966) Phloem-transport in rust-infected plants and cytokinin-directed long distance movement of nutrients. *Phytopathologische Zeitschrift* **56**, 297–309.

Samborski D.J. & Shaw M. (1956) The physiology of host-parasite relations. II. The effect of *Puccinia graminis tritici* Eriks. and Henn. on the respiration of the first leaf of resistant and susceptible species of wheat. *Canadian Journal of Botany* **34**, 601–19.

Scheffer R.P. & Pringle R.B. (1963) Respiratory effects of the selective toxin of *Helminthosporium victoriae. Phytopathology* **53**, 465–8.

Scott K.J. (1965) Respiratory enzymic activities in the host and pathogen of barley leaves infected with *Erysiphe graminis. Phytopathology* **55**, 438–41.

Scott K.J. & Smillie R.M. (1966) Metabolic regulation in diseased leaves. I. The respiratory rise in barley leaves infected with powdery mildew. *Plant Physiology* **41**, 289–97.

Shaw M. & Samborski D.J. (1956) The physiology of host-parasite relations. I. The accumulation of radioactive substances at infections of facultative and obligate parasites including tobacco mosaic virus. *Canadian Journal of Botany* **34**, 389–405.

Shaw M. & Samborski D.J. (1957) The physiology of host-parasite relations. III. The pattern of respiration in rusted and mildewed cereal leaves. *Canadian Journal of Botany* **35**, 389–407.

Smedegaard-Petersen V. (1977) Respiratory changes of barley leaves infected with *Pyrenophora teres* or affected by isolated toxins of this fungus. *Physiological Plant Pathology* **10**, 213–20.

Smedegaard-Petersen V. (1980a) Enzyme-inhibitor studies of respiratory pathways in barley leaves infected with *Pyrenophora teres* and affected by its isolated toxins. *Royal Veterinary and Agricultural University, Yearbook* 1980, 23–35.

Smedegaard-Petersen V. (1980b) Increased demand for respiratory energy of barley leaves reacting hypersensitively against *Erysiphe graminis, Pyrenophora teres* and *Pyrenophora graminea. Phytopathologische Zeitschrift* **99**, 54–62.

Smedegaard-Petersen V. & Stolen O. (1981) Effect of energy-requiring defense reactions on yield and grain quality in a powdery mildew-resistant barley cultivar. *Phytopathology* **71**, 396–9.

Smedegaard-Petersen V., Heide M. & Rajagopal R. (1984) Changes in permeability and lipid composition of plasmamembranes of barley leaves infected with *Pyrenophora teres* or treated with *P. teres* toxin A. (Submitted 1984.)

Tolstrup K. & Smedegaard-Petersen V. (1984) Saprophytic leaf fungi on barley and their effect on leaf senescence and grain yield. *Växtskyddsnotiser* 1984. (In press.)

7 Effects of pathogens on plant transport systems

J.F. FARRAR
School of Plant Biology, University College of North Wales, Bangor, Gwynedd LL57 2UW, UK

Introduction

An understanding of the response of plant transport systems to pathogens must be sought in the nature of the transport systems and the nature of pathogenicity. Since neither is understood independently, their interaction is obviously an area of considerable ignorance. The transport systems of a higher plant will be described first, and then how a pathogen might perturb them.

A simple classification of plant transport systems is given in Figure 7.1. Of the four systems, vapour-filled intercellular space is of most importance in the gas exchange of illuminated leaves and so is covered by Habeshaw (this volume). Transport in the apoplast is largely outside the control of the plant, being mainly a function of water potential gradients imposed by the condition of the soil and atmosphere; pathogen attack can clearly change water status and lead to changes in liquid and vapour phase resistances, as discussed by Ayres (this volume). Only the consequences of such changes for the transport of nutrients in the xylem will be considered; the bulk of this chapter will concern transport in the symplast and across membranes that separate the symplast from the apoplast and from organelles.

For long distance transport it is possible to consider the plant to have a primitive circulation system, with ascent of water in the apoplast (xylem) and descent in the symplast (phloem), coupled to a net upward transfer of water (Figure 7.1). Membrane transport is required for phloem loading. The total amount of a substance within a leaf is simply the integral of differences between import and export in the long-distance transport system.

Each of these transport systems can be altered by pathogens or other factors in one of three ways: physical damage, changes in mass transfer, or change as a result of non-nutritional chemical effects. The first two are simple and perhaps the most probable. The third might involve plant growth regulators (PGRs), phytoalexins, and toxins, which I shall here group and call 'modulators'. This might appear a likely mechanism, but perhaps only because our ignorance of the mechanism of action of PGRs is such that almost any role can be attributed to them. Caution is needed here.

What types of evidence are likely to increase our understanding of the effects of

Wood R.K.S. & Jellis G.J. (1984) *Plant Diseases: infection, damage and loss.* Blackwell Scientific Publications, Oxford.

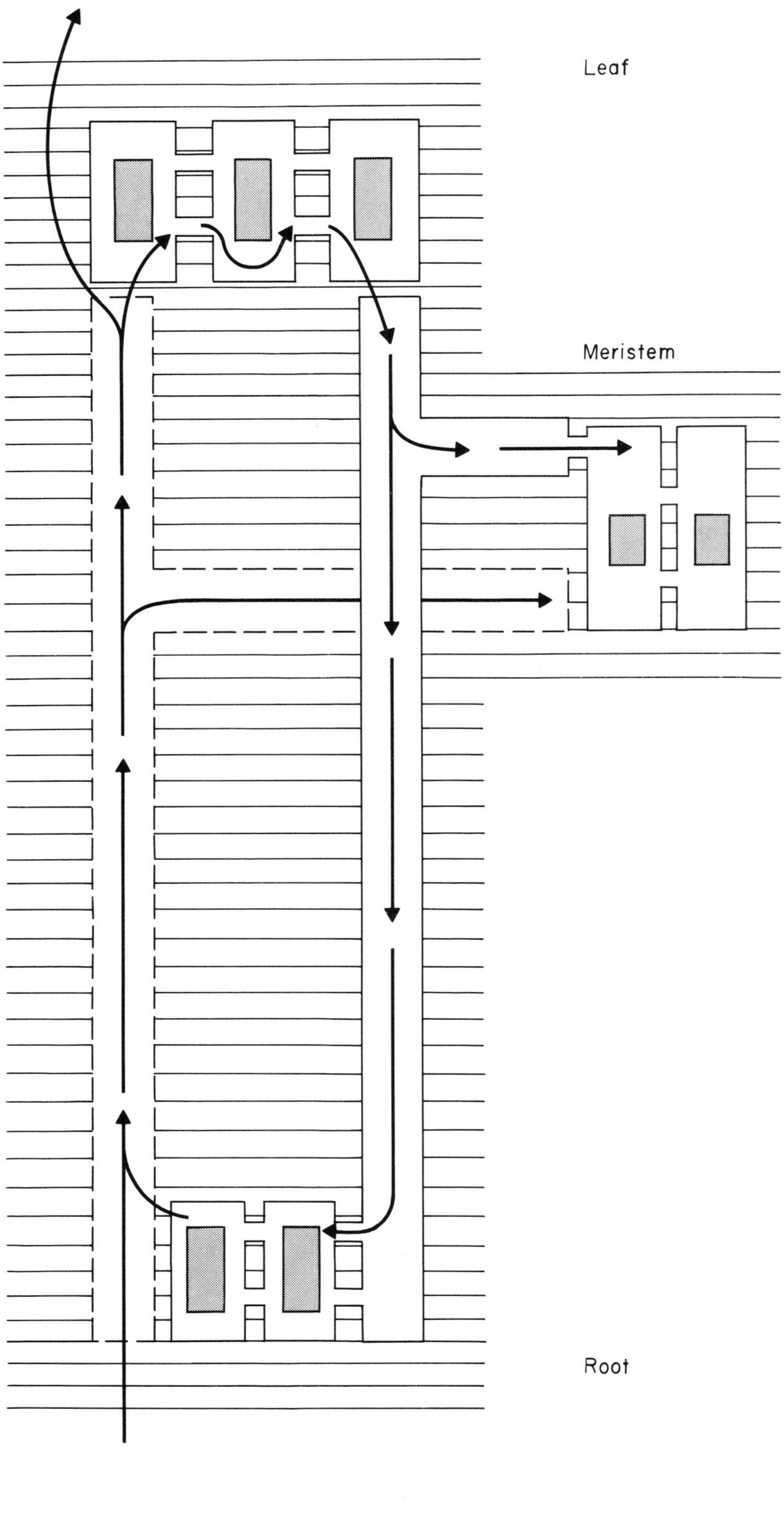
Leaf
Meristem
Root

pathogens on transport? Two criteria, it seems, must be met:
(a) Experiments must have a short time course. In a plant 1 m high, a molecule of Pi could move from top to bottom and back in about 2 h; an experiment on transport with its first harvest 24 h after the start will not even allow the transport system involved to be identified.
(b) Experiments must be truly quantitative. It is necessary to partition changes in transport between the different ways in which disease affects it. Gross, rather than net, fluxes need to be measured. Tracer work alone is not truly quantitative as it does not allow for dilution of the isotope. Also, there may be advantages in measuring a number of fluxes simultaneously for a single system, since each will place constraints on the interpretation of the others.

Unfortunately these two basic criteria are met in rather few published experiments so claims, for example, that changed cytokinin levels cause the altered translocation patterns that follow disease should be treated with caution. Some of the available evidence will now be considered with a view to evaluating how disease might affect transport in apoplast and symplast and across membranes. The emphasis will be on leaf-inhabiting biotrophic pathogens, for which much relevant evidence is available, but with occasional reference to necrotrophs.

The nutrient content of organs: long distance transport

Many reports state that there is a rise in the concentration of a number of nutrients in infected leaves (Yarwood & Jacobsen, 1955; Huber, 1978). Leaving aside for the moment the possibility that the pathogen accumulates nutrients, it seems clear that the nutrients must enter the infected leaf at a higher, or leave it at a lower, rate than in healthy plants.

Import in the xylem

The flux of xylem sap into a leaf will increase if transpiration increases, possibly due to cuticle rupture. If the concentration of nutrients in the xylem is unchanged,

Figure 7.1 (*opposite*). A simplified representation of the transport systems of higher plants. The four main systems are:

1. Symplast (clear), separated by plasma membranes from apoplast and vacuoles (shaded), and with transport driven by thermal diffusion, cyclosis and possibly pressure flow in the phloem.
2. Apoplast (horizontally hatched) consisting of all walls and a high-conductance pathway in the xylem (enclosed by dashed lines). Transport is by thermal diffusion and bulk flow along water potential gradients.
3. Membranes (shown by continuous lines) which form barriers delimiting the symplast. Transport is by thermal diffusion (permeability), facilitated diffusion and active transport.
4. Vapour-filled intercellular spaces (not shown), of central importance for gaseous exchange in leaves and bulky organs.

Water flux is shown by arrows.

the flux of nutrients into a diseased leaf will increase proportionately.

Data for barley infected with brown rust (Owera *et al.*, 1981: Ahmad *et al.*, 1982) show that on the day of sporulation, leaf diffusive resistance is lowered in diseased leaves whereas the composition of xylem sap is unchanged except for an increase in nitrate (Table 7.1). Thus an increase in import of these nutrients would be predicted. For good, quantitative measurements a knowledge of the diurnal course of both xylem sap concentration and transpiration rate is needed, as at least some foliar diseases affect root behaviour (Ayres, this volume) and some seem to lower the sensitivity of stomata to internal and environmental conditions (Ayres, 1981a).

Export in the phloem

The simplest evidence showing reduced phloem export from diseased leaves comes from applying isotopes to diseased leaves. The subsequent loss of isotope to the rest of the plant is far lower than from uninfected leaves. For more precise work, carbon budgets can be used to show the export of organic carbon from infected leaves by measuring phloem translocation as carbon gained in photosynthesis less carbon stored in leaf plus carbon lost in respiration (Owera *et al.*, 1981). If the composition of phloem sap is relatively unchanged, any phloem-mobile element will show reduced export from an infected leaf and thus a rise in amount in the leaf. So if this mechanism were the major cause of nutrient accumulation, phloem-mobile elements such as K should accumulate more than phloem immobile ones such as Ca and Mg. Rate of export can be calculated for any nutrient by comparing the rate at which it and Ca increase in amount in the leaf (Ahmad *et al.*, 1982). In this way it can be shown that the rise in content of K in rust-infected barley leaves can be completely explained by reduced bulk flow of phloem from the leaf (Table 7.2). Ca and Mg contents are not higher than in controls, since they are virtually immobile in the phloem. However, the reduction in retranslocation of N and P accounts for only

Table 7.1. Diffusive resistance (s/m) and xylem sap concentrations (mol/m^3) for barley first leaves 8 days (at sporulation) after infection with brown rust.

	Control	Rusted
Leaf diffusive resistance	763	470*
Mg^{2+}	2.6	2.3
Ca^{2+}	1.3	1.2
K^+	45.3	51.9
PO_4^{2-}	5.7	4.1
NO_3^-	3.1	5.7*
amino-acids (as leucine)	7.6	10.4

Data from Owera *et al.* (1981) and Ahmad *et al.* (1982).

* Significantly different from control (P = 0.05).

about one-half and one-third respectively of the accumulation of these elements, and an additional explanation is needed for these elements (Table 7.2). The lack of ^{32}P retranslocation from diseased first leaves following ^{32}P feeding to the roots whereas control leaves lost about 25% of their ^{32}P (Figure 7.2), confirms the immobility of P.

Clearly, reduced export from diseased leaves will be a major cause of altered transport patterns in the plant as a whole. Each of the three modes of perturbing transport systems listed above suggests a possible cause for this reduction, namely physical destruction or blockage of phloem tissue, reduced supply of carbohydrate, and effects of modulators on phloem loading and transport. Of these, the second perhaps needs explanation. Whereas there is no agreement on the mechanism of phloem transport, there is a consensus view (but not firm evidence) that the direction and velocity of phloem transport is controlled by sucrose concentration gradients (Moorby, 1977; Cronshaw, 1981); thus if infection results in a reduced supply of sucrose to the phloem, mass transfer in the phloem would be reduced accordingly.

There is evidence for physical damage to phloem tissue. The biotroph *Puccinia poarum* in leaves of *Tussilago farfara* infects phloem tissue, including companion cells but not sieve elements (Al Khesraji *et al.*, 1980); also in virus-infected plants, considerable phloem disruption can occur (Favali *et al.*, 1978). Currently it is impossible to quantify the effects of physical damage to transport systems, but at least the extent of the damage can be determined.

The other two possible ways in which transport systems can be affected are harder to separate. The approach in brown rust of barley has been by analysis of experiments in which $^{14}CO_2$ is fed briefly to leaves and the amount of ^{14}C remaining in the leaf is then recorded continuously, giving data of the kind shown in Figure 7.3. These curves can be used to estimate pool sizes of sugars involved with transport, and the rates at which these pools are mobilized, by the method of compartmental analysis. Of interest here is the pool of readily transported carbon (Figure 7.3). The size of this pool is halved in rusted leaves but the rate at which each unit of this pool is translocated is not changed significantly. This is a strong indication that the

Table 7.2. Export in the phloem (μmol/leaf/d) from first leaves of barley 8 days after infection with brown rust.

Element	Control	Rusted	% reduction due to rust
C	479	326*	32
K	4.5	3.6*	19
N	3.3	0.7*	79
P	0.57	0.01*	98

Data for carbon from Owera *et al.* (1981), and for other nutrients from Ahmad *et al.* (1982).

* Significantly different from control ($P = 0.05$).

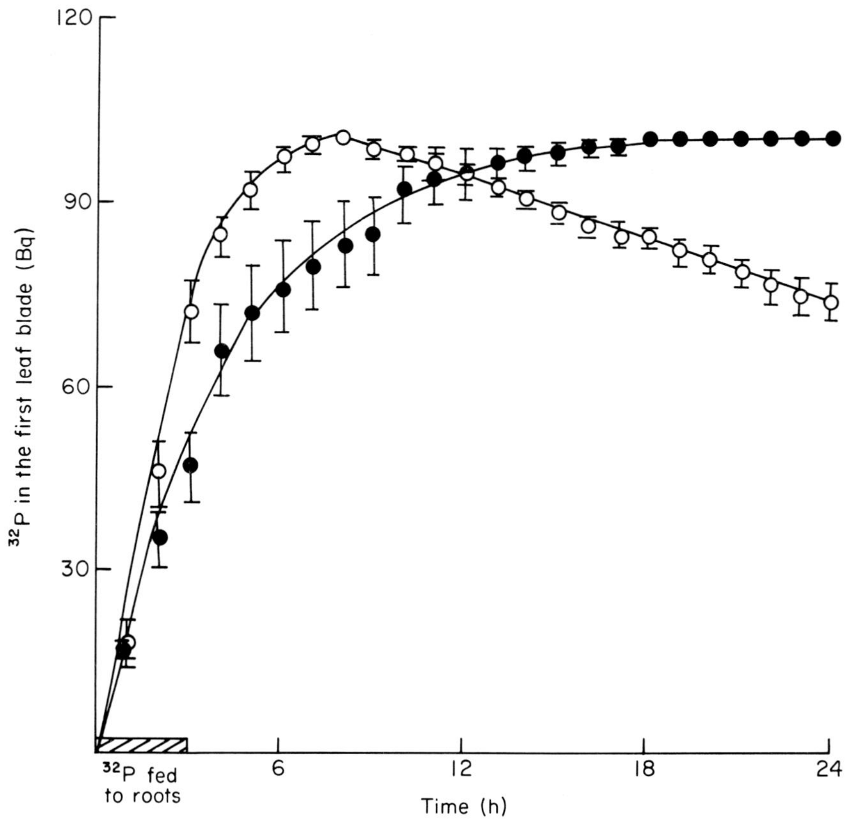

Figure 7.2. The reduction of ^{32}P export from rust-infected barley leaves. On the day of sporulation, barley plants with (●) and without (○) their first leaves infected with brown rust were supplied with 80 GBq of $KH_2\,^{32}PO_4$ to their root system for 3 h, after which the roots were washed and returned to Long Ashton solution for a further 21 h. The ^{32}P status of the first leaf blade was monitored continuously with a Geiger-Muller tube applied to its undersurface.
Unpublished data of I. Ahmad, J.F. Farrar & R. Whitbread

rust affects translocation by reducing sugar supply but not by interfering with phloem loading.

On evidence currently available, fungal effects on translocation seem to be confined to physical disruption of phloem and a reduction in sucrose supply. That modulators seem not to be involved accords with lack of effects of PGRs on translocation from wheat leaves (Dale *et al.*, 1981). However, compounds such as fusicoccin which act on membrane ATPases (Daly, 1981) would be expected to affect phloem loading profoundly.

Figure 7.3 (*opposite*). Carbon fluxes in first leaves of barley infected with brown rust.
(a) The kinetics of ^{14}C efflux from healthy (○) and rust-infected (●) barley first leaves, following a 20 min pulse of $^{14}CO_2$.
(b) The flux of carbon through a transport pool, derived by a compartmental analysis of the data in (a). The amount of sucrose in the transport pool is given in mg/m^2 leaf area, and the rate constants for loss from this compartment in h. The product of pool size and rate constant gives the carbohydrate flux (Owera, Farrar & Whitbread, 1983).

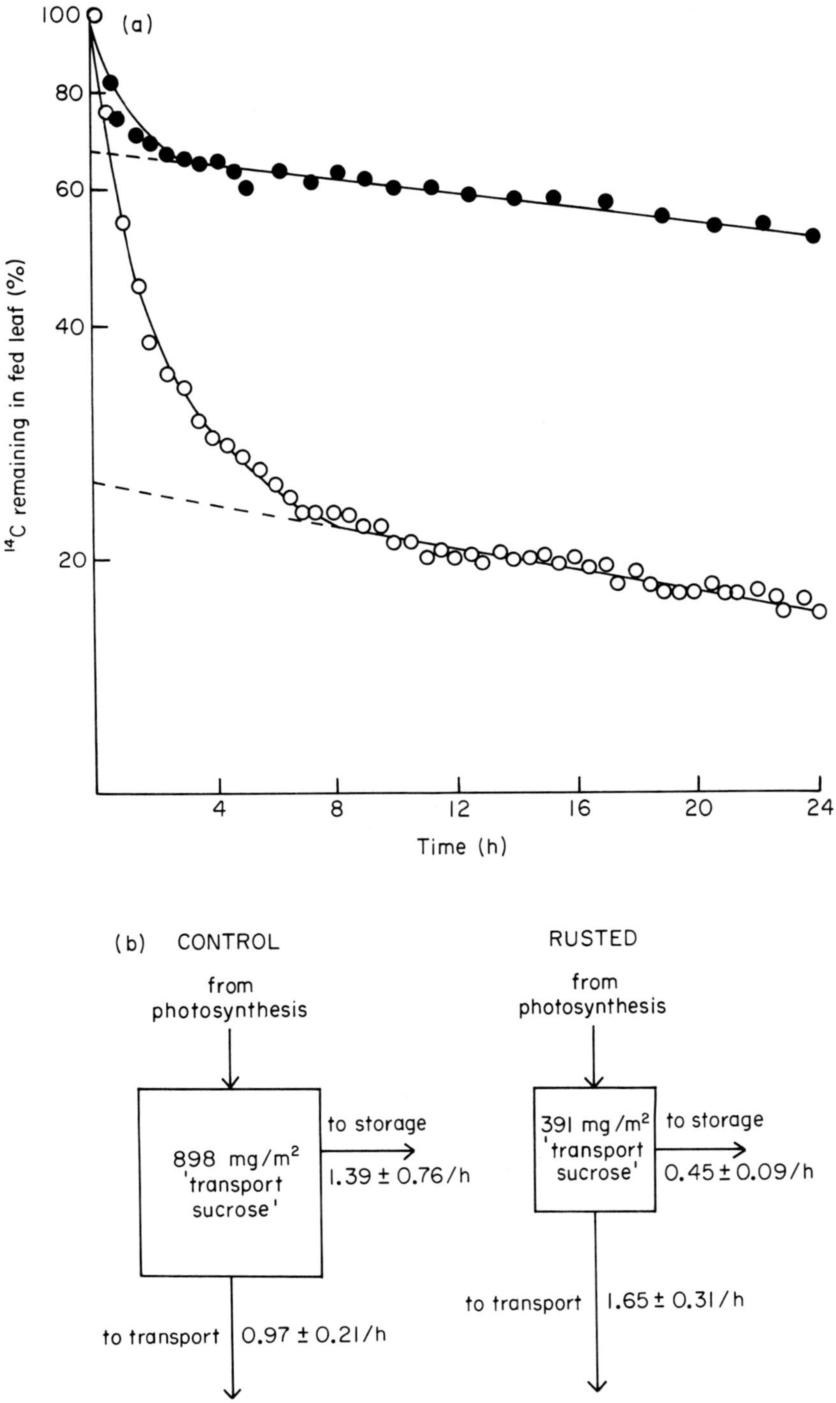
(a)
100
80
60
40
20
^{14}C remaining in fed leaf (%)
4
8
12
16
20
24
Time (h)
(b) CONTROL
from photosynthesis
898 mg/m²
'transport sucrose'
to storage
1.39 ± 0.76/h
to transport
0.97 ± 0.21/h
RUSTED
from photosynthesis
391 mg/m²
'transport sucrose'
to storage
0.45 ± 0.09/h
to transport
1.65 ± 0.31/h

Import in the phloem

From time to time it has been suggested that a net import of material into phloem is necessary to account for the accumulation of material at infection sites, although this argument has not been supported by sufficiently precise experiments.

In grasses and cereals, there is little evidence from any system that suggests phloem import into mature leaves; although looked for (Edwards, 1971) it seems not to occur. In further work with barley two kinds of evidence suggest that this is the case. Firstly, the percentage reduction in ^{14}C export from a diseased leaf is very similar to the percentage reduction in export calculated from a carbon budget which assumes only export in the phloem, and there is enough C left in the leaf to account for fungal growth and starch synthesis (Owera *et al.*, 1981, 1983). Secondly, when $^{14}CO_2$ is fed to a non-infected leaf, very little ^{14}C appears in the infected leaf

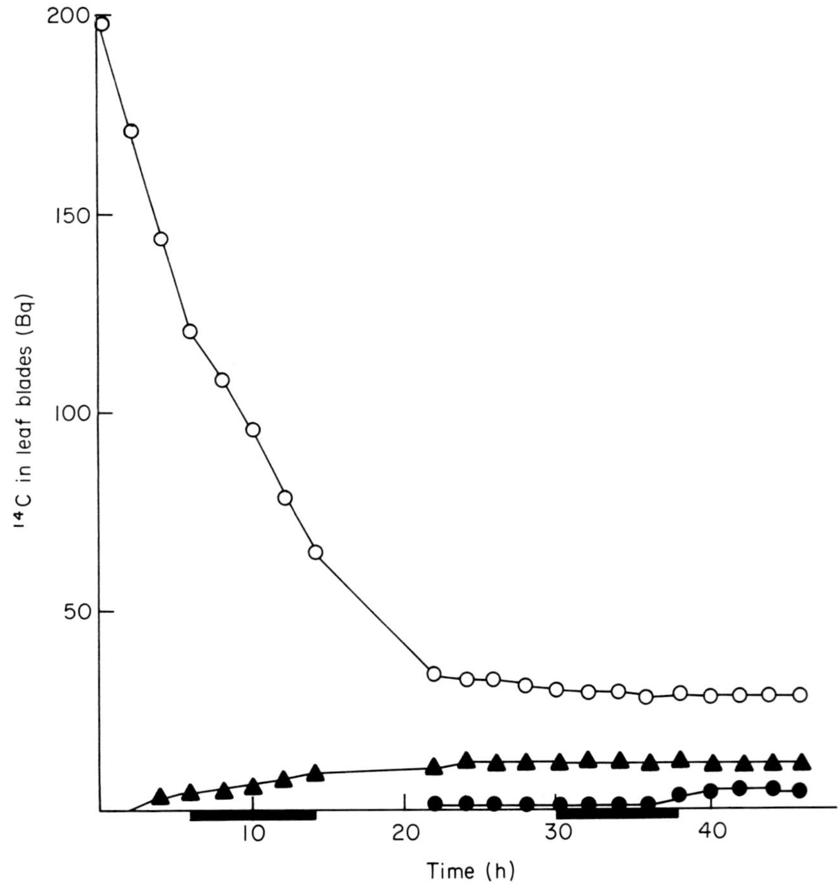

Figure 7.4. Partitioning of ^{14}C from a healthy second leaf (○) of a barley plant in which the first leaf (●) was already infected with brown rust and the third leaf (▲) was just developing. Following a 20 min pulse of $^{14}CO_2$, the three leaf blades were each monitored for 15 minutes in each hour by Geiger-Muller tubes attached to an automatic switching unit. Conditions: 20°C, and a photon fluence rate of 220 μmol/m^2/s for 16 h/d (night periods are indicated by bars on the X-axis).

(Figure 7.4), and this could be due to import of ^{14}C-amino acids to the infected leaf in the xylem, following their production in the root.

In dicotyledons, phloem transport may be more readily reversible, but the major question is still one of carbon balance. Starch which might be presumed available for remobilization to drive carbon export in the phloem at night could make import in the dark unlikely. However, there is evidence that the high concentrations of glucans at infection sites are not subject to diurnal fluctuations (Whipps & Lewis, 1981) and so these glucans may behave quite differently from starch in healthy leaves. For rust diseases of both bean (Livne & Daly, 1966; Durbin, 1967) and coltsfoot (Long *et al.*, 1975) net import has been proposed, with supporting evidence from isotope movement. Re-examination of these systems, using carbon budgets and truly quantitative use of isotope, could be rewarding since they may provide a rare example of a leaf simultaneously importing and exporting carbohydrate.

Phloem transport

Any effect on phloem transport which is independent of physical damage, impaired loading and reduced sucrose supply, is likely to be caused by modulators such as PGRs or toxins. Our lack of understanding of the mechanism of phloem transport does not help decide if pathogens can act on the process; there is no evidence to suggest that they do. Currently the pressure-flow theory of Munch is returning to favour (Fisher, 1978; Cronshaw, 1981; Moorby, 1981) and if this does explain the major source of energy for driving phloem transport, the only place that disease could exert an effect is on the integrity of the sieve-element plasma membrane. As noted above, modulators such as fusicoccin which can affect membrane ATPases could act here. In passing, it is worth noting that any transport into leaves in the phloem would be reduced by impaired phloem transport, with adverse consequences for nutrient supply to the fungus.

Consequences for plant growth

The direct fungal effects on long distance transport systems could have profound effects on plant growth. So could production and transport of modulators consequent on infection. For biotrophic pathogens, the evidence suggests that the former is more important. Thus for brown rust on barley, plant dry-weight loss is entirely accounted for by the reduced carbohydrate supply from the rusted leaf (Owera *et al.*, 1981). Further, Ayres (1981b) has shown compensatory increases in photosynthesis by healthy leaves of pea plants partially infected with powdery mildew. This is a normal plant response to reduced assimilate supply from a part of its leaf system. However, some effects demand a more subtle explanation such as the involvement of changed PGR status. Thus effects of leaf-inhabiting fungi on root processes (Walters & Ayres, 1980; Ahmad *et al.*, 1982) may be explicable in terms of the disruption of normal control processes (Ben Zioni *et al.*, 1971; Luttge & Higinbotham, 1979) but could be more complex, involving the transport and distant action of modulators. However, since leaf-inhabiting biotrophs greatly reduce phloem transport, they may have no effective long-distance transport system for distributing modulators. It is far easier to see how diseases such as vascular wilts can directly inject transport systems with toxins, whatever their physiological significance (Pegg, 1981).

Nutrient distribution within organs: short distance transport

The above effects change the nutrient status of an organ such as a leaf. But commonly the distribution of substances within an organ is far from uniform with many accumulating at or around an infection site. Such redistribution of nutrients, possibly involving uptake by the fungus, is the outcome of short-distance transport within the leaf. That movement to infection sites can occur quite independently of long-distance transport is well shown for calcium by Durbin (1967) and for nitrogen by Bushnell (1970) and Bushnell & Gay (1978), but this is an inadequately investigated area.

Movement to infection sites

Transfer in the apoplast and symplast are best considered separately. Nutrients in the apoplast can move towards the fungal surface either by simple thermal diffusion or by bulk flow. The latter will depend on water potential gradients favouring water flow to the fungus, and these could be caused in two ways. Firstly, nutrient concentrations either outside or inside the fungal plasma membrane could be raised, thereby lowering solute potential. Concentrations will almost certainly be higher inside than outside the fungus (Allen *et al.*, 1979); water flux into a fungus which has not ruptured the host cuticle will be at about the rate of fungal fresh weight increase and so quite inadequate to deliver nutrients by bulk flow. Secondly and far more probably, evaporation from fungal hyphae penetrating the host cuticle will cause a bulk flow in the apoplast to the evaporation site (Harvey, 1930) with a consequent flux of apoplastic nutrients to the fungal surface.

Undoubtedly, therefore, there will be transport of substances to the fungal surface in the apoplast. The pertinent question is what proportion of fungal growth is satisfied in this way? Hancock & Huisman (1981) calculate that all the carbon necessary for growth of leaf-inhabiting bacteria and fungi could be supplied from the apoplast and suggest that other nutrients could be supplied similarly, as long as the metabolite concentration in the apoplast is maintained by normal rates of leakage from the symplast. These workers, and Littlefield & Heath (1979), also point to the growth of fungi entirely intercellularly as proof of apoplastic supply but they do not eliminate the possibility of an enhanced flux of nutrients to the apoplast, as a result of increased membrane leakiness caused by infection. Membrane transport is considered below but its critical importance in controlling supply of nutrients to the

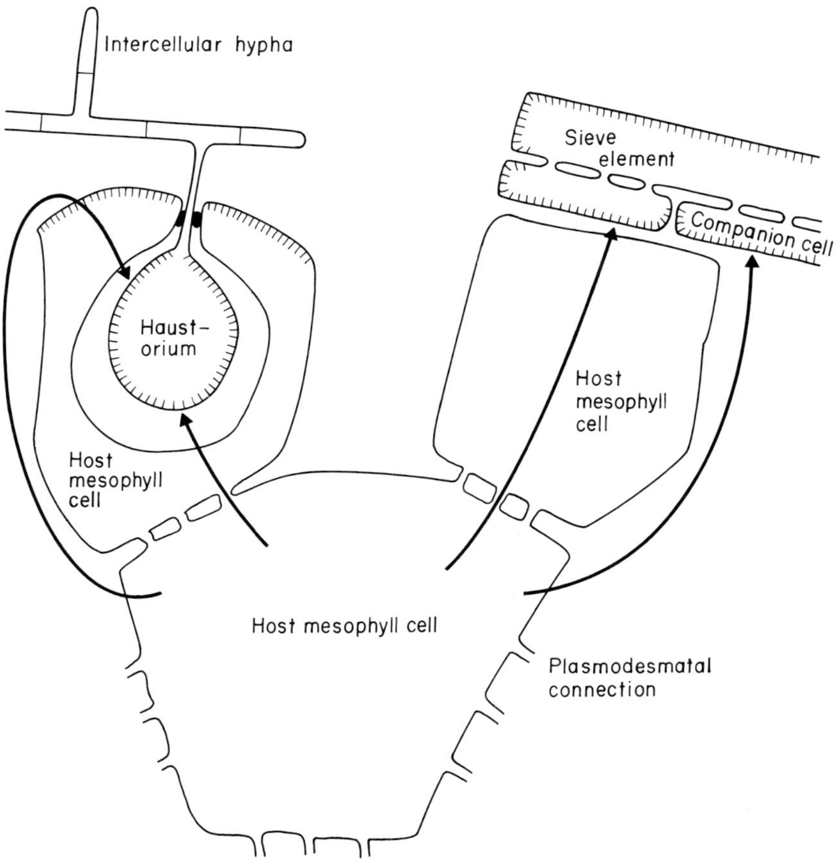

Figure 7.5. A diagrammatic representation of the pathways for transport from a leaf mesophyll cell to phloem and fungal mycelium. Continuous lines represent plasma membranes, and it can be seen that both phloem and fungal tissue must be loading from the apoplast. Hatching on the membranes indicates the localization of ATPases (Spencer-Phillips & Gay, 1981; Cronshaw, 1980), and arrows show the possible routes for assimilate transport to fungus and phloem.

apoplast should be noted here. It may be much less of a barrier in necrotrophic than in biotrophic infections.

Transport in the symplast must also occur. Host symplast in the leaf may have two, non-connected, components, the phloem sieve-tube: companion cell complex and the mesophyll plus epidermis. The phloem has been discussed above. There is an analogy betweed the phloem and the fungal mycelium in their relation to the mesophyll symplast (Figure 7.5). Clearly, a net transport of assimilates and nutrients can occur in response to concentration gradients across cells and across plasmodesmata. Since movement within cells may be effected partly by cyclosis (Allen & Allen, 1978) any pathogen-induced reduction of such metabolically driven rotational streaming would be expected to reduce nutrient transfer towards the pathogen.

It has been considered that cytokinin-like substances might be involved in controlling this short-distance transport; many analogies between colonies of biotrophic pathogens and localized cytokinin application have been made. Two difficulties attend the cytokinin hypothesis. One is how cytokinins trigger transport. The second is whether the other effects caused by cytokinins such as increasing stomatal aperture and stimulating mesophyll growth also occur in fungal colonies. Neither of these difficulties can be resolved with certainty, and the primary cause of modified short-distance transport remains to be determined. Real advances in understanding may await a far more detailed picture of how substances are distributed in diseased tissue. The most pertinent questions concern the proportions of material in apoplast, host symplast and fungal symplast, and whether concentration gradients exist within pustules. Application of techniques such as freeze substitution and quantitative autoradiography may improve our understanding, as would a clear idea of routes of water movement in infected tissues. Unpublished work on phosphorus distribution in freeze-substituted barley leaves infected with brown rust suggests strongly that the bulk of the excess P within infected leaves is held within the fungus (Ahmad, Farrar & Whitbread, unpublished). The simplest hypothesis is that short-distance transport is altered simply by the pathogen taking up nutrients and acting as a sink; where concentration gradients do not favour passive uptake, active transport at its plasma membrane or modification of the absorbed substances may occur.

Uptake by the pathogen

The way in which pathogens can obtain nutrients delivered to their surface by long and short-distance transport will be constrained by the anatomy of the host pathogen interface. This is invariably apoplastic (Hancock & Huisman, 1981) but can range from situations where the pathogen is more or less uniformly surrounded by host tissue, as in bacterial infections, to where most of the interface is between the host and haustoria of powdery mildews, with most of the fungal mycelium outside the host cuticle although with continuity of apoplast. Two questions arise. Firstly, is the growth rate of the pathogen limited by its ability to absorb nutrients — perhaps because of limited surface area — rather than by delivery of nutrients to its surface? (Another possibility is that it is limited by internal constraints of the type proposed for *Neurospora* by Alberghina *et al.*, 1979). The information needed to answer this question is not available. The second question is confined to fungi with organs apparently specialized for uptake. Do these organs, such as haustoria, absorb the bulk of the nutrients needed for fungal growth? It has been argued that in lichens they do not because they seem neither necessary nor sufficiently frequent (Collins & Farrar, 1978). Further, the calculations of Hancock & Huisman (1981) mentioned above show that apoplastic supply of nutrients to the mycelium is adequate for sustaining fungal growth. Conversely, for the powdery mildews we have the elegant and persuasive arguments of Gay and his co-workers (Bushnell & Gay, 1978: Spencer-Philips & Gay, 1980, 1981; Manners & Gay, 1978, 1980, 1982; Gay & Manners, 1981) that the haustoria are the main agents of carbohydrate uptake by the fungus,

Table 7.3. Calculations relating carbohydrate needed for fungal growth to fluxes sustainable into haustoria and intercellular hyphae.

Parameter	Value	Origin of value
1. colony density (m^{-1} leaf surface)	5×10^5	Owera *et al.* (1981)
2. colony radius (μm)	500	measured
3. colony radial growth rate (μm/h)	2	linear growth to 500 μm over 10 days
4. colony dry matter (μg)	4.5	From *1* and *2*, assuming fungus is 0.15 of leaf weight
5. increase in colony mass (μg/d)	0.9	Owera *et al.*, 1981, and colony density in proportion to increase in colony 'area'
6. spore production rate (μg/d/colony)	0.6	Ahmad *et al.* (1982), and colony density
7. ratio: (carbohydrate incorporated into spore and colony mass)/ (carbohydrate taken up by mycelium)	2	Bull & Bushnell (1976)
8. carbohydrate flux into colony to account for growth (μg/d	3	(*5* + *6*) × *7*
9. hexose flux into colony to account for growth (nmol/h/colony)	0.7	(8×10^3)/(180 × 24)
10. hyphal radius (μm)	2.5	From Lane & Shaw (1972)
11. hyphal length per colony (μm)	1.5×10^6	From *4* & *10* assuming a density of 0.15 g dry matter cm^{-3} mycelium; an answer of the same order is obtained by estimating hyphal length and number from Lane & Shaw (1972)
12. haustorial frequency (μm/hypha)	0.01	From Lane & Shaw (1972) and Evans (1907)
13. haustorial radius (μm)	3	From Lane & Shaw (1972) and Evans (1907)
14. haustorial surface area per colony (μm^2)	1.7×10^6	$4\pi(13)^2 \times 11 \times 0.01$
15. hyphal surface area per colony (μm^2)	23.6×10^6	$11 \times 2\pi 10$
16. maximal flux of hexose sustainable across membranes ($nmol/m^2/s$)	500	Sovonick *et al.* (1974); From Jennings (1976)
17. potential flux through haustoria (nmol/h/colony)	3.06	$16 \times 3600 \times 14 \times 10^{-12}$
18. potential flux through hyphal surface (nmol/h/colony)	42.5	$16 \times 3600 \times 15 \times 10^{-12}$

Notes: 1. In the column 'Origin of Value' an italicized number refers to the relevant parameter, and the word 'From' indicates my interpretation of data or figures shown in the papers cited.

2. Compare the flux necessary for growth (*9*) with those sustainable (*17*, *18*); the latter will be lower if carbohydrate concentrations at the membrane surface are low, or if the concentration of specific carrier sites on the membrane is low.

with apoplastic delivery of nutrients to the cells invaded by haustoria. Active uptake by these host cells may then precede the transfer of carbohydrates to the apoplast surrounding the fungal haustoria (Figure 7.5).

The situation for rust fungi is less clear both because the haustoria are less accessible experimentally (Mendgen, 1981; Lane & Shaw, 1972) and because intercellular hyphae are also potentially capable of considerable nutrient uptake. The micro-autoradiographic studies which have shown that rust haustoria in bean leaves may accumulate nutrients are neither conclusive nor do they exclude uptake by intercellular hyphae (Mendgen, 1981). One way to proceed is to calculate the number, or better, the surface area, of haustoria necessary to sustain fungal growth, assuming all material enters through the haustoria, and to compare this with measured haustorial numbers. Such a calculation is presented in Table 7.3. It can be seen that uptake either into haustorial or into intercellular hyphae could account for fungal growth, although far better data are needed from a single host-parasite system. This calculation could be seen as supporting the suggestion that since haustoria may not be necessary for nutrient uptake, they may have a different and specialized role in some pathosystems (Littlefield & Heath, 1979).

Membrane transport

The importance of membrane transport has already been mentioned, notably in phloem loading and in the loss of material from symplast to apoplast. Its more general importance is in controlling fluxes of materials and in facilitating the build-up of membrane potentials and concentration gradients central to continued cellular functioning. It is necessary here to distinguish carefully between permeability as passive diffusion across a membrane and active transport.

Several reports suggest that increased permeability results from pathogen attack. Increased permeability implies a degree of membrane dysfunction and impairment resulting in increased passive leakage; it could also imply decreased active transport across the membrane. Unfortunately, these reports are mostly of enhanced leakage of electrolytes out of infected tissue. They could indeed be explained by increased membrane permeability but equally well by a redistribution of nutrients such that more are in pools available for leakage. The only full kinetic analysis of membrane properties seems to be that of Keck & Hodges (1973) for toxins produced in necrotrophic infections, where increased membrane permeability did result in some cases. Hanchey & Wheeler (1979) give a critical account of peremability phenomena in such diseases and Daly (1981) discusses effects of toxins on membranes.

Several lines of evidence indicate that membranes are not damaged in host plants infected by biotrophic fungi. Cells containing haustoria of powdery mildews have active phosphatases at their surface (Figure 7.5; Spencer-Phillips & Gay, 1981). It was argued above that rust infection does not affect phloem loading and thus the properties of the membranes around the sieve element-companion cell complex in rust-infected barley leaves. Data of various types for four other membrane systems from this pathosystem show that all appear to be functioning unimpaired (Table 7.4). The plasma membrane and tonoplast in infected leaves have been examined using isotope efflux to separate effects at the two membranes (Figure 7.6a) Either alone, or preferably followed by a compartmental analysis (Figure 7.6b), these data indicate no overall change in membrane permeability consequent on rust infection, for plasma membrane or tonoplast. Two types of membrane in the chloroplast have also been examined. Firstly, the chloroplast inner membrane, in which the rate at which the phosphate translocator operates is higher in diseased than in healthy leaves, as each chloroplast is fixing more CO_2 (there are, of course, far fewer chloroplasts). Secondly, there are gradients of H^+ and Mg^{2+} across the chloroplast thylakoid lamellae during photosynthesis which would be dissipated were these membranes leaking; *in vivo* chlorophyll fluorescence kinetics has shown them to be fully functional in diseased leaves (Ahmad, Farrar & Whitbread, unpublished).

Conclusions

It is always dangerous to adduce evidence to support one view of a complex system, and yet it is frequently difficult to do anything else. With this in mind, the

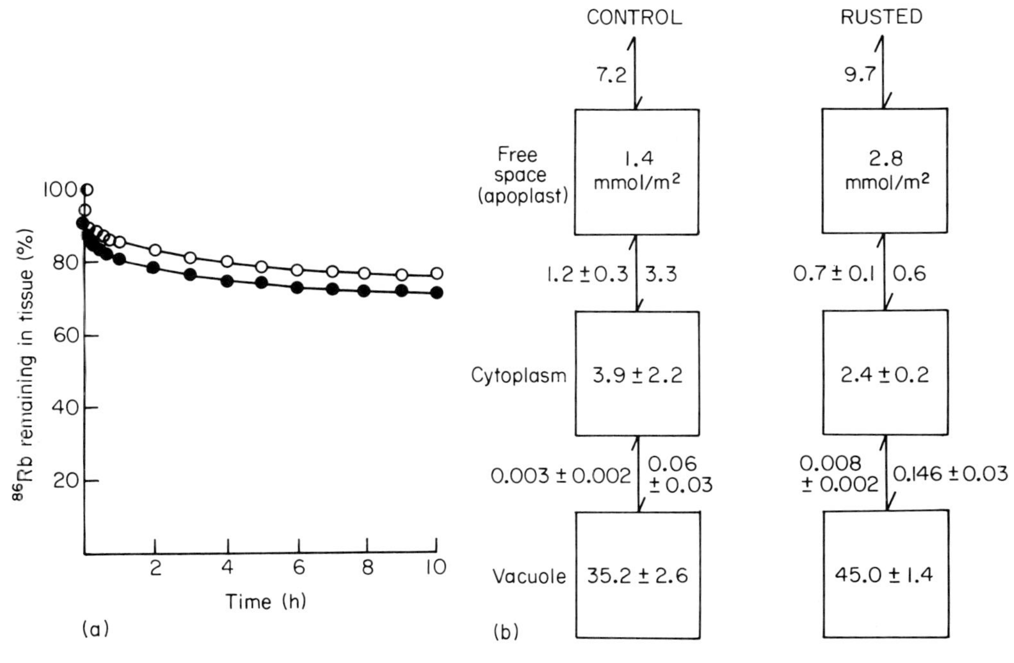

Figure 7.6. Potassium fluxes in leaf discs cut from first leaves of barley infected with brown rust. (a) Leakage of ^{36}Rb into KC1 following 24 h loading with ^{36}RbCl, for control (○) and rusted (●) leaves. (b) Compartmental analysis of fluxes of K in the leaf discs using the data of (a). Amounts of K in the three types of pool are given in mmol/m^2 leaf, and rate constants for transfer between pools in h; no differences are statistically significant. (Ahmad, Farrar & Whitbread, unpublished).

Table 7.4. The integrity of four membrane systems in the leaves of barley infected with brown rust

Membrane system	Parameter and Method	Control	Rusted	Units and Notes
Chloroplast thylakoid lamellae	Quenching capacity; *in vivo* fluorescence	0.29 ± 0.02	0.72 ± 0.02	arbitrary units; a high value denotes functional lamellae
Chloroplast inner membrane	Flux through phosphate translocator; calculation from rates of photosynthesis	0.54	0.73	μmol Pi/m^2 membrane/s. Assumes a 1:1 exchange of Pi : DHAP
Plasma membrane	K^+ influx; ^{86}Rb efflux kinetics and compartmental analysis	1.56 ± 0.29	1.62 ± 0.12	μmol/m^2 leaf/h
Tonoplast	K^+ efflux from vacuole; ^{86}Rb efflux kinetics and compartmental analysis	97.5 ± 51.0	366.0 ± 91.2	μmol/m^2 leaf/h

Unpublished data of Ahmad, Farrar & Whitbread.

thrust of this chapter can be summarized as follows. Biotrophic pathogens inhabit tissue, simply taking up nutrients that reach their surface. Other effects which they cause follow from this; they neither produce, nor stimulate the host to produce, modulators (toxins or PGRs) of significance to transport processes. The many phenomena that may determine compatibility or incompatibility of a host pathogen combination have little or no relevance to the mass transfer of nutrients to the fungus. That host material, such as chloroplasts and phloem, function nearly unimpaired has as a consequence the continued flow of nutrient to the growing and sporulating mycelium. By contrast, necrotrophs may depend in part on impairment of host membranes for their supply of nutrients; this may be achieved by toxins that act mainly on short-distance transport.

References

Ahmad I., Owera S.A.P., Farrar J.F. & Whitbread R. (1982) The distribution of five major nutrients in barley plants infected with brown rust. *Physiological Plant Pathology* **21**, 335–46.

Alberghina L., Sturani E., Costanini M.G., Martegani E. & Zippel R. (1979) Regulation of macromolecular composition during growth of *Neurospora crassa*. In: *Fungal Walls and Hyphal Growth* (Ed. by J.H. Burnett & A.J.P. Trinci), pp. 295–318. Cambridge University Press, Cambridge.

Al Khesraji T.O., Losel D.M. & Gay J.L. (1980) The infection of vascular tissue in leaves of *Tussilago farfara* L. by pycnial-aecial stages of *Puccinia poarum* Nich. *Physiological Plant Pathology* **17**, 193–7.

Allen F.H.E., Coffey M.D. & Heath M.C. (1979) Plasmolysis of rusted flax: a fine-structural study of the host-pathogen interface. *Canadian Journal of Botany* **57**, 1528.

Allen N.S. & Allen R.D. (1978) Cytoplasmic streaming in green plants. *Annual Review of Biophysics and Bioengineering* **7**, 497–526.

Ayres P.G. (1981a) Effects of disease on plant water relations. In: *Effects of Disease on the Physiology of the Growing Plant* (Ed. by P.G. Ayres), pp. 131–48. Cambridge University Press, Cambridge.

Ayres P.G. (1981b) Powdery mildew stimulates photosynthesis in uninfected leaves of pea plants. *Phytopathologische Zeitschrift* **100**, 312–8.

Ben-Zioni A., Vaadia Y. & Lips W. (1971) Nitrate uptake by roots as regulated by nitrate reduction products of the shoot. *Plant Physiology* **24**, 288–90.

Bull A.T. & Bushnell M.E. (1976) Environmental control of fungal growth. In: *The Filamentous Fungi*, Vol. II (Ed. by J.E. Smith & D.R. Berry), pp. 1–13. Edward Arnold, London.

Bushnell W.R. (1970) Patterns in the growth, oxygen uptake and nitrogen content of single colonies of wheat stem rust on wheat leaves. *Phytopathology* **60**, 92–9.

Bushnell W.R. & Gay J.L. (1978) Accumulation of solutes in relation to the structure and function of haustoria in powdery mildews. In: *The Powdery Mildews* (Ed. by D.M. Spencer), pp. 183–236. Academic Press, London.

Collins C.R. & Farrar J.F. (1978) Structural resistances to mass transfer in the lichen *Xanthoria parieina*. *New Phytologist* **81**, 71–83.

Cronshaw J. (1980) Histochemical localization of enzymes in the phloem. *Berichte der Deutschen Botanischen Gesellschaft* **93**, 123–9.

Cronshaw J. (1981) Phloem structure and function. *Annual Review of Plant Physiology* **32**, 465–84.

Dale J.E., Baeurmeister A., & Williams E.J. (1981) The use of compartmental analysis to examine effects of plant growth regulating substances on transport of assimilates in wheat leaves. In: *Mathematics and Plant Physiology* (Ed. by D.A. Rose & D.A. Charles-Edwards), pp. 79–90, Academic Press, London.

Daly J.M. (1981) Mechanisms of action. In: *Toxins in Plant Disease* (Ed. by R.D. Durbin), pp. 331–94. Academic Press, New York.

Durbin R.D. (1967) Obligate parasites: effect on the movement of solutes and water. In: *The Dynamic Role of Molecular Constituents in Plant-Parasite Interaction* (Ed. by C.J. Mirocha & I. Uritani), pp. 80–99. Bruce Publishing Company, St. Paul, Minnesota.

Edwards H.H. (1971) Translocation of carbon in powdery mildewed barley. *Plant Physiology* **47**, 324–8.

Evans I.B.P. (1907) The cereal rusts. I. The development of their uredo mycelia. *Annals of Botany* **21**, 441–6.

Favali M.A., Conti G.G. & Bassi M. (1978) Modifications of the vascular bundle ultrastructure in the resistant zone around necrotic lesions induced by tobacco mosaic virus. *Physiological Plant Pathology* **13**, 247–51.

Fisher D.B. (1978) An evaluation of the Munch hypothesis for phloem transport in soybean. *Plant Physiology* **139**, 25–8.

Gay J.L. & Manners J.M. (1981) Transport of host assimilates to the pathogen. In: *Effects of Disease on the Physiology of the Growing Plant* (Ed. by P.G. Ayres), pp. 85–100. Cambridge University Press, Cambridge.

Hanchey P. & Wheeler H. (1979) The role of host cell membranes. In: *Recognition and Specificity in Plant Host-Parasite Interactions* (Ed. by J.M. Daly & I. Uritani), pp. 193–210. Japan Scientific Societies, Tokyo.

Hancock J.G. & Huisman O.C. (1981) Nutrient movement in host-pathogen systems. *Annual Review of Phytopathology* **19**, 309–31.

Harvey R.B. (1930) The relative transpiration rate at infection sites. *Phytopathology* **20**, 359–62.

Huber D.M. (1978) Disturbed mineral nutrition. In: *Plant Disease*, Vol. III (Ed. by J.G. Horsfall & E.B. Cowling), pp. 163–82. Academic Press, New York.

Jennings D.H. (1976) Transport and translocation in filamentous fungi. In: *The Filamentous Fungi*, Vol. II (Ed. by J.E. Smith & D.R. Berry), pp. 32–64. Edward Arnold, London.

Keck R.W. & Hodges T.K. (1973) Membrane permeability in plants: changes induced by host-specific pathotoxins. *Phytopathology* **63**, 226–30.

Lane W.D. & Shaw M. (1972) Axenic culture of flax rust isolated from cotyledons by cell-wall digestion. *Canadian Journal of Botany* **50**, 2601–3.

Littlefield L.J. & Heath M.C. (1979) *Ultrastructure of Rust Fungi*. Academic Press, New York.

Livne A. & Daly J.M. (1966) Translocation in healthy and rust affected beans. *Phytopathology* **56**, 170–5.

Long D.E., Fung A.K., McGee E.E.M., Cooke R.C. & Lewis D.H. (1975) The activity of invertase and its relevance to the accumulation of storage polysaccharides in leaves infected by biotrophic fungi. *New Phytologist* **74**, 173–82.

Luttge U. & Higinbotham N. (1979) *Transport in Plants*. Springer, Berlin.

Manners J.M. & Gay J.L. (1978) Uptake of ^{14}C photosynthates from *Pisum sativum* by haustoria of *Erysiphe pisi. Physiological Plant Pathology* **12**, 199–209.

Manners J.M. & Gay J.L. (1980) Autoradiography of haustoria of *Erysiphe pisi. Journal of General Microbiology* **116**, 529–33.

Manners J.M. & Gay J.L. (1982) Transport, translocation and metabolism of ^{14}C-photosynthates at the host-parasite interface of *Pisum sativum* and *Erysiphe pisi. New Phytologist* **91**, 221–44.

Mendgen K. (1981) Nutrient uptake in rust fungi. *Phytopathology* **71**, 983–9.

Moorby J. (1977) Integration and regulation of translocation within the whole plant. *Symposium of the Society for Experimental Biology* **21**, 425–54.

Moorby J. (1981) *Transport Systems in Plants.* Longman, London.

Owera S.A.P., Farrar J.F. & Whitbread R. (1981) Growth and photosynthesis in barley infected with brown rust. *Physiological Plant Pathology* **18**, 79–90.

Owera S.A.P., Farrar J.F. & Whitbread R. (1983) Translocation from leaves of barley infected with brown rust. *New Phytologist* **94**, 111–23.

Pegg G.F. (1981) Biochemistry and physiology of pathogenesis. In: *Fungal Wilt Diseases of Plants* (Ed. by M.E. Mace, A.A. Bell and C.H. Beckman), pp. 193–254. Academic Press, New York.

Sovonick S.A., Geiger D.R. & Fellows R.J. (1974) Evidence for active phloem loading in the minor veins of sugar beet. *Plant Physiology* **54**, 886–91.

Spencer-Phillips P.T.N. & Gay J.L. (1980) Electron microscope autoradiography of ^{14}C-photosynthate distribution at the haustorium-host interface in powdery mildew of *Pisum sativum. Protoplasma* **103**, 131–54.

Spencer-Phillips P.T.N. & Gay J.L. (1981) Domains of ATPase in plasma membranes and transport through infected plant cells. *New Phytologist* **89**, 393–400.

Walters D.R. & Ayres P.G. (1980) Effects of powdery mildew disease on uptake and metabolism of nitrogen by roots of infected barley. *Physiological Plant Pathology* **17**, 369–79.

Whipps J.M. & Lewis D.H. (1981) Patterns of translocation, storage and interconversion of carbohydrates. In: *Effects of Disease on the Physiology of the Growing Plant* (Ed. by P.G. Ayres), pp. 47–83. Cambridge University Press, Cambridge.

Yarwood C.E. & Jacobsen L. (1955) Accumulation of chemicals in diseased areas of leaves. *Phytopathology* **45**, 43–8.

8 Effects of infection on root growth and function; consequences for plant nutrient and water relations

P.G. AYRES

Department of Biological Sciences, University of Lancaster,
Lancaster LA1 4YQ, UK

Introduction

Roots take up water and inorganic ions and transport these to the shoot together with certain plant growth regulators that they synthesize. Infection can alter the supply of these materials to the shoot by changing the efficiency with which each unit of root functions and by reducing root growth. Altered efficiency may be caused directly by invasion and disruption of root tissue by the pathogen. Since the network of roots has a single connection with the shoot (though several vascular traces may pass through the connecting zone) lesions on roots which are close to the shoot tend to have a greater effect on transport than do more distant lesions. Altered efficiency can be caused indirectly by a pathogen which invades the shoot and reduces movement to the root of photoassimilates needed for active processes. Prolonged or severe infections reduce root growth but relationships between the amount or duration of infection and the reduction in root growth are usually ignored because small or rapid changes in root growth are very difficult to measure in practice.

The functional efficiency of each unit of root changes

Water transport

Water moves through the soil-plant-air continuum in response to differences in water potential (ψ) between component parts of the system. By analogy with Ohm's Law, in a steady state

$$\text{flux of water} = \frac{\psi_s - \psi_l}{r_p} = \frac{\psi_l - \psi_a}{r_g}$$

where ψ_s is the water potential at the soil-root interface, ψ_l is the leaf water potential, ψ_a the water potential of the ambient air; r_p is the sum of a number of component resistances in root, stem and leaf, and represents the total resistance to liquid flow through the plant, and r_g represents the resistance to water movement in the gas phase.

Wood R.K.S. & Jellis G.J. (1984) *Plant Diseases: infection, damage and loss.* Blackwell Scientific Publications, Oxford.

Resistance to the flow of liquid water through roots has both an axial and a radial component. Axial flow, through xylem vessels, approximates to laminar flow and, as such, is governed by Poiseuille's Law,

$$\text{flux of water} = \triangle P \times \frac{\pi r^4}{8L\eta}$$

where $\triangle P$ is the driving force (difference in hydrostatic pressure across the system), L and r are the length and radius of the vessel and η is the viscosity of the medium transported. Halving the radius of a vessel reduces the flux by a factor of 16. Vessels of infected plants may be wholly or partially occluded by cells or products of the pathogen, fragments of host cell walls, tyloses or embolisms (Hall & MacHardy, 1981; Ayres, 1982a). Zimmerman & McDonough (1978) suggest that embolisms may be the most common cause of total occlusion since the xylem sap is under a negative hydrostatic pressure (tension) and, thus, air may often enter when vessel walls are punctured by an invading pathogen.

A normal water balance is maintained in tissues of healthy plants because r_g is at least as great as the 'upstream' r_p. In some foliar infections r_g may be reduced because the pathogen induces excessive stomatal opening or ruptures the host cuticle (Ayres, 1978, 1982a; also Table 3 of Farrar (this volume)). In plants infected by root pathogens the balance is disturbed because r_p increases and wilting results. Thus, in soybean infected by *Phytophthora megasperma* var. *sojae* the resistance of unit root was increased from 1.73 to 6.98 MPa.sec.g/cm^3 10 days after inoculation. Four days later, ψ_l of infected plants was -0.16 MPa (0.1 MPa = 1 bar) compared with −0.10 MPa in controls (Kittle & Gray, 1980). In safflower infected by *Phytophthora cryptogea* (*drechsleri*), ψ_l was also lowered by infection and wilting was seen 5 days after inoculation. The reason was that whole-root resistance of infected plants rose from approximately 0.3×10^5 to 2.4×10^5 MPa.sec.g/cm^3, whereas the resistance of the leaves remained at 0.3×10^5 MPa.sec.g/cm^3 cm (Duniway, 1977). (In the second example resistances were expressed on a leaf area basis for ease of comparison; the published data indicate clearly that infection also increased resistance per unit root). Only a small part of the increase in root resistance could be attributed to the reduction in mean vessel diameter which occurred in infected plants.

A reduction in vessel diameter of 25% was also noted in roots of barley infected by powdery mildew (*Erysiphe graminis* f.sp. *hordei*) (Walters, 1981). However, although this would have caused the resistance to axial flow to increase, the measured resistance of the whole root system (axial plus radial) was reduced by infection from 1.75×10^5 to 1.48×10^5 MPa.sec.g/cm^3 (Walters & Ayres, 1982a). Infection must have caused a reduction in the resistance to radial flow; possibly, this reduction was caused by other changes in the anatomy of the root, for example, the reduction by 18% in the diameters of the root and the endodermal cylinder. Similar changes in anatomy have been noted in seminal roots of wheat infected by yellow rust (*Puccinia striiformis*) (Martin & Hendrix, 1974), and may well be common in plants infected by foliar pathogens, but their effects on water transport remain unexplored.

Ion transport

Investigations of several fungi that invade roots of wheat show that the position and depth of lesions dictate the effects that arise (Fitt & Hornby, 1978). The most detailed studies have concerned take-all of wheat caused by *Gaeumannomyces graminis*. Where dark-brown lesions were apparent on roots, but hyphae had not yet penetrated the stele, ion uptake by intact root systems kept for 24 h in nutrient solution was little different from that in healthy controls (for Ca,103% of control; for P,123% and for Fe,93%). When single seminal axes with well-developed lesions located midway along their length were supplied with ^{32}P above the lesion or ^{85}Sr (a tracer for Ca) below the lesion it was found that the uptake of Ca was reduced and transport toward the shoot was inhibited, but uptake and transport of P was unaffected by infection (Table 8.1) (Clarkson *et al.*, 1975). Invasion of the phloem eventually inhibited the translocation of carbohydrates and phloem-mobile solutes towards root tips. Root elongation ceased 2–3 days after invasion of the phloem and transport of xylem-mobile ions stopped as roots exhausted their metabolic reserves. While Rovira (1979) reported that take-all reduced the mineral (K, P and Ca) content of shoots of wheat grown in the field, Fitt & Hornby (1978) found that when plants were grown in sand and were well-supplied with nutrients, infection reduced the K contents of shoots but increased their P and Ca contents (Figure 8.1). An increased P content is difficult to account for, but it may have occurred because older parts of roots increased their uptake and transport of P (Table 8.1) to an extent that overcompensated for the loss of function in younger parts distal to lesions. (P can be taken up along the whole length of roots). An increased Ca content probably occurred because Ca is immobile once deposited in leaves; this can lead to the accumulation of high levels when infected leaves fail to expand as normal.

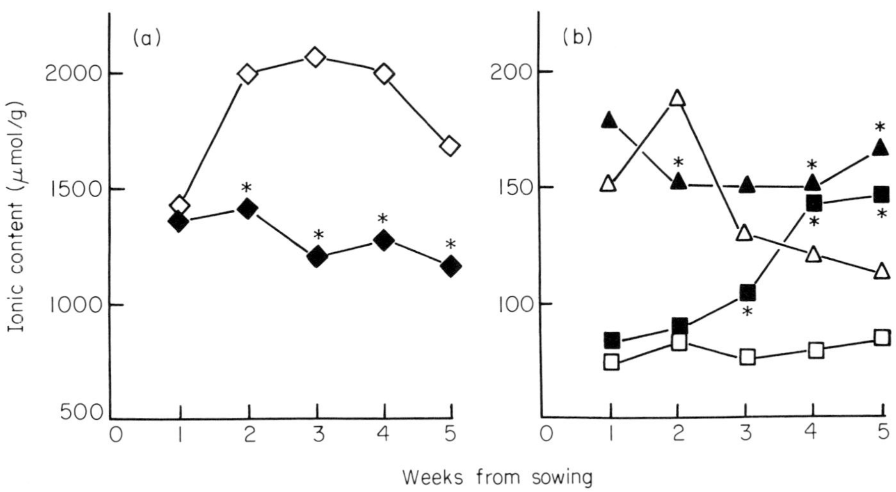

Figure 8.1. Changes in (a) shoot K (□) and (b) shoot Ca (□) and P (△) in wheat healthy (open symbols) or infected by *Gaeumannomyces graminis* (closed symbols). *Significantly different from control, $P \leq 0.05$. (After Fitt & Hornby, 1978; Hornby & Fitt, 1982.)

Table 8.1. The effect of take-all lesions, located midway along the seminal axis, on the uptake and distribution of calcium and phosphate by segments of intact wheat roots.

	Calcium (Supplied below lesion)		Phosphate (Supplied above lesion)	
	Control	Infected	Control	Infected
	nmol/mm^3 segment/day ± S.E.			
Shoot	1.74 ± 0.74	0	0.08 ± 0.02	0.12 ± 0.06
Root	2.93 ± 0.90	0.51 ± 0.09†	0.13 ± 0.14	0.12 ± 0.04
Treated segment	3.19 ± 0.25	2.54 ± 0.50	1.66 ± 0.61	2.83 ± 0.58
Total uptake	7.86 ± 0.83	3.05 ± 0.68	1.87 ± 0.61	3.06 ± 0.67
% Translocated from segment	58 ± 6	14 ± 4	14 ± 4	6 ± 1

† Nearly all of this would have been located between the point of labelling and the lesion; none reached the shoot. Data from Clarkson *et al.* (1975).

Whereas root pathogens may inhibit translocation of photoassimilates to parts of root axes distal to sites of infection, foliar pathogens may affect the whole root system. Nitrate uptake in healthy plants is very sensitive to a reduction in energy supply and, not surprisingly, uptake is reduced in barley infected by *E. graminis* f.sp. *hordei* (Walters & Ayres, 1980); ammonium uptake, which is less sensitive to the energy supply, is unaffected. Ammonium accumulates in roots and leaves because there is a decrease in assimilation through glutamate dehydrogenase and glutamine synthetase/GOGAT pathways, as well as enhanced degradation of glutamic acid and asparagine.

Phosphate uptake is also dependent on energy metabolism. However, its uptake, unlike that of nitrate, increased per unit of root tissue in barley infected with powdery mildew (Figure 8.2a) and net transport to the shoot was stimulated at rates of water flux corresponding to those found in normally transpiring plants in the light (Figure 8.2b) (Walters & Ayres, 1981). The effect of a decreased supply of energy may have been offset by abnormally high amounts of ammonium, since ammonium is known to stimulate P uptake in roots of healthy barley (Klemm, 1967).

Assimilate partitioning in the root

Infection of the shoot inhibits photosynthesis (Habeshaw, this volume) but has more complex effects on assimilate availability in roots than have other inhibitory factors such as shading or grazing by herbivores because it may also reduce the percentage of assimilates that is transported to roots. For example, percentage transport to roots is reduced in wheat infected by yellow rust (*Puccinia striiformis*) (Doodson *et al.*, 1965) or loose smut (*Ustilago nuda*) (Gaunt & Manners, 1971b), maize infected by smut (*Ustilago maydis*) (Billett & Burnett, 1978) and barley infected by powdery mildew (Ayres, 1982b).

Roots of young plants have few insoluble reserves. Most imported assimilates are rapidly utilized for growth, respiration or osmoregulation. In young sunflowers,

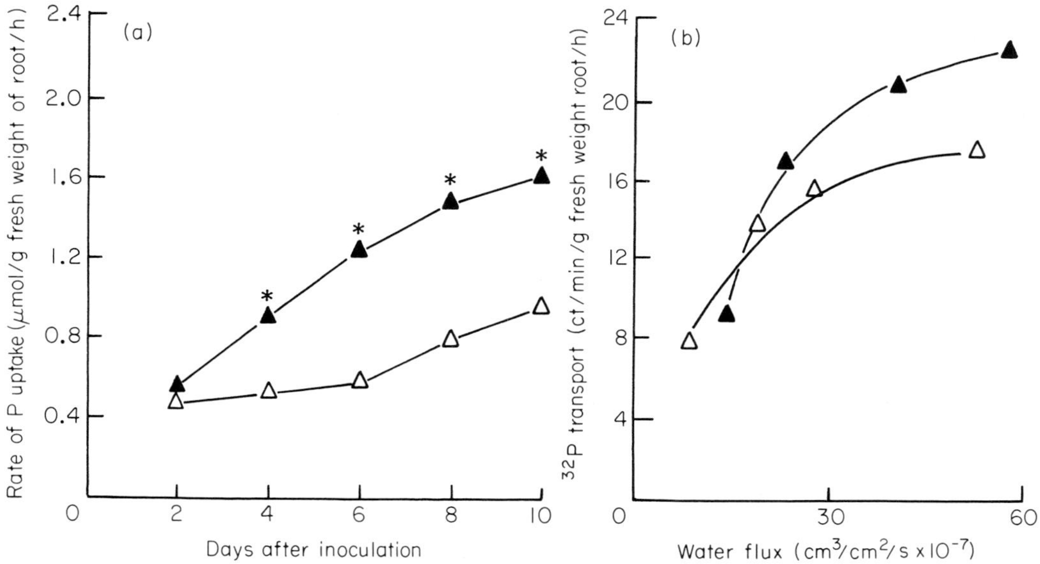

Figure 8.2. (a) Phosphate uptake by intact barley plants healthy (△) or infected by *Erysiphe graminis* f.sp. *hordei* (▲). *Significantly different from healthy, $P \leq 0.05$. (From Walters & Ayres, 1981).
(b) Effect of water flux, driven by an externally applied hydrostatic pressure, on transport of ^{32}P-labelled phosphate across root systems excised from barley plants healthy (△) or infected 7 days by *E. graminis* f.sp. *hordei* (▲). Data from Walters & Ayres (1982a).

two out of every three sucrose molecules reaching roots are immediately respired and no reserves are maintained (Hatrick & Bowling, 1973). Farrar (1981) estimates that the pool of soluble carbohydrates in roots of barley is so small that it is turned over every 30 min. In roots of barley infected by powdery mildew, incorporation of imported assimilates into the soluble pool of carbohydrates was affected before incorporation into storage or structural pools (Walters & Ayres, 1982b). Processes consuming assimilates are in competition with each other to some extent. The allocation of substrates to different respiratory processes varies according to the environment of the plant. In healthy barley, from which all but five crown roots had been excised 2 days prior to measurements, Veen (1981) found that 13% of respiration supported ion transport, 79% supported growth and 8% was maintenance respiration. Five days later, when ion uptake was less limited by root size, the relative proportions were 48, 38 and 14%. The unusual conditions of Veen's experiment (excision followed by regrowth) may have led to overestimation of the amount of energy normally needed for active ion transport. Luttge & Pitman (1976) calculated that steady state fluxes in barley roots consume 4.3 J/g fresh weight of root, i.e. 5.7% out of an available total of 75.0 J/g fresh weight provided by respiration in 24 h. Bowling (1976) reached the similar conclusion that ion transport consumed relatively little energy; he calculated that potassium uptake in sunflower needed only 0.3 J in 24 h compared with a total energy expenditure by roots of 785 J.

In spite of requiring little energy, ion transport often appears sensitive to a reduction in energy supply. Excision or darkening of the shoot of sunflower brought about a rapid decrease in the phosphate-transport-linked electrogenic component of the membrane potential of root cortical cells in sunflower (Graham & Bowling, 1977) and white clover (Bowling & Dunlop, 1978). Veen (1981) found that respiration linked to nitrate transport in barley was more sensitive than was growth or maintenance respiration to a reduction in energy supply caused by darkening the shoot. In contrast, Crapo & Ketallapper (1981) found that shading the shoot of barley, wheat or tomato had a greater effect on growth than on ion (K) uptake. Farrar (1981) has concluded that respiratory activity in roots is not a simple function of supply from the shoot, although it may be controlled by some factor produced or modified in the shoot, e.g. amounts of reduced nitrogen. This is an area in which studies of plants infected by foliar pathogens might help to resolve a basic problem in plant physiology.

Growth of the root system is reduced

A reduction in the size of the root system and in the volume of soil explored by that system may assume great importance when infected plants compete with healthy plants of the same or other species for limited amounts of soil water and nutrients. This last possibility has rarely been explored at a physiological level, but it is predictable that any reduction in the rate at which roots extend, or in the functional activity of root tips, will be of critical importance in situations where there is competition for dwindling resources.

Water transport

When plants are growing in soil, steep gradients of water potential may develop in the perirhizal zone because the hydraulic conductivity of soil, which is normally lower than that of the root under dry conditions (Karamanos, 1981), may restrict water movement or, and judged by Weatherly (1979) to be more important, the hydraulic conductance (reciprocal of resistance) at the root-soil interface may become extremely low because the root shrinks away from soil particles as its volume decreases under stress. Perirhizal and interfacial gradients will be greatest wherever the requirements for water uptake are greatest.

When Sterne *et al.* (1978) investigated the causes of wilt of avocado resulting from infection by *Phytophthora cinnamomi* they found that healthy plants had ψ_l at night of -0.18 MPa, but ψ_l of well-watered infected plants and water-stressed healthy plants reached only -0.80 and -0.70 MPa respectively. Highest ψ_l (approximately equal to xylem potential) normally occurs at night when evaporative demand is low and stomata are closed. There was a consistent relationship between ψ_l and transpiration in well-watered healthy trees but in infected and water-stressed healthy trees there was a marked hysteresis, i.e. at a given transpiration rate, ψ_l was lower in the morning than in the afternoon (Figure 8.3). Since ψ_l is decreased by a reduction in ψ_{soil}, if neither transpiration rate nor resistance from soil to leaf

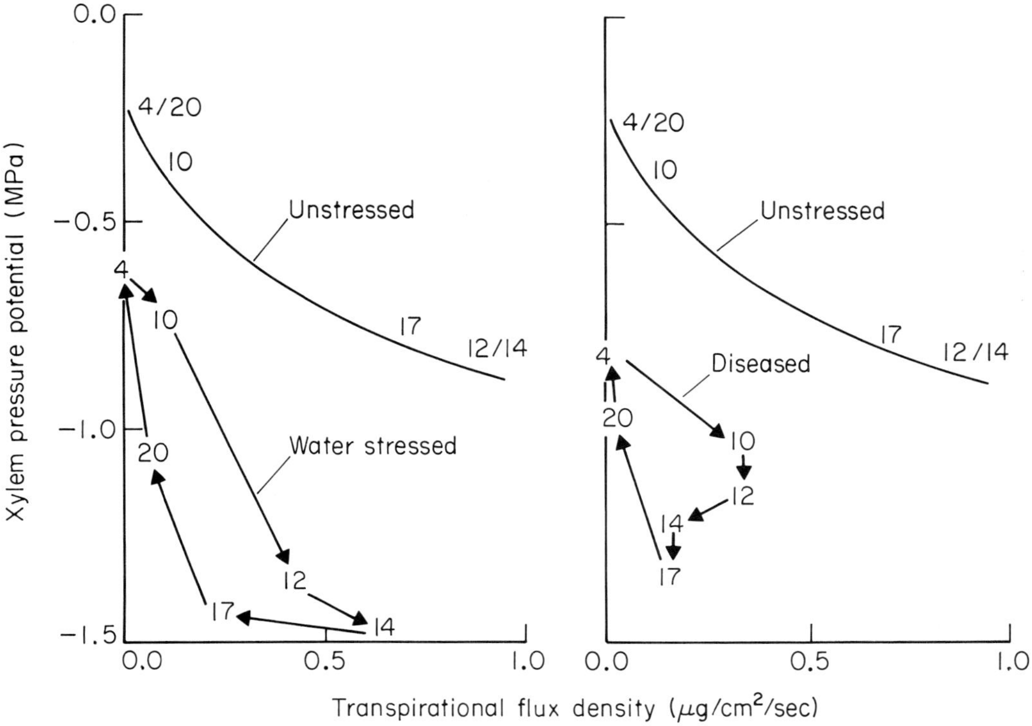

Figure. 8.3. Relationship between water potential and transpiration in leaves of avocado trees, unstressed (irrigated every 7 to 8 days), water-stressed (without irrigation for 30 days) or infected by *Phytophthora cinnamomi*. Numbers indicate time of measurement, e.g. 4 = 0400 hours. Line for unstressed was fitted after regression analysis. Data from Sterne *et al.* (1978).

changes, i.e.

$$\psi_l = \psi_s - \text{flux}\ (r_{\text{soil to leaf}})\ \text{(Kaufmann, 1976)}$$

this suggests that the hydraulic conductivity of the soil decreased around roots of healthy water-stressed and well-watered infected plants. Sterne *et al.* (1978) suggest that in infected plants the demands of the shoot on the smaller root system caused a lowering of perirhizal ψ_{soil} equivalent to that around the larger root system of healthy trees under stress.

The branching pattern of a root system, and the relative growth rates of its component parts, may vary in response to non-uniform distribution of water and nutrients in the soil. Consider the simplest case in which drought develops as soil dries regularly from the surface downward. Water stress will be avoided so long as only shallow roots are lost and the remaining deeper roots are able to transport sufficient water for the shoot's normal requirements. Peas responded to drought in this way despite suffering an absolute reduction in root growth caused by infection by powdery mildew (*Erysiphe pisi*) (Ayres, 1981). Normal water potentials were maintained at the shoot apex of infected plants because, although growth of lateral roots was reduced as soil dried, growth of the tap root was unaffected. *In vitro* experiments showed that root tips of infected plants were unable to accumulate sufficient solutes to maintain turgor and extension growth when the water potential of the external environment was low (ψ_{ext} = -0.8 MPa), but extension growth continued when ψ_{ext} was high (Figure 8.4). The water supply to the shoot of infected plants may also be protected by an increase in the hydraulic conductance of roots (see earlier) (Walters & Ayres, 1982a).

Ion transport

Ions move in aqueous solution from the soil to the xylem. Their precise pathway depends on the ion and the development of the root. Some ions, such as Ca^{2+}, follow the same pathway as water which moves freely through the cell walls and intercellular spaces (free space) of the cortex and only enters the symplast at the endodermis where its radial pathway is blocked by the hydrophobic Casparian strip deposited in radial and transverse walls. This deposition, termed the Stage 0/I transition, occurs 3–7 mm behind the root tip of barley. Later in the development of the root direct access to the endodermal cells is blocked as a deposit of suberin is laid down between plasma membrane and cell wall. This change, termed Stage I/II transition, which occurs 10–20 cm behind the root tip of barley, greatly reduces the radial transport of water and Ca^{2+} (Figure 8.5). Uptake of other ions such

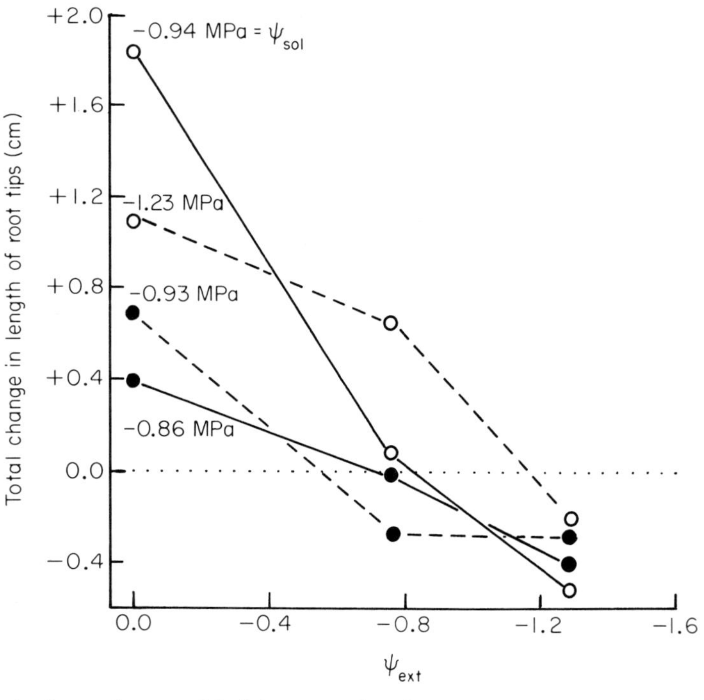

Figure 8.4. Effects of osmotic potential of the external medium (ψ_{ext}) on the extension of root tips (0.5 cm long) excised from peas grown in culture solution. Plants were grown in Hoagland's solution (−0.05 MPa) throughout the experiment (——), or were stressed by transfer to Hoagland's solution plus polyethylene glycol 4000 (-0.3 MPa) 4 days after inoculation (----). Healthy (○), infected (●). For each treatment the extension of 12 root tips was summed after 6 hours. ψ_{sol} is a representative value for the solute potential of the tissue. Data from Ayres (1981).

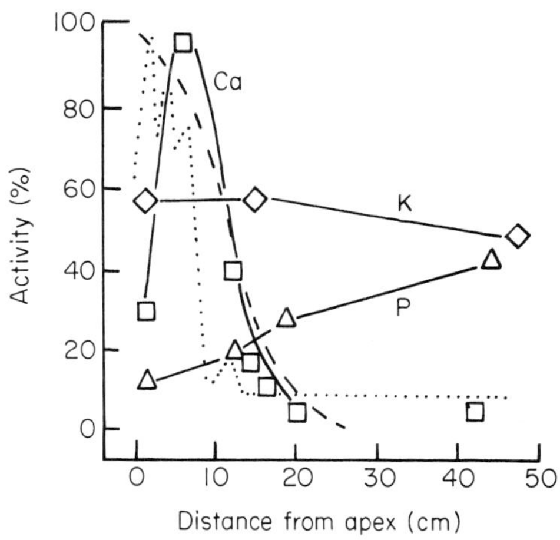

Figure 8.5. Transport of ions and water in the development of the endodermis. The percentage of endodermal cells in Stage I (----) is related to water flow (····) and Ca uptake (□). Uptake of K^+ (◇) and P (△) are independent of the development of Stage II. K data from marrow, other data from barley. Data from Pitman (1981).

as K^- and PO_4^{3-} is unaffected by the Stage I/II transition because these ions are taken up into the symplast while still within the cortex and so their radial passage is unaffected by changes that take place in walls of the endodermal cells.

It is implicit in the above that K^+ and PO_4^{3-} uptake can occur equally well at all points along the length of the root. In soil this does not happen. Some ions have higher diffusion constants than others (Table 8.2). Chloride and nitrate diffuse rapidly because they interact little with the solid phase in the soil, whereas K^+ and, more particularly, PO_4^{3-}, diffuse slowly because they interact strongly with the solid phase. Thus, when PO_4^{3-} is taken up by roots a zone of depletion rapidly builds up around the root. During a season of plant growth PO_4^{3-} is unlikely to move a greater distance than the width of a root hair (Harper, 1977). The pattern is somewhat disturbed by mass flow as water percolates through the soil or moves to replace that taken up by roots. However, mass flow to replace transpirational losses is greatest late in the season when leaf areas and demand are greatest and when most of the annual uptake of nutrients is complete, e.g. in cereals 90% of the total N and P content of mature plants is absorbed before the plant has achieved 25% of its final dry weight. (Incidentally, this is some justification for the over-emphasis on seedling material in experimental investigations). Thus, continued growth into unexploited soil is important if plants are to obtain sufficient P and other ions of low diffusion constant.

It should be noted that high transpiration rates can enhance the transport of most ions under suitable conditions. Thus, increased transpiration in leaves of barley infected by leaf blotch (*Rhynchosporium secalis*) was associated with the increased absorption of a root-absorbed solute, ^{86}Rb, and its accumulation at infection sites (Ayres & Jones, 1975). A rapid flux of water may sweep ions out of the xylem in the root and so favour further uptake (Figure 8.2b).

Table 8.2. Diffusion coefficients for some ions in solution and in moist soils at 20–25°C.

	Diffusion coefficient ($cm^2/s \times 10^5$)	
	Free solution	Soil
Nitrate	1.92	0.5
Chloride	2.03	0.5
Potassium (or Rubidium)	1.98	0.01–0.24
Phosphate	0.89	0.0005–0.001

Data from Scott Russell (1977)

A functional equilibrium between root and shoot

The effects of an absolute reduction in root growth on the acquisition of water and mineral nutrients have been discussed (p. 110), but what are the consequences of a reduction in the ratio root mass:shoot mass? A reduction is often recorded, e.g. in wheat infected by yellow rust (Doodson *et al.*, 1964), loose smut (Gaunt & Manners, 1971a) or speckled leaf blotch (*Septoria tritici*) (Gough & Merkle, 1977), although occasionally increases are noted, e.g. in wheat infected by take-all (Asher, 1972) or powdery mildew (*E. graminis* f.sp. *tritici*) (Lupton & Sutherland, 1973). When the source of water and nutrients is not limiting, as in a complete culture solution, a significant proportion of the roots of healthy plants can be lost without affecting transport to the shoot. Excision of roots does not always reduce plant growth, especially in cereals (Andrews & Newman, 1968), although Humphries (1958) found that dry matter accumulation in barley was closely related to ion uptake from a complete culture solution.

Simultaneous studies of water and ion transport have rarely been made in plants with damaged root systems so it is not possible to be certain whether water or ion transport may be more important in restricting growth. Brouwer (1963) found that excision of roots of bean affected the transport of ions more than that of water. Leaf growth was related to the uptake of P and nitrate-N rather than the uptake of water (Figure 8.6a). Few data are available from infected plants, but it is interesting to note that the mineral composition of shoots of wheat, 18% root axes of which were infected by take-all, did not differ from that of controls in pot experiments (Hornby & Fitt, 1982). Clearly this subject deserves further investigation but any conclusions will depend *inter alia* on the age of the plant because, as described earlier (p. 113), its need for mineral nutrients declines with age (Figure 8.6b).

Deviations from the normal root:shoot ratio caused by mowing or grazing of the shoot or artificial pruning of the root are usually rapidly corrected, either by increased growth of the affected organ or by decreased growth of the other organ (Andrews & Newman, 1968; Brouwer, 1963). Increased growth of nodal roots is known to occur in wheat infected by take-all (Asher, 1972) and similar correction

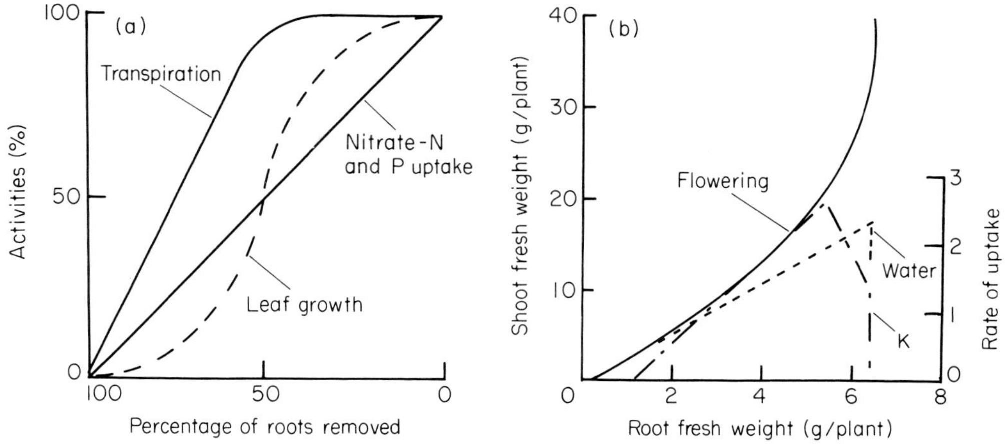

Figure 8.6. (a) Effects of removing increasing proportions of the root system of bean plants on the uptake of water, nitrate-N and P, and on the growth of leaves.
(b) Effects of plant development on the root : shoot fresh weight ratio in pea and on the uptake of water and K. Data from Brouwer (1963).

may occur in other diseases, particularly where infection is limited or overcome. Brouwer (1963, 1977) proposes there is a functional equilibrium between root and shoot. Equilibrium is mediated by water and nutrient ions moving from shoot to root and by photoassimilates moving in the other direction. Undoubtedly it is also mediated by growth regulatory substances, including those produced by roots, e.g. cytokinins and abscisic acid. Some effects that infection has on the production of growth regulators are described by Pegg (this volume).

References

Andrews R.E. & Newman E.I. (1968) The influence of root pruning on the growth and transpiration of wheat under different soil moisture conditions. *New Phytologist* **67**, 617–30.

Asher M.J.C. (1972) Effect of *Ophiobolus graminis* infection on the growth of wheat and barley. *Annals of Applied Biology* **70**, 215–23.

Ayres P.G. (1978) Water relations of diseased plants. In: *Water Deficits and Plant Growth*, Vol. V (Ed. by T.T. Kozlowski), pp. 1–60. Academic Press, New York.

Ayres P.G. (1981) Root growth and solute accumulation in pea in response to water stress and powdery mildew. *Physiological Plant Pathology* **19**, 169–80.

Ayres P.G. (1982a) Effects of disease on plant water relations. In: *Effects of Disease on the Physiology of the Growing Plant* (Ed. by P.G. Ayres), pp. 131–48. Cambridge University Press, Cambridge.

Ayres P.G. (1982b) Water stress modifies the influence of powdery mildew on root growth and assimilate import in barley. *Physiological Plant Pathology* **21**, 283–93.

Ayres P.G. & Jones P. (1975) Increased transpiration and the accumulation of root absorbed ^{86}Rb in barley leaves infected by *Rhynchosporium secalis* (leaf blotch). *Physiological Plant Pathology* **7**, 49–58.

Billett E.E. & Burnett J.H. (1978) The host-parasite physiology of the maize smut fungus, *Ustilago maydis*. II. Translocation of ^{14}C-labelled assimilates in smutted maize plants. *Physiological Plant Pathology* **12**, 103–12.

Bowling D.J.F. (1976) *Transport of Ions by Plant Roots*. Chapman & Hall, London.

Bowling D.J.F. & Dunlop J. (1978) Uptake of phosphate by white clover. I. Evidence for an electrogenic phosphate pump. *Journal of Experimental Botany* **29**, 1139–46.

Brouwer R. (1963) Some aspects of the equilibrium between underground and overground plant parts. *Jaarboek, Institute voor Biologische en Scheikundig, Wageningen* **213**, 31–44.

Brouwer R. (1977) Root functioning. In: *Environmental Effects on Crop Physiology* (Ed. by J.J. Landsberg and C.V. Cutting), pp. 229–45. Academic Press, New York.

Clarkson D.T., Drew M.C., Ferguson I.B. & Sanderson J. (1975) The effect of take-all fungus, *Gaeumnanomyces graminis*, on the transport of ions by wheat plants. *Physiological Plant Pathology* **6**, 75–84.

Crapo N.L. & Ketellapper H.J. (1981) Metabolic priorities with respect to growth and mineral uptake in roots of *Hordeum, Triticum* and *Lycopersicon. American Journal of Botany* **68**, 10–16.

Doodson J.K., Manners J.G. & Myers A. (1964) Some effects of yellow rust (*Puccinia striiformis*) on the growth and yield of spring wheat. *Annals of Botany* **28**, 459–72.

Doodson J.K., Manners J.G. & Myers A. (1965) Some effects of yellow rust (*Puccinia striiformis*) on ^{14}carbon assimilation and translocation in wheat. *Journal of Experimental Botany* **16**, 304–17.

Duniway J.M. (1977) Changes in resistance to water transport in safflower during development of *Phytophthora* root rot. *Phytopathology* **67**, 331–7.

Farrar J.F. (1981) Respiration rate of barley roots: its relation to growth, substrate supply and the illumination of the shoot. *Annals of Botany* **48**, 53–63.

Fitt B.D.L. & Hornby D. (1978) Effects of root-infecting fungi on wheat transport processes and growth. *Physiological Plant Pathology* **13**, 335–46.

Gaunt R.E. & Manners J.G. (1971a) Host–parasite relations in loose-smut of wheat. I. The effect of infection on host growth. *Annals of Botany* **35**, 1131–49.

Gaunt R.E. & Manners J.G. (1971b) Host–parasite relations in loose-smut of wheat. II. The distribution of ^{14}C-labelled assimilates. *Annals of Botany* **35**, 1141–50.

Gough F.J. & Merkle O.G. (1977) The effect of speckled leaf blotch on root and shoot development of wheat. *Plant Disease Reporter* **61**, 597–9.

Graham R.D. & Bowling D.J.F. (1977) Effect of the shoot on the transmembrane potentials of root cortical cells of sunflower. *Journal of Experimental Botany* **28**, 886–93.

Hall R. & MacHardy W.E. (1981) Water relations. In: *Fungal Wilt Diseases of Plants* (Ed. by M.E. Mace, A.A. Bell and C.H. Beckman), pp. 255–98. Academic Press, New York.

Harper J.L. (1977) *Population Biology of Plants*. Academic Press, New York.

Hatrick A.A. & Bowling D.J.F. (1973) A study of the relationship between root and shoot metabolism. *Journal of Experimental Botany* **24**, 607–13.

Hornby D. & Fitt B.D.L. (1982) Effects of root-infecting fungi on structure and function of cereal roots. In: *Effects of Disease on the Physiology of the Growing Plant* (Ed. by P.G. Ayres), pp. 101–30. Cambridge University Press, Cambridge.

Humphries E.C. (1958) Effect of removal of part of the root system on the subsequent growth of the root and shoot. *Annals of Botany* **22**, 251–7.

Karamanos A.J. (1981) The development of water deficits in plants. In: *Water Stress on Plants* (Ed. by G.M. Simpson), pp. 34–88. Praeger, New York.

Kaufmann M.R. (1976) Water transport through plants: current perspectives. In: *Transport and Transfer Processes in Plants* (Ed. by I.F. Wardlaw and J.B. Passioura), pp. 313–27. Academic Press, New York.

Kittle D.R. & Gray L.E. (1980) Effects of infection by *Phytophthora megasperma* var. *sojae* on the water relations of soybean. *Crop Science* **20**, 504–7.

Klemm K. (1967) The effects of nitrogen form on the nutrient uptake by plants. *Bodenkultur* **18**, 210–28.

Lupton F.G.H. & Sutherland J. (1973) The influence of powdery mildew (*Erysiphe graminis*) infection on the development of four spring wheats. *Annals of Applied Biology* **74**, 35–9.

Luttge U. & Pitman M.G. (1976) Transport and energy. In: *Encyclopedia of Plant Physiology* Vol. 2A, *Transport in Plants*, Part A, *Cells* (Ed. by U. Luttge and M.G. Pitman), pp. 251–9. Springer-Verlag, Berlin.

Martin N.E. & Hendrix J.W. (1974) Anatomical and physiological responses of Baart wheat roots affected by stripe rust. *Washington Agricultural Experimental Station, Technical Bulletin* **77**, 16 pp.

Pitman M.G. (1981) Ion uptake. In: *Physiology and Biochemistry of Drought Resistance* (Ed. by L.G. Paleg and D. Aspinall), pp. 71–96. Academic Press, New York.

Rovira A.D. (1979) Biology of the soil–root interface. In: *The Soil-Root Interface* (Ed. by J.L. Harley and R. Scott Russell), pp. 145–60. Academic Press, New York.

Sterne, R.E., Kaufmann, M.R. & Zentmyer, G.A. (1978) Effect of *Phytophthora* root rot on water relations of avocado; interpretation with a water transport model. *Phytopathology* **68**, 595–602.

Scott Russell R. (1977) *Plant Root Systems.* McGraw Hill, London.

Veen B.W. (1981) Relation between root respiration and root activity. *Plant & Soil* **63**, 73–6.

Walters D.R. (1981) Root function in mildewed barley. Ph.D. thesis, University of Lancaster.

Walters D.R. & Ayres P.G. (1980) Effects of powdery mildew disease on uptake and metabolism of nitrogen by roots of infected barley. *Physiological Plant Pathology* **17**, 369–79.

Walters D.R. & Ayres P.G. (1981) Phosphate uptake and transport by roots of powdery mildew infected barley. *Physiological Plant Pathology* **18**, 195–205.

Walters D.R. & Ayres P.G. (1982a) Water movement through root systems excised from healthy and mildewed barley; relationships with phosphate transport. *Physiological Plant Pathology* **20**, 275–84.

Walters D.R. & Ayres P.G. (1982b) Translocation and partitioning of ^{14}C-labelled assimilates in roots of mildewed barley: effects on mitotic index. *Plant Pathology* **31**, 307–13.

Weatherly P.E. (1979) The hydraulic resistance of the soil-root interface — a cause of water stress in plants. In: *The Soil-Root Interface* (Ed. by J.L. Harley and R. Scott Russell), pp. 275–86. Academic Press, New York.

Zimmerman M.H. & McDonough J. (1978) Dysfunction of the flow of food. In: *Plant Disease*, Vol. III (Ed. by J.G. Horsfall and E.B. Cowling), pp. 117–40. Academic Press, New York.

9 Tolerance of parasitic infection in plants

D.D. CLARKE
Department of Botany, Glasgow University,
Glasgow G12 8QQ, UK

Introduction

Resistant cultivars can provide the cheapest and most effective way of avoiding crop losses due to parasitic infection. However, for many crops adequate resistance to important pathogens is not available, e.g. take-all disease of wheat and barley caused by *Gaeumannomyces graminis* (Scott, 1981), and in such instances tolerance could be an additional breeding objective leading to the development of cultivars which produce a yield acceptable both in quality and quantity even when heavily infected. As yet the potential value of tolerance cannot be assessed properly because little is known about its physiological basis or of the degree of tolerance that may be achieved.

Advance in our knowledge of tolerance has been partly hindered by the differing ways in which the term has been defined and used by various workers. It has often been loosely used and even misused to describe host–parasite interactions where terms implying some form of resistance or disease escape would have been better. Dictionary definitions of *tolerant* and *tolerance* include 'the ability to endure' or 'capable of enduring without showing serious effects'. The operative word in both definitions is *endure*, which means 'to put up with' or 'to bear with patience'; it certainly does not imply resistance. Thus *tolerance* should only be used in relation to host properties or reactions which enable the host to *endure* the parasite and never in relation to any property or reaction which limits in any way the growth and development of the parasite. Some misuse of the term has probably come from the failure of certain definitions to stress sufficiently the importance of the amount of infection. Thus two commonly used definitions are quoted in a guide to the use of terms in plant pathology (Anon., 1973) as follows: *tolerant*, 'able to endure infection by a particular pathogen without showing severe disease', and, with slight modification, *tolerance*, 'the ability of the affected organism to endure invasion by a pathogenic organism or virus with little or no reaction, as shown by the more or less complete absence of 'symptom expression and damage'. The guide goes on to point out that the definitions are inadequate because neither draws attention to the importance of relating tolerance to the degree of infection to which

Wood R.K.S. & Jellis G.J. (1984) *Plant Diseases: infection, damage and loss.* Blackwell Scientific Publications, Oxford.

the host tissues are subjected and states that a host can properly be termed tolerant only if it can be extensively colonized by a parasitic organism or virus without symptom development. Thus to establish tolerance in the terms of the definitions it must be shown that the tolerant host can be infected to an extent which would cause significant damage in non-tolerant hosts.

By putting so much emphasis on the lack of a host reaction, or at most a limited host reaction in response to infection, the definitions clearly only encompass mechanisms of tolerance based on the relative insensitivity of the host to the parasite or its metabolic products, i.e. tolerance of the parasite, and ignore the possibility that plants may also be tolerant of any damage parasites may cause, i.e. be tolerant of disease. Tolerance of disease would be shown by plants which, although sensitive to the parasite and its metabolic products, can compensate for the tissue damage so that growth and yield are less affected than those of plants with similar levels of disease but without the same compensatory capacity.

In contrast to the definitions given in the guide (Anon., 1973) some authors have defined tolerance to encompass tolerance of disease and ignored the possibility of tolerance of the parasite. Thus Caldwell *et al.* (1958) defined tolerance as 'the ability of plants to endure severe (rust) disease without severe losses in yield or quality'. This definition or modifications of it have been widely employed and a number of reviews and articles have been produced under the general heading of 'tolerance to disease' (Schafer, 1971; Mussell, 1980; Gaunt, 1981). However all, including Caldwell *et al.* (1958), appear to interpret the definition very loosely because the examples they quote, especially of biotrophic infections, include many which involve as much tolerance of the parasite as of the disease. Both Schafer (1971) and Gaunt (1981) in fact refer to the possibility of tolerance of the pathogen (parasite) as distinct from tolerance of disease and Schafer (1971) offers a definition of tolerance which encompasses both forms thus: 'that capacity of a cultivar resulting in less yield or quality loss relative to disease severity or pathogen development when compared with other cultivars or crops'.

Mussell (1980), in his review includes some interesting examples of both forms of tolerance as defined above but unfortunately goes on to include examples which clearly involve forms of resistance or infection escape. In fact he uses the term for a wide range of host–parasite interactions from the level of the cell to that of the population.

In contrast Parbery (1978) considers that *tolerance to disease* is a meaningless term and that all forms of tolerance are in reality of the parasite.

This review will not deal with tolerance in the all-embracing sense employed by Mussell (1980), nor with the restricted view adopted by Parbery (1978), but with the two forms of tolerance, tolerance of the parasite and tolerance of disease, described above. It will also only be concerned with tolerance of the individual plant and its components and although both forms of tolerance may occur in the same plant they will be treated separately because each is clearly determined by different sets of factors.

Tolerance of the parasite

Parasites may stress or damage their host by various mechanisms during infection. Firstly, they may divert host metabolites to support their own growth and development. Secondly, they may produce metabolites which either directly or indirectly affect normal functioning of the cells and tissues of the host. Thirdly, they may disrupt the structure and integrity of host tissues. Some of this damage may be unavoidable in the sense that it is bound to occur when the parasite infects its host and completes its life cycle. Other damage may not be essential for parasite development and such damage is avoidable.

Damage caused by the diversion of host metabolites into the production of parasite structures

Viral, bacterial and fungal parasites of the aerial parts of plants will obtain all their requirements from their host but some fungal parasites of subterranean structures, whether roots or modified stems, with hyphae within both host and soil may obtain some of their requirements from the soil.

Damage caused by the diversion of host metabolites will depend upon how they are diverted and how much is taken. Necrotrophs, once infection is established, kill host tissues and can destroy more host biomass than they can utilize. In contrast biotrophs grow in association with living cells and gain access to host metabolites in more subtle ways with the quantities diverted being related not only to the efficiency with which they are converted into parasite structures but also to the total biomass of parasite produced. Alexander (1961) considered that fungi were much more efficient converters than bacteria *in vitro* but more recent studies indicate that bacteria may convert about 50% of the assimilated carbohydrate into cell structures whereas in general, the conversion efficiencies of fungi are nearer 40% (Callow, 1977).

Little is known about rates of accumulation of parasite biomass or its total production in relation to host biomass in any host — parasite interaction. Most information relates to infections by bacteria and viruses because of the particulate nature of these pathogens. Bacterial biomass may be determined by macerating infected tissues and dilution-plating the macerate, while for viruses a variety of techniques are available including particle counting or chemical and serological analyses. In contrast exact measurement of the amounts of fungus in infected tissues has rarely been attempted. Hyphae have been measured directly (Clifford, 1972; Kassim, 1976) but the method is very tedious and while chemical analysis for particular components such as chitin (Ride & Drysdale, 1972), mannan and enzymes such as invertase (Whipps *et al.*, 1982) is a possible indirect approach it may be difficult to relate amounts found to fungal biomass. Thus Sharma *et al.* (1977) found that the amount of chitin per unit of fungal hypha may vary with age, with conditions of culture or with other environmental factors.

Some calculations of the amounts of energy or carbon likely to be used in the production of given biomasses of some viral and bacterial parasites are given by Kosuge (1978). Based on known daily yields of 2.5 mg of tobacco mosaic virus/g fresh weight of host tissue and starting from CO_2, total energy requirement for the production of the 2.375 mg protein and 0.125 mg of viral nucleic acid required would be about 243 μmols of host adenosine triphosphate (ATP)/day/g fresh weight of leaf. It was concluded that this, together with the host's requirements for maintenance and growth would not stress a plant system which is capable of producing up to 1.5 mmols of ATP/g fresh weight/h by photophosphorylation. Similarly he calculated that bacterial infections, with populations of up to 10^{10} cells (2 mg biomass)/g dry weight of host tissue and assuming a low conversion efficiency of 20%, would require about 10 mg of host carbohydrate which is no more than 1 to 2% of total host carbohydrate.

Intercellular biotrophs probably require few special adaptations for obtaining nutrients from their host tissues although the haustoria of fungal parasites may function in this respect. Thus it is well known that plant cells are leaky and that even under mild leaching conditions a wide range of organic and inorganic materials can be leached out of roots and leaves (Tukey, 1970). Such materials arrive at the surface by diffusion through the apoplast and while in this space would provide a freely available source of nutrients for intercellular biotrophs. Hancock & Huisman (1981) have reviewed information about the amounts of nutrients within the apoplast and conclude that, with normal rates of replenishment, there would be sufficient to support known populations of intercellular parasites. Plants appear to be able to lose significant amounts of metabolites to the external environment by leaching, particularly to the rhizosphere, without ill effect and so there is no reason to believe that the host should suffer from loss due to diversion into the parasite in the plant. In any case plants have considerable capacity to compensate for such losses by increased photosynthesis. Thus there is little evidence to indicate that the diversion of metabolites to the parasite should stress the infected plant.

Damage caused by structural components of the parasite or by metabolites secreted into the host

Bacteria and fungi secrete a wide range of metabolites into culture media with both the types and amounts of compounds produced being determined by the composition of the media and the conditions of incubation (Berry, 1975). Bacteria and fungi also produce a range of metabolites in their hosts including toxins, growth regulators and enzymes (Turner, this volume; Pegg, this volume; Cooper, this volume) and it is reasonable to suppose that their production will also be affected by the chemical and physical environment within the host. These metabolites include the determinants of the changes in host metabolism, in transport systems and in growth which are such marked features of many infections. Equally important are constituents of the parasite, for example components of the fungal hyphal wall (Keen, 1982) or of the bacterial cell wall and its surrounding polysaccharide layers (Kelman, 1979). Virus infections also cause marked changes in the metabolic and growth processes of their host but the viral components involved have not been clearly identified. It appears that the coat protein plays little part and that some form of viral or viroid nucleic acid is involved (Zaitlin, 1979).

The active structural components or metabolites of parasites may be classified into two groups. Firstly, there are those whose activities are essential for parasitism such as enzymes, growth regulators and toxins (Wheeler, 1975). To the extent that their production is essential for the establishment of the parasitic relationship, any damage they cause will clearly be unavoidable. However, damage caused by overproduction is potentially avoidable. The second group includes structural components or metabolites which are not involved in the establishment of a parasitic relationship and if they damage the host then that damage would be potentially avoidable.

Damage caused to the structure and integrity of the host's tissues

Parasites may damage the structure and integrity of host tissues in various ways. Necrotrophs kill tissues in advance of colonization and since some damage appears to be essential for the development of the parasite it must be unavoidable although overkill would be avoidable. On the other hand some bacteria and fungi produce toxins which diffuse well ahead of the parasite causing damage which could well be considered avoidable. Thus *Helminthosporium maydis* produces a host–specific toxin, T-toxin, which has high activity towards maize cultivars with Texas male sterile (Tms) cytoplasm but little if any activity towards other maize cultivars, although it does invade and colonize both types of cultivar (Ullstrup, 1970). Thus the extreme action of the toxin in cultivars with Tms cytoplasm does not appear to be essential for the establishment of the parasite relationship. Turner (this volume) has also reported that toxin-less mutants of some bacteria colonize their hosts to the same extent as toxin-producing strains.

Intercellular parasites generally do not cause direct physical damage to host tissues during vegetative growth because they can be readily accommodated within the intercellular spaces of the host. On the other hand many fungi, such as rusts, cause considerable damage at the time of sporulation. Thus sori develop subepidermally and as spores mature their bulk tears apart the covering epidermis and cuticle. The host then no longer has overall control of water loss and the tissues may become water-stressed. This damage, however, is clearly unavoidable.

Some parasites, by producing growth regulators or by stimulating host tissues to do so (Pegg, this volume; Dixon, this volume), induce growth abnormalities such as galls. They cause damage in part by disrupting the normal arrangements of the cells so that the tissues no longer longer function normally. Thus in galls caused by *Plasmodiophora brassicae*, the vascular connections may be so disrupted that translocation in either direction through the galled region is appreciably impaired

(Agrios, 1978). It is unlikely that such damage is essential for the development of the parasite and so it may well be avoidable.

In conclusion, the major damage caused by parasites appears to be due, not to the diversion of host metabolites to form their structure, but to the activity of their metabolites or to the physical damage caused to the host at various stages of their development.

Tolerance of disease

The extent to which lesion development may affect plant growth will depend on the amount of tissue affected and its position in the plant. Thus a small lesion may girdle a stem and kill the plant whereas a similar amount of diseased tissue on leaves or roots may be insignificant in relation to total leaf area or the whole of the root system. There is evidence that some plants have a greater capacity for photosynthesis than is expressed in growth and yield (Ellis, 1954), and Grainger (1959) has shown that removing some of the foliage of plants at various stages of growth does not generally reduce yield in proportion. Thus plants may be able to tolerate the loss of some tissue by disease if the remaining tissue can compensate for that lost. The maximum tolerance for any given amount of disease is likely to occur in those plants in which the effects of the parasite are confined to the lesion with no production of translocatable toxins which could affect the metabolism of tissues outside the lesion.

A clear example of tolerance of disease is given by cereal take-all caused by *Gaeumannomyces graminis* (Manners & Myers, 1981). Infections restricted to the root cortex have no obvious effect on the growth and functioning of the host plant. Real damage occurs only when hyphae breach the endodermis and colonize the vascular tissue because this rapidly blocks the transport of water and nutrients past the lesion and root tissue distal to the lesion will then be cut off from the rest of the plant. The maximum damage caused to the plant will then be related to the amount of root tissue lost and the plant can compensate to some extent for this loss by producing adventitious roots. The capacity of infected barley plants to produce adventitious roots is greater than that of wheat plants so that given levels of root infection reduce the yield of barley less than that of wheat. Thus barley has greater tolerance of take-all than wheat has.

Evolution of tolerance by natural selection

In general the greater the damage caused by a given level of infection, the lower the reproductive success of the host is likely to be. Natural selection should therefore favour the host–parasite interactions which are least damaging and so avoidable damage would be expected to be eliminated and the fittest plants would be those with the maximum tolerance of any unavoidable damage. It is common observation that some wild plants can become heavily infected without showing significant disease (Tarr, 1972) and there is some experimental evidence to indicate the nature of some of the unavoidable damage that may occur in one wild plant, the common annual weed *Senecio vulgaris*, in response to infection by a powdery mildew (Ben-Kalio & Clarke, 1979). This work is briefly discussed below.

Erosion of tolerance during crop development

In contrast to the observations on wild plants, relatively low levels of infection in crop plants can lead to significant disease and marked reductions in growth and yield (Chester, 1950; James, 1974). It is to be expected that the wild ancestors of crop plants were as tolerant of most of their parasites as wild plants are of theirs today and so presumably this tolerance has been lost during the development of crop plants.

The exploitation of tolerance

It should be possible to re-select for either tolerance of the parasite or tolerance of disease in crop plants. Clearly it will not be possible to eliminate unavoidable damage and so a knowledge of the likely nature and extent of this form of damage would help to identify the range and extent of the avoidable damage which could then be selected against.

It is likely that the type of unavoidable damage produced will vary with the parasite and its mode of parasitism. Thus rusts or other endoparasites and ectoparasitic powdery mildews are likely to damage their hosts in different ways. Some indication of the unavoidable damage crop plants may suffer during infection by particular parasites may be obtained by studying the reactions of wild plants to invasion by closely related species. Thus studies of interactions between the common annual weed *Senecio vulgaris* and *Erysiphe fischeri* (Ben-Kalio & Clarke, 1979; Harry, 1980) may indicate the type of unavoidable damage powdery mildews are likely to cause. These studies have shown that even heavy infections when 75 to 100% of the leaf area is colonized do not affect leaf chlorophyll levels, the rate of dry matter production/unit area of leaf tissue, or the distribution of photosynthates. However, total plant growth is reduced, apparently as the result of a reduction in leaf expansion which may result from water stress caused by higher rates of water loss from infected leaves. The greater water loss could thus be a consequence of unavoidable damage. However some of the effects of mildew infection in crop plants e.g. the reductions in chlorophyll levels (Allen, 1942), in rates of photosynthesis (Allen, 1942; Last, 1963), the altered translocation patterns (Frič, 1975) and the reduction in the root:shoot ratio (Last, 1962) were not found and so these effects may constitute avoidable damage.

Studies such as these should indicate whether tolerance has a future in crop protection and if so how it may best be exploited. It is unlikely that it could be used alone because heavy infection clearly reduces plant growth but it could be useful in association with levels of partial resistance which allowed some tissue colonization.

References

Agrios G.N. (1978) *Plant Pathology*, 2nd edition. Academic Press, New York.

Alexander M. (1961) *Introduction to Soil Microbiology*. John Wiley & Sons, New York.

Allen P.J. (1942) Changes in the metabolism of wheat leaves induced by infection with powdery mildew. *American Journal of Botany* **29**, 425–35.

Anon (1973) *A Guide to the Use of Terms in Plant Pathology*. Phytopathological Papers No. **17**, Commonwealth Mycological Institute, Kew.

Ben-Kalio V.D. & Clarke D.D. (1979) Studies on tolerance in wild plants: effects of *Erysiphe fischeri* on the growth and development of *Senecio vulgaris*. *Physiological Plant Pathology* **14**, 203–11.

Berry D.R. (1975) The environmental control of the physiology of filamentous fungi. In: *The Filamentous Fungi*, Volume 1 (Ed. by J.E. Smith & D.R. Berry), pp. 16–32. Edward Arnold, London.

Caldwell R.M., Schafer J.F., Compton L.E. & Patterson F.L. (1958) Tolerance to cereal leaf rusts. Science **128**, 714–5.

Callow P. (1977) Conversion efficiencies in heterotrophic organisms. *Biological Reviews* **52**, 385–409.

Chester K.S. (1950) Plant disease losses: their appraisal and interpretation. *Plant Disease Reporter Supplement* **193**, 190–362.

Clifford B.C. (1972) The histology of race non-specific resistance to *Puccinia hordei* Orth. in barley. In: *Proceedings of the Third European and Mediterranean Cereal Rusts Conference, Prague* **1**, pp. 75–8.

Ellis R.T. (1954) Tolerance to the maize rust *Puccinia polysora* Underw. *Nature* (London) **174**, 102.

Frič F. (1975) Translocation of ^{14}C-labelled assimilates in barley plants infected with powdery mildew (*Erysiphe graminis* f.sp. *hordei* Marchal). *Phytopathologische Zeitschrift* **84**, 88–95.

Gaunt R.E. (1981) Disease tolerance — an indicator of thresholds? *Phytopathology* **71**, 915–6.

Grainger J. (1959) Effects of diseases on crop plants. *Outlook on Agriculture* **2**, 114–26.

Hancock J.G. & Huisman O.C. (1981) Nutrient movement in host–pathogen systems. *Annual Review of Phytopathology* **19**, 309–31.

Harry I.B. (1980) Studies on the *Senecio vulgaris* L. *Erysiphe fischeri* Blumer. Plant pathosystem. Ph.D. Thesis, University of Glasgow.

James W.C. (1974) Assessment of plant diseases and losses. *Annual Review of Phytopathology* **12**, 27–48.

Kassim M.Y.A. (1976) Potato tissue resistance to the growth of *Phytophthora infestans* (Mont.) de Bary. Ph.D. Thesis, University of Glasgow.

Keen N. (1982) Specific recognition in gene-for-gene host-parasite systems. *Advances in Plant Pathology* **1**, 35–82.

Kelman A. (1979) How bacteria induce disease. In: *Plant Disease, An Advanced Treatise*, Vol. IV (Ed. by J.G. Horsfall & E.B. Cowling), pp. 181–202. Academic Press, New York.

Kosuge T. (1978) The capture and use of energy by diseased plants. In: *Plant Disease, An Advanced Treatise*, Vol. III (Ed. by J.G. Horsfall & E.B. Cowling), pp. 85–116. Academic Press, New York.

Last F.T. (1962) Analysis of the effects of *Erysiphe graminis* D.C. on the growth of barley. *Annals of Botany N.S. London* **26**, 279–89.

Last F.T. (1963) Metabolism of barley leaves inoculated with *Erysiphe graminis* Marchal. *Annals of Botany N.S. London* **27**, 685–90.

Manners J.G. & Myers A. (1981) Effects on host growth and physiology. In: *Biology and Control of Take-all* (Ed. by M.J.C. Asher & P.J. Shipton), pp. 237–48. Academic Press, London.

Mussell H. (1980) Tolerance to disease. In: *Plant Disease, An Advanced Treatise*, Vol. V (Ed. by J.G. Horsfall & E.B. Cowling), pp. 39–52. Academic Press, New York.

Parbery D.G. (1978) A consideration of tolerance to parasites in plants. *Annals of Applied Biology*, **89**, 373–7.

Ride J.P. & Drysdale R.B. (1972) A rapid method for the chemical estimation of filamentous fungi in plant tissue. *Physiological Plant Pathology* **2**, 7–15.

Schafer J.F. (1971) Tolerance to plant disease. *Annual Review of Phytopathology* **9**, 235–52.

Scott P.R. (1981) Variation in host susceptibility. In: *Biology and Control of Take-all* (Ed. by M.J.C. Asher & P.J. Shipton), pp. 219–36. Academic Press, London.

Sharma P.D., Fisher P.J. & Webster J. (1977) Critique of the chitin assay technique for estimating fungal biomass. *Transactions of the British Mycological Society* **69**, 479–83.

Tarr S.A.J. (1972) *Principles of Plant Pathology*. Macmillan, London.

Tukey H.B.J. (1970) The leaching of substances from plants. *Annual Review of Plant Physiology* **21**, 305–24.

Ullstrup A.J. (1970) History of southern corn leaf blight. *Plant Disease Reporter* **54**, 1100–02.

Wheeler H. (1975) *Plant Pathogenesis*. Springer-Verlag, Berlin.

Whipps J.M., Haselwandter, K., McGee, E.E.M. & Lewis, D.H. (1982) Use of biochemical markers to determine growth development and biomass of fungi in infected tissues with particular reference to antagonistic and mutualistic biotrophs. *Transactions of the British Mycological Society* **79**, 385–400.

Zaitlin M. (1979) How viruses and viroids induce disease. In: *Plant Disease, An Advanced Treatise*, Vol. IV (Ed. by J.G. Horsfall & E.B. Cowling), pp. 257–71. Academic Press, New York.

Section 3
Infection and host damage

10 Infection processes, symptomatology and host damage

B.E.J. WHEELER
Department of Pure and Applied Biology,
Imperial College, Silwood Park, Ascot, Berks. SL5 7PY, UK

Introduction

This chapter has two aims: one is to provide a general background for the topics reviewed in this section of the book; the other is to provide a link between the processes of infection which, so far, have been considered mainly at the cellular level, and the relationships between host damage and crop losses.

Pathogenesis at the cellular level can be viewed as a series of reactions between two components, basic functions of the host cell and activities of the pathogen, viz:

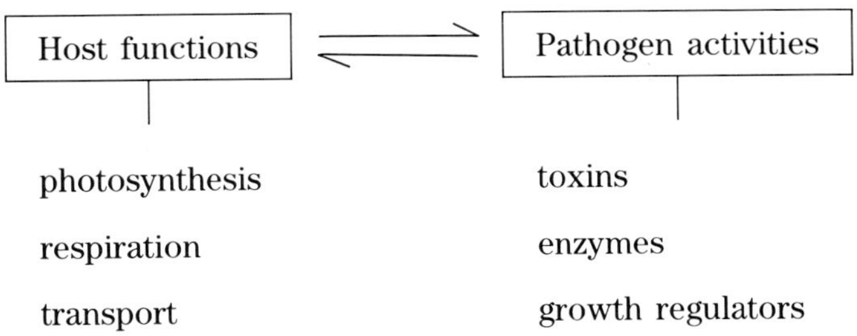

In contrast, for individual plants, it is useful to consider three components in the system:

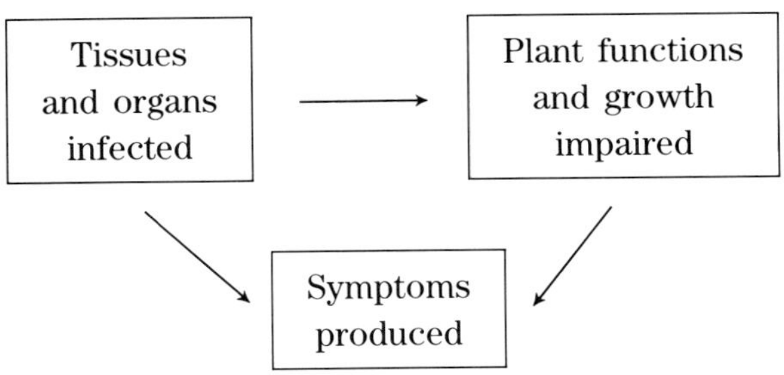

Infection of plant tissues and organs

In terms of the types of elements and the corresponding tissues they comprise three are of primary interest to the plant pathologist, vessels of the xylem, sieve tubes of

Wood R.K.S. & Jellis G.J. (1984) *Plant Diseases: infection, damage and loss.* Blackwell Scientific Publications, Oxford.

the phloem and parenchymatous cells which constitute the bulk of the ground tissue (parenchyma) of plants. To these must be added the meristems from which they are derived. Three features of these elements are especially important in relation to colonization of tissues by plant pathogens — whether they are living or not, their function and their accessibility.

Vessels are non-living and serve primarily in the transport of water and mineral salts from roots to the developing shoots; they terminate near, but not at, the ends of rootlets or at leaf margins and leaf tips. In terms of nutrients they thus present a rather peculiar milieu for the development of plant pathogens and one to which access is not readily available. But, once entered, then the nature of the transport system (or transpiration stream) facilitates the movement of either the pathogen itself or its products and thus usually results in effects far removed from the site of entry. It is hardly surprising, therefore, that pathogens colonizing vessels are restricted to a small group of fungi and bacteria which cause characteristic effects to which the disease name 'vascular wilt' is commonly applied.

Apart from their physiological effects on the host, which quite rightly have received much attention, the modes of entry of vascular wilt pathogens and their initial growth in plant tissues are of most interest and present some problems. Some, like *Pseudomonas solanacearum*, invade through wounds in the roots, possibly at sites where secondary roots appear, through abrasions or feeding punctures made by nematodes. In these instances it is not entirely clear whether the wound must be deep enough to expose a vessel or whether the bacterium is able to grow in and degrade the exposed parenchyma as it does in the later stages of the disease cycle. This type of problem is even more apparent in the entry of wilt-inducing species of *Fusarium* and *Verticillium*. These typically invade the root cortex and become established first in the small, spirally-thickened vessels and then progress upwards within the large xylem elements. However, in the root cortex they cause little obvious damage and this phase remains one of the least investigated and least understood aspects of pathogenesis by these fungi.

There are similar problems connected with the entry of wilt-inducing pathogens into shoots. When a cucumber beetle (*Diabrotica* sp.) carrying *Erwinia tracheiphila* feeds on a cucumber leaf there is an obvious mechanism by which a vessel is exposed and the bacterium is introduced into it. When a cell of *Xanthomonas campestris* enters through a hydathode at the margin of a cabbage leaf it is not directly in contact with the vessel or tracheid at the end of a leaf vein. It is separated by a layer of parenchyma cells with fine intercellular spaces through which it must pass. Presumably its motility facilitates this process and its subsequent colonization of the vessels against the transpiration flow. However, it is far less easy to understand how colonization by *Ceratocystis ulmi* follows from the introduction of a non-motile spore by the feeding of a *Scolytus* sp. beetle on a young elm twig, although this fungus produces conidia which are capable of being transported upwards through vessels.

The sieve tubes of the phloem, in contrast to the xylem vessels, constitute a conducting system of living cells for the transfer of carbohydrates from their source to places where they can be utilized or stored and it is accepted that within different sieve tubes transport may be in different directions depending on the location of carbohydrate source and sink. For pathogens, therefore, sieve tubes present an environment rich in nutrients, perhaps even too rich for most, with living barriers (the cytoplasm and its organelles) and physical barriers (the sieve plates) to their colonization and an inaccessibility at least comparable with, if not greater than that of the vessels. Therefore it should not be too surprising that pathogens colonizing this special environment themselves possess particular features and gain access through highly evolved mechanisms. They are certain types of viruses, viroids, mycoplasmas and rickettsias some of which, like beet curly top virus, are known to be introduced by insects feeding directly on the phloem tissue. But there are others, like the mycoplasma associated with lethal yellowing of coconut for which the mode of entry has not been established unequivocally. Once within sieve tubes and other phloem cells, however, they cause effects beyond this tissue because of their effects on its function of transport.

To these two specialized habitats can be added a third, the meristems from which plant growth is initiated. In this respect they are areas of high activity but their other special features of significance to colonization by plant pathogens are not obvious. Clearly there are some, because in many infected hosts with abundant virus particles within cells throughout the plant, few or no particles are found within the apical meristems of shoots and this feature has been utilized extensively in the development of virus-free stock through meristem culture. Also, very few other plant pathogens exist and grow within meristems. Indeed those that do are an odd collection with little in common except that, in a certain phase of their disease cycle, they are biotrophic. Examples are the downy mildew *Peronospora viciae* on pea, the smut *Ustilago nuda* on barley and the agaric *Crinipellis perniciosa* on cocoa. Infection by the species of *Peronospora* and *Crinipellis* induces a more rapid meristematic activity which is reflected in a swelling of the shoot but *U. nuda* causes no such effect and, externally, infected and healthy plants appear similar until heading. These pathogens thus present further problems in fungal biotrophy.

It would be an over-generalization to say that the fourth element of plant tissues, parenchyma cells, are not special in any way because obviously within this wide group some are modified for particular purposes such as photosynthesis, storage and stomatal regulation. Nevertheless they constitute the ground-tissue within ready access of pathogens entering by direct penetration, e.g. of the cuticle, or entering through stomata and through wounds. As such this is the type of tissue which most plant pathogens colonize. They do so by physical insinuation between cells as with *Erwinia amylovora* on apple or pear or *Puccinia hordei* in a barley leaf, or by enzymic degradation of the middle lamella as with *Erwinia carotovora* on potato and *Botrytis fabae* on faba bean; and they derive nutrients either from leakage through damaged host membranes or through haustoria.

To complete this review of infection processes in plants the aggregation of tissues into specific plant organs — the roots, stem, leaves, flower and seed — must also be considered, if for no other than the strictly practical reason that infection of and damage to these organs are the concern of the grower and the pathologist who advises him. Clearly some pathogens have developed the capability of infecting one particular type of organ, for example *Gaeumannomyces graminis* the roots and *Claviceps purpurea* the flowers of some cereals and grasses. To some extent such examples might reflect separation of susceptible tissue from source of inoculum but this is certainly not always so. Zoosporangia of *Phytophthora infestans* washed from lesions on potato leaves and stems into the soil infect the modified shoots (tubers) but do not apparently infect roots.

However, the development of plant organs does profoundly affect the extent of infection in ways which, relatively, are easier to understand. This applies both to the development of a single organ or the sequential development of the same or different organs. For example, the hypocotyl and young root of many seedlings are infected by *Pythium* spp. only in their early stages of development after which they become resistant; the same is true of onion cotyledons in relation to infection by the smut fungus *Urocystis cepulae*. The sequence of leaf development in cereals facilitates successive infections to the ear and its awns of pathogens such as *Puccinia striiformis* and, similarly, with oilseed rape, infections of leaves lead to invasion of the pod by *Alternaria brassicae*.

Impairment of plant functions and growth

Inevitably, any review of the infection of plant tissues and organs involves consideration of their functions and to that extent pre-empts what might be said about the second component of the system — the impairment of plant functions and growth. We can, however, extend the analysis by relating types of disease (rather than types of pathogen) to the observed establishment and growth of plants, a system for the classification of plant diseases originally proposed by McNew (1950, 1960) and which forms the basis on which the papers in this section are defined. Thus we can recognize seven plant functions and seven corresponding groups of diseases the causal pathogens of which interfere primarily with each function (Table 10.1). To this list might be added reproduction and its associated diseases, head smuts and ergot. Some of these groups are fairly well defined because they also relate to tissues with particular functions e.g. the vascular wilts, or to particular organs, e.g. root and foot rots. Perhaps the least satisfactory grouping is that for photosynthesis which is affected by many different types of pathogen and in different ways.

Symptoms

Infection and impairment of plant functions lead to alterations in cells, tissues and organs, many of which result in gross morphological changes which can be seen

Table 10.1. Plant functions and groups of diseases affecting them.

Mobilization of stored food	:	damping-off and seedling blights
Absorption of water and minerals	:	root and foot rots
Water transport	:	vascular wilts
Meristematic activity	:	leaf curl, witches' broom, club-root, galls
Photosynthesis	:	leaf spots, anthracnose, blight, mildew, rusts, leaf smuts, viruses
Translocation	:	some diseases caused by viruses, viroids, mycoplasmas, rickettsias
Storage	:	post harvest diseases of fruits and storage organs

After McNew (1950) and McNew (1960)

with the naked eye or microscopically. These changes are symptoms which provide evidence of the pathogenic processes involved. In this respect symptoms can be divided into three groups: those that result from the disruption of cellular processes and lead to death of cells (necrotic symptoms or necroses); those that in some way suppress growth or differentiation (hypoplastic symptoms or hypoplases); and those that result in a greater and abnormal development (hyperplastic symptoms or hyperplases). Most diseases have more than one symptom and it is the total symptom picture or syndrome which is the basis of disease diagnosis and is fundamental to the work involved in disease clinics, plant disease surveys and eradication campaigns. Many advisory pathologists regard disease diagnosis almost as an art and there is no doubt that in these matters experience counts a great deal. Several plant pathologists from earlier generations, notably H.H. Whetzel of Cornell University, were much taken with the science of symptomatology and attempted to combine an understanding of the underlying processes causing symptoms with the production of keys to the various symptoms and definitions of the terms used to describe them. A modification of Whetzel's key is to be found in the text by Roberts & Boothroyd (1972).

Unfortunately Whetzel's text on symptomatology (Whetzel, 1935) was never formally published in its entirety but many of his terms used to describe symptoms have been widely adopted. This is evident from the *Guide to the Use of Terms in Plant Pathology* prepared under the aegis of the Federation of British Plant Pathologists (1973). There are 41 terms common to both lists, although occasionally the definitions of individual terms differ slightly (Table 10.2). Six terms in the Guide do not appear in Whetzel's list, five of them relating particularly to symptoms associated with virus diseases. Then there are 28 terms in Whetzel's list which find no place in the Guide but about 14 of these are either self-explanatory (bleeding, bronzing, cracking, odors) or are used commonly in describing diseases (abscission, blast, fern leaf, firing, leak, mummification, pitting, proliferation, russetting, silvering). Others are unlikely to gain wide acceptance either because they entail a

Table 10.2. Terms for symptoms of plant disease.

(a) Common to text of Whetzel (1935) and FBPP Guide (1973)			
albinism	enation	leaf curl	scab
anthracnose	epinasty	leaf roll	scald
blight	etiolation	leaf scorch	scorch
blotch	fasciation	necrosis	shot-hole
canker	gall	plesionecrotic	spot
callus	gummosis	phyllody	streak
chlorosis	holonecrotic	resinosis	stripe
damping-off	hyperplasia	ringspot	stunting
die-back	hypertrophy	rosetting	virescence
dwarfing	hypoplasia	rot	wilt
			witches' broom

(b) Whetzel (1935) text only		(c) FBPP Guide (1973) only
abscission	leak	enation
achromatosis	mummification	epinasty
anthocyanescence	odors	mosaic
blast	pitting	mottle
bleeding	prolepsis	vein banding
bronzing	proliferation	yellows
chloronemia	restoration	
cracking	russetting	
fasciculation	sarcody	
fern leaf	savoying	
firing	shelling	
heterotophy	silvering	
hydrosis	suppression	
intumescence	tumefaction	

knowledge of the pathogenesis involved which limits their practical use (e.g. chloronemia = yellowing due to degeneration or disorganization of the chlorophyll or chloroplasts) or because they are not pleasing to the ear (e.g. albication = complete failure of chlorophyll formation). It is easy to be critical of a list so exhaustive as Whetzel's. Its great merit was to encourage a discerning appraisal of symptoms, an aspect which, regrettably, is sometimes neglected now in teaching plant pathology.

Host damage

It is doubtful whether there will ever be general agreement on the use of terms for plant disease symptoms and discussing terms can become an end in itself. There are more important aspects to consider. As well as diagnoses of causes, symptoms provide the basis for assessing amounts of disease and appraising the amount of damage caused. It is the relationship between symptoms and host damage which is most difficult to unravel — for three major reasons. One is that symptoms and the associated damage take time to develop. Infection by plant pathogens and impairment of plant functions inevitably lead to host damage. Essentially the difference between biotrophs and non-biotrophs is the time taken for damage to occur. *Phytophthora infestans* rapidly causes a necrotic lesion on a potato leaf; many downy mildews, e.g. *Plasmopara viticola* on a vine leaf and *Bremia lactucae* on a lettuce leaf, only do so when, following massive sporulation by the fungus, the underlying tissues collapse and turn brown. In the former situation the loss in photosynthetic area is swift and obvious and can be related directly to tuber development; in the latter the relationship is much more difficult to appraise. The loss of photosynthates may be disproportionately large in relation to the area damaged because of the nature of the pathogen involved.

The second reason is that location of symptoms often determines the impact of the damage caused. A large lesion on a leaf, e.g. of *Alternaria alternata* on tobacco, is limited in the total damage that it does to the plant even for crops like tobacco where the leaf is the end product. But a lesion at the stem base which girdles the main stem, e.g. that caused by *Sclerotinia sclerotiorum* on oilseed rape, can result in damage far in excess of its size, although that may be considerable. Similarly a lesion caused by *Venturia inaequalis* on an apple leaf is relatively unimportant compared with a similar lesion and the resulting scab on the fruit where considerations of quality are paramount.

The third reason is that individual plants have a remarkable ability to compensate for damage and to this must be added the compensation between individuals in a plant population, i.e. in crops. For example, the rate at which many root rots cause damage is determined not only by the rate at which a particular fungus grows in the root cortext but also by the ability of the plant to produce new roots above those that are infected. Similarly, the killing of one potato plant by girdling of the sprouts by *Rhizoctonia solani* often allows adjacent plants in the row to grow more vigorously and thus compensate for yield (Jellis & Boulton, this volume).

References

Federation of British Plant Pathologists (1973) *A Guide to the Use of Terms in Plant Pathology*. Phytopathological Paper No. **17**, Commonwealth Mycological Institute, Kew.

McNew G.L. (1950) Outline of a new approach in teaching plant Pathology. *Plant Disease Reporter* **34**, 106–10.

McNew G.L. (1960) The nature, origin and evolution of parasitism. In: *Plant Pathology, An Advanced Treatise* Vol. 3 (Ed. by J.G. Horsfall & A.E. Dimond), pp. 19–69. Academic Press, New York.

Roberts D.A. & Boothroyd C.W. (1972) *Fundamentals of Plant Pathology*. W.H. Freeman & Co., San Francisco.

Whetzel H.H. (1935) Symptomatology. Mimeographed paper, Plant Pathology Department, Cornell University, New York.

11 Infection by foot and root rot pathogens and subsequent damage

P.J. SHIPTON
North of Scotland College of Agriculture,
Aberdeen AB9 1UD, UK*

Introduction

Foot and root rots are enormously diverse groups of disease caused mainly by pathogens generally regarded as soil-borne. The ecological groupings of the range of soil-inhabiting and root-infecting fungi have been defined by Garrett (1956) thus:

- Root-infecting fungi
  - Many mycorrhizal fungi
  - Root disease fungi
    - Specialized parasites
    - Unspecialized parasites
- Obligate soil saprophytes

Root-inhabiting fungi: Many mycorrhizal fungi; Specialized parasites

Soil-inhabiting fungi: Unspecialized parasites; Obligate soil saprophytes

These groupings are important in understanding the characteristics of the behaviour of the root pathogens to be discussed here.

The pathogens causing root rots and foot rots can collectively be grouped as root disease fungi. These comprise the specialized parasites such as *Phymatotrichum*, *Armillaria* and *Gaeumannomyces* which are root-inhabiting fungi with a relatively restricted host range, have a low growth rate even in the parasitic phase, rarely producing metabolic toxins and subsequently suffering a declining saprophytic phase after the death of the host. In contrast, the unspecialized parasites such as *Pythium*, *Rhizoctonia*, *Phytophthora* and *Sclerotinia* etc. are soil-inhabiting fungi which have a much wider host range, aided by having strong competitive saprophytic abilities in the soil, with a high growth rate on the host and often producing toxins.

* Present address: Butterywells, Potterton, Aberdeenshire AB4 OUY.

Wood R.K.S. & Jellis G.J. (1984) *Plant Diseases: infection, damage and loss.* Blackwell Scientific Publications, Oxford.

Seedling blights and damping-off pathogens, vascular wilts and pathogens such as crown gall causing hypertrophy will not be considered in this chapter as they are discussed elsewhere in this volume, although their survival is associated with the soil environment and they enter plants often at or below ground level.

A review of foot and root rots needs to be concerned both with pathogenic activity in the host and saprophytic survival in the soil; in particular, with two distinct aspects of the latter environment which govern the activity of inoculum. The first is related to its survival phase in the bulk of soil in the absence of the host and the second to its pathogenic phase in the rhizosphere around the root, within which the inoculum is likely to be involved actively with infection of the host root and the progression of the disease.

The 'theatre of operations' involving the host, its pathogen and other micro-organisms is very largely confined to the surface layers of the soil. According to Kreutzer (1960), 60% of all soil-borne pathogens such as *Pythium*, *Rhizoctonia*, *Phytophthora*, *Fusarium*, *Sclerotinia* and *Helminthosporium* are confined to the top 7.6 cm, where they are subject to greater fluctuations in soil moisture, temperature, oxygen, other physical and chemical conditions, and biological activity. Some 30% of pathogens, for example, *Fusarium solani*, *Fusarium oxysporum* and *Verticillium albo-atrum* are found between 7.6 and 30.5 cm. Only 10%, such as *Phymatotrichum*, *Armillaria* and *Fomes*, occur much below 30.5 cm.; at these depths conditions are usually more stable, although it is more likely that oxygen, carbon dioxide and other constituents of the soil atmosphere will occur at concentrations critical to growth of host and pathogen.

Host–pathogen interactions in the soil environment

Within the soil the fundamental interaction between host and pathogen that can ultimately result in disease is substantially influenced by the soil microbes. These are present in all field soils as natural inhabitants but respond rapidly to the presence of host and pathogen to such an extent that they become part of a triangle of interactions, being involved with host exudates, the production of antibiotics and antagonism (in all its broadest senses). These interactions are represented diagrammatically in Figure 11.1.

The soil environment comprises an interrelated complex of physical and chemical factors (dependent on the soil type and even climate) which affect the activities of the host plant and the micro-organisms in the system more intimately than in the aerial environment.

The outcome of this three-way interaction, at each stage in a sequence of events, determines firstly the processes of infection (described elsewhere in this volume), then secondly the progress of disease and development of the syndrome (discussed in this section) and finally the expression of disease as seen by its effect on the crop (described in Section 4).

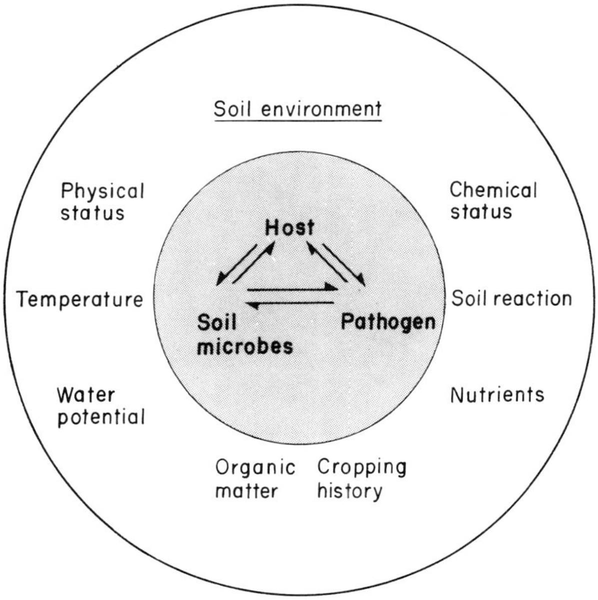

Figure 11.1. Diagramatic representation of the interaction between the physical and chemical factors and the biological and microbiological components in the soil environment. (From Park, 1963.)

The pathological importance of water potential

The concept of water potential needs to be considered here because of its importance for the activities of the host, its pathogens and other soil micro-organisms. Any organism in contact with pure, free water in a saturated soil does not need to do work to obtain that water. If, however, the same soil becomes unsaturated, then the organism must work to overcome the forces holding the water in the soil. The difference in work between the two situations is total soil water potential, which by definition decreases with decreased soil water content. Water potential in this context is a negative force measured in bars, one bar being almost equal to 1 atmosphere. At field capacity, water potential is about -0.1 to -0.5 bar, whereas at permanent wilting it is around -15 bar, corresponding to 99% relative humidity (Griffin, 1972). Large changes in water potential are coincident with quite small changes in soil atmospheric relative humidity (RH), caused for example by transpiration of crops. Thus -15 bar results from a 1% reduction in RH from 100 to 99%, while -40 bar is equivalent to about 97% RH.

Table 11.1 shows the reduction from optimum, half optimum to zero growth rate for four well-known soil-borne pathogens as water potential (expressed as bar matric potential) is reduced.

For example *Rhizoctonia solani* does not grow at -32 bar (equal to about 96% RH) but *Fusarium roseum* 'Culmorum' continues to grow down to about -75 bar or 94% RH (Cook & Papendick, 1972). It is significant that growth of pathogens can proceed well below the permanent wilting point of most host plants.

Table 11.1. Water potential and the effect on growth of pathogens.

Pathogen	Growth permitted (-bar matric potential)		
	Optimal	Reduced by half	Prevented
Rhizoctonia solani	>3	15	<32
Phytophthora cinnamomi	>5	20	35
Gaeumannomyces graminis	>1	20	50
Fusarium roseum f.sp. *cerealis* 'Culmorum'	0–5	45	75

Data from Cook & Papendick (1972)

At the same time, very small reductions in water potential have an even more drastic and immediate impact on bacterial activity, because of the need of motile species for a complete film of water over soil particles and roots to allow motility. The continuity and persistence of this film is one of the principal factors which determines the effectiveness of the microbial antagonism to root pathogens. As a saturated soil starts to drain off water, all the bigger voids empty rapidly, followed progressively by the pores between the soil particles. Even by -1 bar, spaces of 1 to 1.5 microns will be emptying, thus effectively inhibiting bacterial activity and much of the consequent antagonism. According to soil type this critical point is reached at -0.5 bar or so in coarse sandy soils and at -5 bar in fine silty soils (Griffin, 1972).

In general, bacterial activity is much reduced by -3 bar and almost extinguished at -10 bar. Microbial antagonism of soil-borne fungal pathogens is thus minimized at levels where both infection by pathogens and plant growth can still proceed.

Crop management and disease expression

Dryland foot rot of wheat caused by *F. roseum* 'Culmorum' is a well understood example of short-term crop management practices substantially altering the relationships between water potential and the host–pathogen–microflora triangle. In the Pacific North West of the USA the disease is naturally worse with (a) less than 30 cm winter rainfall, and (b) hot dry summers which may leave only 20 cm water reserves in the soil profile (Papendick & Cook, 1974).

Stress factors which have been proved to aggravate disease development are drought and high temperatures which lead to rapid exhaustion of the limited supply of water, high rates of nitrogen application and a high plant density (i.e. 1 plant/cm row). The effects of nitrogen rate and row spacing on water potential in the leaves and on the severity of foot rot are shown in Table 11.2.

The most severe disease developed earliest on plants at higher nitrogen rates and closer spacing; the extreme stress is shown by leaf water potentials of more than -30 bar. Wilting was more or less irreversible at -35 bar when water potentials

Table 11.2. The effect of nitrogen rate and row spacing on leaf water potential and severity of dryland foot rot in wheat.

Nitrogen rate (kg/ha)	Row spacing (cm)	Leaf water potential (- bar)	Disease severity index (0–10)
0	30	23	1.1
	90	29	3.3
67	30	32	7.4
	90	27	3.2
134	30	39	7.1
	90	32	2.2

Data from Papendick & Cook (1974)

of the surface soil were also at -35 bar. The explanation is that the initially vigorous and dense crop increases the rate of uptake and depletion of the limited water reserves, resulting in high plant and soil water potentials. Severe plant stress predisposes the crown of the wheat to rapid attack by the pathogen which was established in plants in the previous autumn as a slow-decay parasite. Soil microbial antagonism can protect the wheat from infection in wetter soils at -10 bar by causing lysis and formation of new chlamydospore germlings. In soils progressively drying-out to -50 and -60 bar, far below that required for bacterial activity, chlamydospores germinate and grow freely to cause infection without any inhibition of their activity.

Soil suppressiveness and conduciveness

Another disease in which soil microbial activity is determined by crop and management practices is fusarium root rot of beans. Severe attacks develop in conducive ('root-rot') fields, in newly irrigated areas of Washington State, USA. In some adjacent suppressive ('resistant') fields under long-term irrigation, the disease is consistently almost absent.

To investigate the nature of this phenomenon, Burke (1965a) progressively diluted infested root-rot soil with suppressive soil or with uncropped soil from a nearby native sagebrush site. In non-treated natural ('virgin') soils there was less disease with suppressive than with virgin soil, especially at low levels of inoculum (Table 11.3).

Sterilization hardly altered the effect of virgin soil but apparently destroyed the effect of the suppressive soil, which suggested that the suppression was caused by a microbial factor. Suppressiveness could not be transferred to, or artificially induced in, a conducive root-rot soil or the comparable virgin soil but this failure may be due to the way in which this effect is produced. Paradoxically, suppressiveness was found to be due to a highly stimulatory effect of the suppressive soil which caused chlamydospores to germinate half-abortively before producing fewer,

Table 11.3. Severity of fusarium root rot of beans with different soil treatments.

Dilution of root-rot soil	Disease severity index (0–100)			
	Suppressive soil		Virgin uncropped soil	
	Sterilized	Untreated	Sterilized	Untreated
1 : 1	83	71	91	95
1 : 10	76	60	89	82
1 : 100	79	38	58	74
1 : 1000	54	22	52	35
Unmixed controls	0	18	24	19

Data from Burke (1965a)

smaller, new chlamydospores with the result that the soil had lower inoculum than did the root-rot soil.

Even on heavily infested root-rot sites, plants could grow well if physical soil conditions were ideal. Growth at soil bulk densities of about 1.2 g/cm^3 was good but root-rot severity was greatly increased when soil compaction resulted in a high soil density of 1.5 g/cm^2. Disease was further increased in dry soil because of small reductions in soil water potential from -0.2 bar to -0.8 bar. Any stress due to drought and root restriction leading to disease and loss in yield could readily be eliminated by irrigation and subsoiling (Burke, 1965b).

A potentially valuable new development in disease forecasting has been the recognition and mapping of soil series or soil associations in the same area containing both conducive and suppressive soils. For example, *Pythium ultimum* is a major pathogen in the intensively cropped areas of the San Joaquin Valley in California. More than 87% of the sites on the moist finely textured soils in the Oxalis and Lethen soil series were suppressive, whereas 70% of the drier, more coarsely textured, Panoche series were conducive to disease development.

Experimentally, the suppressive properties of the more finely textured soils could be transferred by addition to the conducive soils, clearly implicating a microbial component (Hancock, 1979). Various wilts induced by *Fusarium* spp. provide another example of an association of suppressiveness and conduciveness with certain specific soil series (Alabouvette *et al.*, 1979).

One of the most thoroughly investigated examples (Broadbent & Baker, 1975) of disease-suppressive soils is that of some sub-tropical soils in Queensland and New South Wales which can limit the pathogenicity of *Phytophthora cinnamomi* on avocado (Weste, this volume). In practice, the methods for enhancing suppressiveness in the field have been evolved empirically by the avocado growers to excellent effect by using an intensive crop and soil management practice which greatly increases soil organic matter, nitrogen and calcium levels. In the virgin native rain forest soil, the pathogen has rarely if ever been recorded but when these soils are cleared and cultivated with avocado or pineapple, devastating attacks of root rot can occur. In some adjacent avocado groves, however, the disease is absent or negligible in spite of the presence of inoculum as indicated by bioassays. These groves are cultivated with cover crops of legumes and grasses which are ploughed-in with the addition of grass, chicken manure, cereal straw and dolomitic limestone to give a massive composting mulch. In contrast conducive soils have much lower organic matter and soil nutrient levels. The vigorous suppressiveness produced in the soil beneath the trees has been shown to involve extensive lysis of the hyphae of *P. cinnamomi*; Broadbent & Baker (1975) reported that as much as half the mycelium was destroyed in 72 h by various species of *Pseudomonas*, *Bacillus*, and *Streptomyces*.

Disease progression

It is important to study disease progression and development during the life of the crop and in successive crops when trying to forecast the effect of disease on crop production.

Tomatoes, for example, are regularly grown year after year in monoculture under glass and brown (corky) root rot (*Pyrenochaeta lycopersici*) becomes an increasing problem. It was found by Ebben & Last (1975) that yield losses were inevitable if the incidence of root rot exceeded 20% within 12 weeks of planting.

The predicted rate of increase in disease from the first years was not in fact constant, because there was a peak incidence of over 60% at 24 weeks after planting after 7–8 years continuous cropping and thereafter a slower rate of increase (Table 11.4). Comparisons with disease levels at 8 and 16 weeks after planting showed that the peak had occurred up to 2–3 years earlier. Yield losses, however, continued to increase in the eighth and ninth crops, suggesting that monoculture was influencing the host's response to infection. The behaviour of disease initially followed the pattern of increase expected for non-motile, soil-borne inocula. It was suggested that the limitation of disease might have resulted from disease which was so severe so early that the root system failed to allow the inoculum to be fully self-perpetuating, or because of a reduced virulence of the pathogen.

Ectotrophic growth habit

The specialized root pathogens with an ectotrophic habit merit attention because of their economic significance. Infection only becomes established after initial superficial colonization of the root surface by the mycelium. According to Garrett (1956) the degree of aggregation of the mycelia into strands and rhizomorphs has important effects on their behaviour and efficiency of growth ectotrophically. For example, *Helicobasidium purpureum*, spreading by mycelial strands, may grow ectotrophically only a few millimetres on many herbaceous plants, whereas *Armillaria mellea*, spreading by rhizomorphs, can extend itself many metres on some of its tree hosts. The time scale for the development of disease and progression of symptoms is longer for root pathogens with an ectotrophic growth habit

Table 11.4. Changes within and between years in the incidence of tomato brown root rot over nine successive years.

Year of cropping	Incidence of brown root rot (%)					
	Predicted			Actual		
	weeks after planting			weeks after planting		
	8	16	24	8	16	24
1st	3	8	18	3	8	18
3rd	11	23	38	11	23	38
5th	15	42	71	15	35	54
7th	23	62	90	16	38	62
9th	34	82	100	14	36	63

Data from Ebben & Last (1975)

than for many other root pathogens. Hence there are more opportunities for controlling the disease, either directly by cultural practices aimed directly at enhancing host resistance, or indirectly, by activating the antagonistic activity in the rhizosphere.

Many physical, chemical and microbial changes are produced in arable soils following direct drilling cultivation. Among many effects seen during the early work on direct drilling of cereals, one of the most striking was the pronounced reduction in the incidence of take-all (*Gaeumannomyces graminis*) after several successive direct-drilled crops of winter wheat compared to similar, conventionally-sown crops (Brooks & Dawson, 1968). The incidence of take-all was reduced by 50% after direct drilling even though soil inoculum levels were comparable to that of conventional crops. There was no differential effect on the establishment of infection and the differences in infection on mature crop plants appeared to depend on factors which limited and restrained the ectotrophic progress of the pathogen after establishment.

In Malaya, white root disease caused by *Rigidoporus (Fomes) lignosus* causes more losses in young rubber trees up to 10 years old than all other diseases and pests put together (Fox, 1977). Once the disease is established, pathogenesis is dependent on epiphytic growth of the strands, but if they can be checked, host wound healing reactions can stop any substantial further progress. Epiphytic growth along the root can be up to 30 cm a month, but in favourable conditions there is also extensive random mycelial growth in all directions. Disease caused by *R. lignosus* is most severe in near-saturated soils but is almost stopped when the relative humidity falls below 99%. Paradoxically, mycelial growth is highest at 40% soil water holding capacity (WHC) and lowest at above 80%. However, this is more than offset by the number and width of mycelial strands which are greatest at 80–90% WHC, thus ensuring maximum inoculum potential in the rhizosphere. To this end, in the Ivory Coast, cover plants of the family *Compositae* are sown to dry out the top-soil by evapo-transpiration.

Strategies for the control of white root disease in Malaya (described below) exemplify the practical integration of a variety of control measures which according to Fox (1977) can reduce loss by one half. Regardless of apparent infection every tree to be replaced is poisoned with sodium arsenite or 2, 4, 5-T herbicide well before felling in order to promote death and saprophytic invasion of roots by competitive saprophytes discouraging to *R. lignosus* development. Trees are planted in bare soil with wide strips of leguminous cover crops such as *Pueraria, Calapogonium* and *Centrosema* spp. in between which encourage rhizosphere actinomycetes antagonistic to *R. lignosus*. Some disease reduction occurs through loss of inoculum following production of large numbers of fruiting bodies which deplete food reserves rapidly. If the cover plants are not kept clear of the base of the tree, collar infection is promoted leading to the death of the tree. The litter layer and soil below the cover is highly antagonistic to *R. lignosus* with such intense mycostatic and mycolytic effects that mycelial destruction can be total within 16 days, whereas little if any lysis can be seen in soil kept bare. In antagonistic soils even epiphytic strands disappear in a matter of months so that the fungus appears to be moribund or in equilibrium with the microflora. However, it can be readily reactivated by any physical disturbance of the soil, so no mechanical cultivation is permitted between the trees. Young rubber trees are inspected post-planting for early symptoms which reveals primary root-tip infections. On suspect trees, the collar and roots are exposed to a depth and radius of 0.6 m so that advancing mycelial strands or rhizomorphs can be stripped off back towards their origin. Even if the tree has more than half the collar infected, death is not inevitable. Spontaneous healing can occur due to wound barrier formation and a 50–75% recovery has been reported in Java and Sumatra. Furthermore some reduction and destruction of old roots infected by *R. lignosus* by active feeding roots of healthy rubber trees can occur.

In spite of the suggestion by Fox (1977) that the mycelial strands of *R. lignosus* have rhizosphere effects detrimental to their own survival by encouraging their own antagonists, the concept of competition and displacement by other micro-organisms is not valid. This is because it is an unusually tenacious fungus within infected roots from which pure cultures of the pathogen can be obtained, although the roots are very old and well decayed.

Conclusion

This brief and selective summary demonstrates the importance of microbial activity and soil water relations in the complex of interactions determining the development of foot and root rots. More study is still needed because many practical agricultural operations can successfully alter the balance of the soil environment against the pathogen and in favour of the crop.

References

Alabouvette C. Rouxel F. & Louvet J. (1979) Characteristics of *Fusarium* wilt-suppressive soils and prospects for their utilization in biological control. In: *Soil-borne Plant Pathogens* (Ed. by B. Schippers & W. Gams), pp. 165–82. Academic Press, New York.

Broadbent P. & Baker K.F. (1975) Soils suppressive to *Phytophthora* root rot in eastern Australia. In: *Biology and Control of Soilborne Plant Pathogens* (Ed. by G.W. Bruehl), pp. 152–7. American Phytopathological Society. St. Paul, Minnesota.

Brooks D.H. & Dawson M.G. (1968) Influence of direct-drilling of winter wheat on incidence of take-all and eyespot. *Annals of Applied Biology* **61**, 57–64.

Burke D.W. (1965a) *Fusarium* root rot of beans and behaviour of the pathogen in different soils. *Phytopathology* **55**, 112–6.

Burke D.W. (1965b) Plant spacing and *Fusarium* root rot of beans. *Phytopathology* **55**, 757–9.

Cook R.J. & Papendick R.I. (1972) Influence of water potential of soils and plants on root disease. *Annual Review of Phytopathology* **10**, 349–74.

Ebben M.H. & Last F.T. (1975) Incidence of root rots — their prediction and relation to yield losses: a glasshouse study. In: *Biology and Control of Soil-borne Plant Pathogens* (Ed. by G.W. Bruehl), pp. 6–10 American Phytopathological Society. St. Paul, Minnesota.

Fox R.A. (1977) The impact of ecological, cultural and biological factors on the strategy and costs of controlling root diseases in tropical plantation crops as exemplified by *Hevea brasiliensis*. *Journal Rubber Research Institute, Sri Lanka* **54**, 329–62.

Garrett S.D. (1956) *Biology of Root-Infecting Fungi*. Cambridge University Press, Cambridge.

Griffin D.M. (1972) *Ecology of Soil Fungi*. Chapman & Hall, London.

Hancock J.G. (1979) Occurrence of soils suppressive to *Pythium ultimum*. In: *Soil-Borne Plant Pathogens* (Ed. by B. Schippers & W. Gams), pp. 183–9. Academic Press, New York.

Kreutzer W.A. (1960) Soil treatment. In: *Plant Pathology, an advanced treatise*, Vol. III (Ed. by J.G. Horsfall & A.E. Dimond), pp. 431–76. Academic Press, New York.

Papendick R.I. & Cook R.J. (1974) Plant water stress and development of root rot in wheat subjected to different cultural practices. *Phytopathology* **64**, 358–63.

Park D. (1963) The presence of *Fusarium oxysporum* in soils. *Transactions of the British Mycological Society* **44**, 119–22.

12 Foliar diseases: the damage caused and its effect on yield

ELLIS GRIFFITHS
Department of Agricultural Botany, University College of Wales,
Aberystwyth, Dyfed SY23 3DD, UK

Introduction

Before considering the damage caused by foliar pathogens and the consequential effects of such damage on yield, it is necessary to define normal leaf function in relation to the growth and development of the plant as a whole. In most plants leaves are the major sites of photosynthesis, capable of generating surplus carbohydrate which is translocated to other parts of the plant, providing both the basic materials and the energy for growth and development. At the end of a growing season, the dry matter produced by a plant represents the summation of surplus carbohydrate production (mainly from leaves) over the whole of that season.

When viewed in this way the performance of a leaf would, ideally, be measured by the amount of carbohydrate exported during its life. It follows that damage by a foliar pathogen would then be measured by the reduction in exported carbohydrate. In practice these measurements cannot be made directly and other related features of leaf function (light interception, green area, carbon dioxide uptake) have to be used instead. The usefulness of these measurements, as shown later, varies greatly for different host pathogen systems.

Assuming that a meaningful measure of leaf function (or dysfunction) can be obtained there remains the problem of relating this to yield. With the exception of forage plants, total dry matter production is not of primary importance to the agriculturalist. He is usually concerned with one component (e.g. seeds or tubers) which constitutes the commercial yield of the crop. It is important, therefore, to establish whether measurements of leaf function appropriate to dry matter production are also appropriate to seed or tuber production. For some crops the answer seems to be yes. Many studies have shown that leaf function, as measured by total light interception over the growing season, is closely related to total dry matter production (Monteith, 1977). In barley (Gallagher & Biscoe, 1978) and potatoes (Allen & Scott, 1980), there is an equally good relationship between total light interception and yield of seed and tubers, respectively. In these crops the fraction of the total dry matter which is harvested (harvest index) is a relatively stable character of the crop cultivar. For foliar diseases in such crops, accurate assessment of leaf dysfunction should define the effect on yield. There is, however,

Wood R.K.S. & Jellis G.J. (1984) *Plant Diseases: infection, damage and loss.* Blackwell Scientific Publications, Oxford.

one important proviso: measurements have to be integrated over the whole or at least the greater part of the growing season. Point estimates will, to varying degrees, give erroneous results.

Unfortunately not all crops have a stable harvest index. In faba beans (*Vicia faba*), for example, yield of seed is often poorly related to total dry matter production; other factors, notably those affecting pod setting, are more important (Lawes, 1974). It is also evident that much of the foliage in currently used cultivars does not contribute to seed yield (McEwen, 1972; Tamaki *et al.*, 1972; Ishag, 1973). This means that relating damage by foliar pathogens to reduction in seed yield is more complex.

Foliar diseases of barley

Barley offers a good system for examining relationships between leaf damage and yield because dry matter production and grain yield are closely related and there are several foliar pathogens, *Erysiphe graminis* (powdery mildew), *Rhynchosporium secalis* (leaf blotch) and *Puccinia hordei* (brown rust), which are largely confined to leaf laminae. Complications arising from infection of other plant parts are, therefore, minimal.

Powdery mildew

Virulent races of *E. graminis* f.sp. *hordei* produce the familiar white, powdery colonies on the leaf surface and a visible consequence of this type of infection is rapid senescence of affected leaves. Several workers have suggested a relationship between green leaf area (GLA) and yield loss (Last, 1962; Rea & Scott, 1973; Jenkyn, 1977) but this has only recently been examined in detail (Carver & Griffiths, 1981, 1982). Plants of the barley cultivar Julia were grown in the glasshouse or in the field at a spacing equivalent to 300 plants/m^2 and were exposed to infection by powdery mildew from growth stage (GS) 2 (Feekes scale) until maturity. Treatments with ethirimol were used to provide five different epidemic patterns, viz. no mildew (CC), early mildew (EM), late mildew (LM), continuous mildew (NF) and continuous mildew but delayed by seed treatment (F).

Mildew (as per cent GLA affected) and total GLA of primary shoots and tillers were recorded at regular intervals. These data were plotted graphically (Figure 12.1, 12.2) and areas under curves determined. Similar results were obtained in all experiments and analysis of the data showed consistently strong correlations ($r > -0.93$) between areas under mildew and GLA curves. It was also apparent that reduction in GLA lagged behind disease spread.

Grain yield was strongly correlated ($r > 0.94$) with the total area under the GLA curves plotted from GS 2 to GS 10.5 (Figure 12.3) Detailed evaluation of the data indicated that leaf laminae were not contributing significantly to grain yield after this time but this does not mean that assimilates produced by other structures

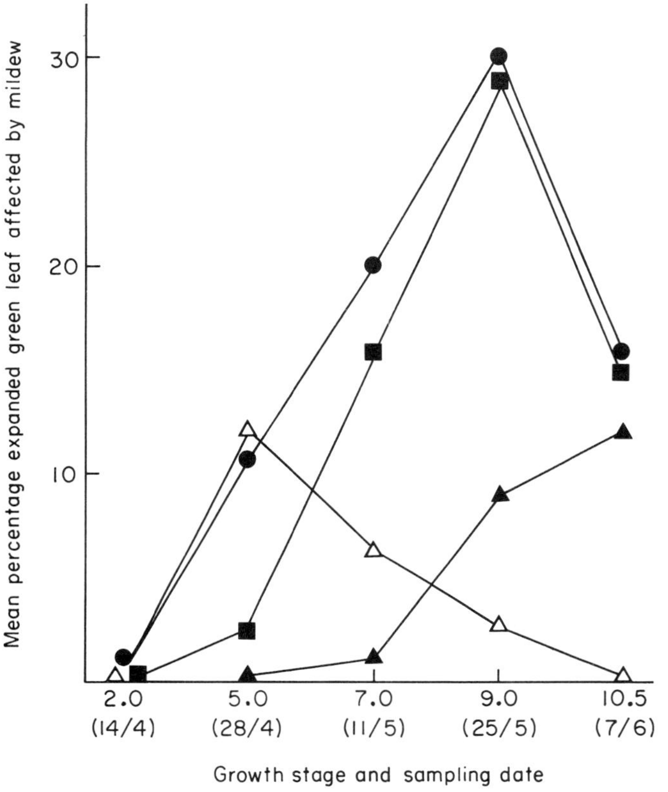

Figure 12.1. Mildew development on leaves of primary shoots in a glasshouse experiment (1980).
●——● = no fungicide (NF); ■——■ = ethirimol seed dressing (F); △——△ = early mildew (EM);
▲——▲ = late mildew (LM).
Data from Carver & Griffiths (1981)

(head, culm, leaf sheaths) were making no contribution. In barley these structures are relatively unaffected by mildew and their contribution does not affect the *variation* in yield associated with the disease. This is well shown by some recent data obtained by Slawson (unpublished) who found for barley cv. Golden Promise and Sundance that inclusion of the very substantial leaf sheath areas did not affect the correlation between GLA and grain yield (Table 12.2).

So far only effects of mildew and GLA in relation to total yield per plant have been considered. Effects on the components of yield vary somewhat in different experiments and in view of the complex factors influencing distribution of assimilates both between and within tillers (Kirby & Faris, 1972; Kirby, 1977; Kirby & Jones, 1977) this is not surprising. A lengthy discussion of this aspect would be inappropriate here but it is worth drawing attention to effects of early mildew epidemics which are subsequently well controlled. It has regularly been found (Griffiths *et al.*, 1975; Scott & Griffiths, 1980; Carver & Griffiths, 1981, 1982) that such attacks, in addition to causing some reduction in tiller number, also reduce number of grains per ear or grain size or both (Table 12.1). When these effects were first observed it seemed possible that spikelet differentiation might have been affected

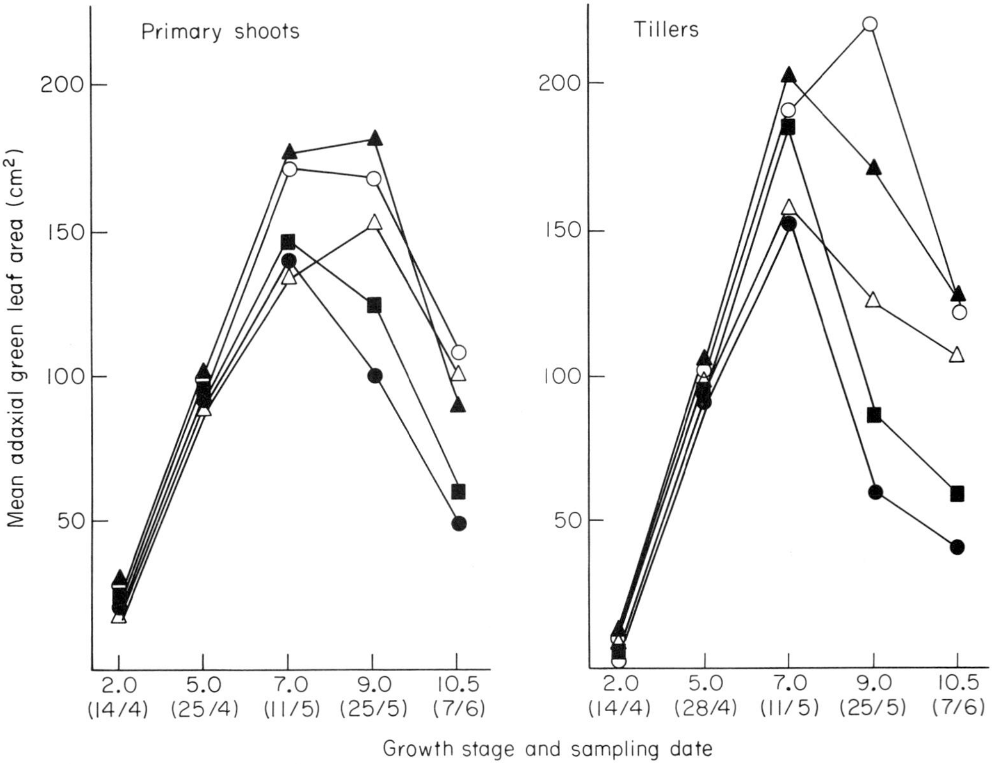

Figure 12.2. Green leaf area of leaves of primary shoots and tillers of barley plants in a glasshouse experiment (1980). ○——○ = no mildew; other symbols as Figure 12.1.
Data from Carver & Griffiths (1981)

but this now seems unlikely. In their experiments, Carver & Griffiths (1981, 1982) carefully recorded numbers of spikelet primordia and the number and sizes of leaves produced and found them to be the same in all treatments. Some alternative explanation for effects of early mildew on grain size and number was thus needed.

The first point to note is that at anthesis (GS 10.5) plants in the early mildew treatment are disease free and have a GLA equivalent to the mildew-free control (Figure 12.1, 12.2). They should, therefore, be at no disadvantage in terms of assimilate supply during the grain-filling period. This is obviously not so, implying that surplus photosynthate generated after anthesis is inadequate for full development of the grain. It is known that reserves of carbohydrate, stored in the culm, can contribute substantially to grain filling (Gallagher *et al.*, 1975) and it was inferred (Carver & Griffiths, 1981) that early mildew attacks might exert their effect on grain development through reduction in reserves. This was subsequently confirmed in an experiment (Carver & Griffiths, 1982) where early mildew was shown to reduce reserve carbohydrate by almost 50%.

It thus seems that for barley mildew, damage can be adequately defined in terms of reduction of GLA integrated over the period of GS 2 to GS 10.5 and that effects

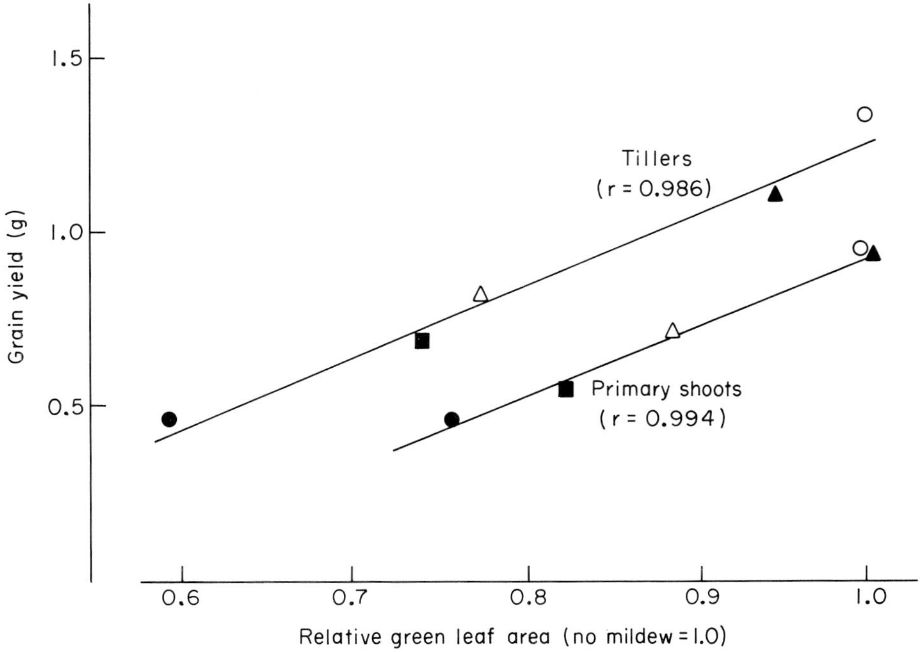

Figure 12.3. Relationship between grain yield of tillers and primary shoots and the relative green leaf areas (areas under curves in Figure 12.2). Symbols denote mildew treatments; legend as Figure 12.1. Data from Carver & Griffiths (1981)

Table 12.1. Mean dry weight of grain, dry weight per grain and number of grains produced on primary shoot ears of barley grown under different mildew treatments.

		Treatments*					
		CC	EM	LM	F	NF	$D_{5\%}$[†]
Glasshouse, 1980	Dry wt. of grain per ear (g)	0.957	0.726	0.951	0.566	0.468	0.175
	Dry wt. per grain (g)	0.043	0.039	0.047	0.031	0.028	0.009
	No. grains per ear	22.25	18.63	20.25	18.25	16.72	1.43
Field, 1981	Dry wt. of grain per ear (g)	0.930	0.770	0.928	0.715	0.661	0.112
	Dry wt. per grain (g)	0.042	0.037	0.041	0.036	0.033	0.004
	No. grains per ear	22.40	21.00	22.40	20.00	19.80	2.30

* CC — no mildew, EM — early mildew, LM — late mildew, F, mildew attach delayed by seed dressing, NF — continuous mildew

[†] Significant differences (P = < 0.05) based on the method of Tukey (Snedecor, 1956)

Data from Carver and Griffiths (1981, 1982)

on yield can be reasonably interpreted in terms of variation in assimilate supply.

Problems, however, remain. Although reduction in GLA is *proportional* to the area of the leaf occupied by mildew it is not equal to that area. *Prima facie* there seems to be no good reason for assuming that the relationships between mildew and GLA and between GLA and yield will be the same for all cultivars and under all

Table 12.2. Correlation coefficients of green leaf area (GLA), integrated from growth stage 2 to 10.5, with dry weight of grain (field experiment, 1981).

Disease	Barley cultivar[†]	GLA (leaf lamina)	GLA (leaf lamina plus sheath)
Powdery mildew	Golden Promise (S)	0.973***	0.975***
	Sundance (R)	0.951***	0.952***
Brown rust	Midas (S)	0.651§	0.553Π
	Triumph (R)	0.759	0.764
Leaf blotch	Georgie (S)	0.747*	0.728*
	Koru (R)	0.714*	0.734

*, ***, denote significance at 5% and 0.1% levels of probability
§ r = 0.851 for Midas when GLA post-anthesis is included in the correlation
Π r = 0.826 for Midas when GLA post-anthesis is included in the correlation
[†] (S) and (R) denote resistant and susceptible respectively
Data from Slawson (unpublished).

conditions. The little information at present available indicates that differences do exist between cultivars. Thus although excellent relationships exist between, say, GLA and yield for any one cultivar, different cultivars do not fit on the same regression line. Environmental factors also influence the relationship in absolute terms as is evident from the comparison of different experiments (Carver & Griffiths, 1982).

Finally, the foregoing has been concerned with effects of virulent races of *E. graminis*; it should not be forgotten that avirulent races may also be damaging, although in a rather different way. Smedegaard-Petersen & Stolon (1981) and Smedegaard-Petersen (this volume) have shown that exposure of barley plants to avirulent races results in much enhanced respiration rates even though visible symptoms of infection are absent. Since this would reduce the surplus carbohydrate available for export from leaves, some reduction in yield would be expected; reductions of about 7% (mainly due to smaller grains) have been obtained when plants were continuously exposed to the pathogen. This effect would be of little or no significance on crops of single cultivars but might be of some importance in cultivar mixtures. These are designed in such a way that the components are resistant to different races; the consequence is a marked retardation of the progress of an epidemic (Wolfe, 1977) because much of the inoculum generated on one cultivar is avirulent on the others. Therefore, a small yield penalty may be exacted.

Leaf blotch

The excellent relationship between severity of mildew infection, reduction in GLA and yield naturally raises the question whether similar relationships exist for other diseases. At first sight leaf blotch, with its well defined necrotic areas, appears a promising candidate but preliminary experiments are not encouraging (Table 12.2). In a field experiment where plants infected with mildew showed an excellent relationship between GLA and yield, a much poorer relationship was found with plants infected with leaf blotch. Only about 50% of the yield variation could be attributed to differences in GLA in the plots infected with leaf blotch compared with over 90% in the plots infected with mildew.

It is not known why the relationship was poorer with leaf blotch. Leaves infected by *R. secalis* frequently display an intricate pattern of necrotic and yellow areas which make it difficult to estimate accurately the remaining green areas. More precise measurement may enable closer correlation between GLA and yield to be established but preliminary studies indicate that the improvement may be only marginal. A more likely explanation is that damage extends beyond areas visibly affected i.e. those necrosed or yellowed. Evidence in support of this has been provided by Habeshaw (1979) who has shown that photosynthesis (measured as O_2 uptake) is reduced in green areas of leaves infected by *R. secalis*. Toxins may be involved as suggested by Auriol *et al.* (1978) although there has subsequently been criticism of this work (Daly, 1981). Whatever the correct explanation it is evident that account has to be taken of this impairment of photosynthetic activity in parts of the leaf which are not visibly damaged. It is important to note that the correlation of GLA with yield would be unaffected if there were a constant relationship between the area visibly damaged and the green area in which photosynthesis was impaired. It is likely, however, that the relationship is variable. For example, if photosynthesis was impaired for a fixed distance around visibly damaged areas then different patterns of visible damage, even of equal area, would obviously have different consequences for photosynthetic activity in the leaf as a whole. If this proved to be the case then accurate assessment of 'functional' GLA would be very difficult without the use of sophisticated scanning devices.

Brown rust

Preliminary studies with brown rust indicate that the nature of the damage caused and its consequential effects on yield differ from that in mildew and leaf blotch. In the last two diseases inclusion of GLA post-anthesis had no effect on the correlation between GLA and yield but with leaf rust a substantial improvement in correlation was achieved (Table 12.2). In this experiment rust was most severe during the later stages of plant growth. However, in view of the data obtained in experiments with powdery mildew, it is difficult to believe that it was loss of GLA at this time that was responsible for yield loss.

In the absence of specific data for this disease any interpretation must be largely speculative. There are, however, a number of points which can be made. First, it should be noted that, unlike mildew and leaf blotch, the area colonized cannot be accurately defined by visual observation. What is normally measured is the area occupied by sporing pustules; this is certainly a gross underestimate of the area occupied by the fungus. Second, rusts may permit leaf cells to remain in an apparently functional state (green island effect) and, for a time, may even enhance CO_2 uptake (Livne & Daly, 1966). In these circumstances reduction in GLA becomes a poor indicator of leaf dysfunction. Third, rusts constitute powerful sinks. In a recent survey of effects of disease on translocation of photosynthetic products, Whipps & Lewis (1981) reported observations made on seven rust diseases. Reduction in export of photosynthate was regularly noted and in four instances infected leaves were found to import photosynthate.

If brown rust provides a powerful sink during the grain filling period its effect on yield could be considerable. However, much of the effect at this time might be due to exploitation of photosynthate, normally available for grain filling (e.g. reserves in the culm) rather than to any curtailment of its production.

These three diseases of barley highlight many of the difficulties in relating severity of foliar infection to damage and loss of yield. To consider damage in terms of a reduction in surplus carbohydrate available to the plant for growth and development seems a useful concept but devising suitable methods for measurement of this key variable is not always easy. Integrated values for GLA are, subject to the limitations already noted, adequate for mildew and refinement in methods of measurement may enable GLA to be used for leaf blotch also. Rust, however, poses a more difficult problem as some means of measuring the ability of the pathogen to import photosynthate is needed.

Foliar diseases of faba beans

Faba beans are affected by a number of foliar pathogens but most attention has been devoted to chocolate spot caused by *Botrytis fabae*. Severe attacks can cause massive defoliation but unless the disease occurs early there may be little or no effect on seed yield as shown by Grainger (1949). The reasons why early attacks are so damaging and why late attacks are of little consequence have emerged from a number of studies examining effects of partial defoliation (McEwen, 1972; Ishag, 1973; Tamaki *et al.*, 1972). The broad conclusion from these studies is that after flowering the only leaves influencing yield are those at flowering nodes; leaves above and below these nodes can be removed with negligible effect on seed yield.

These observations prompted a comparative study of effects of mechanical defoliation and disease on seed production in faba beans (Griffiths & Amin, 1977). Plants of the cultivar Maris Beagle were sown in December out of doors and treatments applied when flowers formed at the ninth flowering node. Three defoliation treatments were applied in all combinations:

(a) removal of all leaves below the first flowering node (zone A),
(b) removal of all leaves at the nine flowering nodes (zone B) and
(c) removal of all leaves above the ninth flowering node by severing the stem (zone C).

The same three zones were inoculated separately and in all combinations with *B. fabae*. Leaves in zone A were rapidly killed but those in zone B were less affected (35% area infected 3 weeks after inoculation). Only a small amount of infection occurred in zone C (8% after 3 weeks). In accordance with expectation, defoliation

Table 12.3. Mean effects of zonal defoliation (D) and inoculation with *Botrytis fabae* (I) on seed yield of faba beans.

Zones* defoliated or inoculated	Total seed weight per plant (g)	
	D	I
A	16.5	14.9
B	12.1	10.0
C	13.8	14.6
A + B	13.7	11.5
A + C	13.1	12.0
B + C	8.9	9.2
A + B + C	6.0	8.8
None	17.2	18.2
$D_{5\%}$ †	2.11	3.78

* A — nodes below first flowering node, B — first nine flowering node, C — nodes above ninth flowering node
† Significant differences (P = < 0.05) based on the method of Tukey (Snedecor, 1956)
Data from Griffiths & Amin (1977)

or inoculation of zone B had the most damaging effects on seed yield (Table 12.3). Treatments applied to other zones were not, however, without effect and the view that only leaves at flowering nodes are important is evidently an over-simplification. Without a much fuller understanding of assimilate supply and its partitioning in faba beans, establishing relationships between disease and seed yield will continue to be a problematical affair.

Conclusions

Leaves may simply be regarded as sites of photosynthesis and providers of surplus product for plant growth. Foliar pathogens reduce growth in proportion to the degree to which they inhibit this function. Devising means of measuring damage may be difficult but, in principle, should be achievable. Relating damage to loss of yield in cases where only a portion of the total plant is harvested, poses different problems; its attainment depends on the reliability of models describing assimilate partitioning within the plant.

Damage to the supply and translocation of assimilates are of primary importance but foliar pathogens also have other effects which may influence growth e.g. on growth regulators (Pegg, 1981, and this volume) and on water relations (Ayres, 1981, and this volume). Ultimately, these also have to be integrated with effects on assimilates to provide an overall appraisal of the relationships between infection, damage and yield loss.

References

Allen E.J. & Scott R.A. (1980) An analysis of growth of the potato crop. *Journal of Agricultural Science, Cambridge* **94**, 583–606.

Auriol P., Strobel G., Pio Beltran J. & Gray G. (1978) Rhynchosporoside, a host-selective toxin produced by *Rhynchosporium secalis*, the causal agent of scald disease of barley. *Proceedings of the National Academy of Sciences USA* **75**, 4339–43.

Ayres P.G. (1981) Effects of disease on plant water relations. In: *Effects of disease on the physiology of the growing plant* (Ed. by P.G. Ayres), pp. 131–48. Society for Experimental Biology — Seminar Series II. Cambridge.

Carver T.L.W. & Griffiths E. (1981) Relationship between powdery mildew infection, green leaf area and grain yield of barley. *Annals of Applied Biology* **99**, 255–66.

Carver T.L.W. & Griffiths E. (1982) Effects of barley mildew on green leaf area and grain yield in field and greenhouse experiments. *Annals of Applied Biology* **101**, 561–72.

Daly J.M. (1981) Mechanisms of Action. In: *Toxins in Plant Disease* (Ed. by R.D. Durbin), pp. 331–94. Academic Press, New York.

Gallagher J.N. & Biscoe P.V. (1978) Radiation absorption, growth and yield of cereals. *Journal of Agricultural Science, Cambridge* **91**, 47–60.

Gallagher J.N., Biscoe P.V. & Scott R.K. (1975) Barley and its environment. V. Stability of grain weight. *Journal of Applied Ecology* **123**, 319–36.

Grainger J. (1949) The tithe of plant disease. *Transactions of the Royal Highland and Agricultural Society of Scotland* **61**, 56–76.

Griffiths E. & Amin S.M. (1977) Effects of *Botrytis fabae* infection and mechanical defoliation on seed yield of field beans (*Vicia faba*). *Annals of Applied Biology* **86**, 359–67.

Griffiths E., Jones D.G. & Valentine M. (1975) Effects of powdery mildew at different growth stages on grain yield of barley. *Annals of Applied Biology* **80**, 343–9.

Habeshaw D. (1979) The effect of foliar pathogens on leaf photosynthetic carbon dioxide uptake of barley. In: *Photosynthesis and Plant Development* (Ed. by R. Marcelle, H. Clijsters & M. Van Poucke), pp. 355–73. W. Junk, The Hague.

Ishag H.M. (1973) Physiology of seed yield in field beans (*Vicia faba* L.). II. Dry matter production. *Journal of Agricultural Science, Cambridge* **80**, 191–9.

Jenkyn J.F. (1977) Effects of mildew (*Erysiphe graminis*) on green leaf area of Zephyr spring barley, 1973. *Annals of Applied Biology* **82**, 485–8.

Kirby E.J.M. (1977) The growth of the shoot apex and the apical dome of barley during ear initiation. *Annals of Botany* **41**, 1297–308.

Kirby E.J.M. & Faris D.G. (1972) The effect of plant density on tiller growth and morphology in barley. *Journal of Agricultural Science, Cambridge* **78**, 282–8.

Kirby E.J.M. & Jones H.G. (1977) The relations between the main shoot and tillers in barley plants. *Journal of Agricultural Science, Cambridge* **88**, 381–9.

Last F.T. (1962) Analysis of the effects of *Erysiphe graminis* D.C. on the growth of barley. *Annals of Botany* **26**, 279–89.

Lawes D.A. (1974) Field beans: improving yield and reliability. *Span* **17**, 21–3.

Livne A. & Daly J.M. (1966) Translocation in healthy and rust-affected beans. *Phytopathology* **56**, 170–5.

McEwen J. (1972) Effects of defoliating different zones on the plant in field beans (*Vicia faba* L.). *Journal of Agricultural Science, Cambridge* **78**, 487–90.

Monteith J.L. (1977) Climate and the efficiency of crop production in Britain. *Philosophical Transactions of the Royal Society of London* **281**, 277–94.

Pegg G.F. (1981) The involvement of growth regulators in the diseased plant. In: *Effects of disease on the physiology of the growing plant* (Ed. by P.G. Ayres), pp. 149–78. Society for Experimental Biology — Seminar Series II. Cambridge.

Rea B.L. & Scott R.K. (1973) The effects of mildew (*Erysiphe graminis*) on leaf growth and yield of spring barley. *Proceedings of the 7th British Insecticide and Fungicide Conference* pp. 29–37.

Scott S.W. & Griffiths E. (1980) Effects of controlled epidemics of powdery mildew on grain yield of spring barley. *Annals of Applied Biology* **94**, 19–31.

Smedegaard-Petersen V. & Stolen O. (1981) Effect of energy-requiring defense reactions on yield and grain quality in a powdery mildew-resistant barley cultivar. *Phytopathology* **71**, 396–9.

Snedecor G.W. (1956) *Statistical Methods*, 5th edn. lowa State University Press, Ames.

Tamaki K., Naka J. & Asanuma K. (1972) Physiological studies on the growing process of broad bean plants. VI. Effects of partial leaf removal in the flowering and maturing stages on the growth and variation of chemical components. *Kagawa Diagaku Nogakuku Grakuzyutu Hokuku* **24**, 1–9. In: *Field Crop Abstracts* **26**, 442.

Whipps J.M. & Lewis D.H. (1981) Patterns of translocation, storage and interconversion of carbohydrates. In: *Effects of Disease on the Physiology of the Growing Plant* (Ed. by P.G. Ayres), pp. 47–84. Society for Experimental Biology — Seminar Series II. Cambridge.

Wolfe M.S. (1977) Yield stability in barley using varietal mixtures and disease control. *Cereal Research Communications* **5**, 119–24.

13 Infection by pathogens affecting reproduction and resulting damage and loss

S. SHAW

Department of Biological Sciences,
North East Surrey College of Technology, Reigate Road,
Ewell, Surrey, KT17 3DS, UK

Introduction

A broad definition of reproductive disease might encompass any harmful deviation from normal functioning of physiological processes that can affect the reproductive capacity of a plant. Reviews on this subject (Ciccarone, 1960; Mathre, 1978) have not only described pathogens that directly affect reproductive capacity through colonization of reproductive parts at the time of flowering, but also those which reduce the vegetative plant's capacity to support the reproductive structures. The health and quality of seeds and fruits colonized by pathogens during the post-fertilization phase may also be considered, especially when a disease predisposes towards a reduction of the reproductive capacity of the next generation of plants.

Seedling diseases associated with seed-borne pathogens are described by White (this volume) and there are several comprehensive reviews on seed pathology (Noble, 1971; Neergaard, 1977; Hewett, 1978, 1979). This paper primarily concerns infection, damage and loss caused by pathogens which, at least for part of their life cycle, are more directly associated with reproduction, particularly in grass and cereal crops in UK.

Infection and host damage

Establishment of infection

The infection court of pathogens causing reproductive disease may be on bud scales and perianth, accessory to reproduction, or in stamens and carpels. Infection at later, post-fertilization, stages may occur through the pericarp and testa of a developing fruit or seed. In loose smut disease of wheat and barley caused by *Ustilago nuda* there is limited growth of hyphae in the scutellum and plumular regions of the embryo, from where hyphae subsequently colonize shoot apices and inflorescence primordia of developing seedlings. In other smut diseases, including common bunt

Wood R.K.S. & Jellis G.J. (1984) *Plant Diseases: infection, damage and loss.* Blackwell Scientific Publications, Oxford.

of wheat caused by *Tilletia caries*, seedlings are infected from seed-borne or soil-borne inoculum. Hyphae eventually enter the differentiating inflorescence, having reached the stem apex early in the seedling stage.

Blossom blight

Many pathogens take advantage of the suitably humid microclimate in flowers and floral buds to infect host plants. Petals, stamens and carpels seldom have a complete or effective cutin layer as the first line of defence against invading bacteria and fungal hyphae; the presence of pollen (Chou & Preece, 1968), stigmatic secretions and nectar may stimulate growth of the pathogen during the pre-penetration period before infection. Infection of some pome fruits, hawthorn and related ornamentals by *Erwinia amylovora* commences as bacteria enter any of the numerous natural openings on flowers. These include stomata on the styles, sepals and fruit epicarp, and hydathode-like structures sometimes located on sepals and at the base of the nectary (Rosen, 1936; Schroth *et al.*, 1974).

In stone fruits, where blossom appears before leaf buds expand, flowers are important sites for early infection by *Monilinia* species. *M. fructicola* and *M. laxa* can infect blossom of stone fruits through petals, stigmas or stamens in high humidity and then blight all floral structures within a few days (Weaver, 1950; Byrde & Willetts, 1977). The ubiquitous fungus *Botrytis cinerea* is typical of facultative parasites that cause flower blights and rots in that the initial stage of infection is often restricted to damaged or senescent tissues, particularly anthers and stigmas which wither after pollination (Ogawa & English, 1960; McClellan & Hewitt, 1973). In cereals, dehisced anthers are susceptible to invasion by fungal parasites and saprophytes including *Fusarium*, *Cladosporium*, *Alternaria* and *Helminthosporium* species which may then spread over the seed surface or colonize the seed testa.

Carpel and embryo infection

Infection of carpels is well illustrated by the pathogens causing ergot, loose smut and blind seed diseases of gramineous hosts. Spores of *Claviceps purpurea*, the cause of ergot disease of certain cereals and grasses, may be deposited on exposed stigmas of host inflorescences. Until recently, it was accepted that *C. purpurea* only penetrated the ovary wall (Campbell, 1958) but light and electron microscope studies of infection of wheat and rye (Shaw & Mantle, 1980) have now demonstrated that primary infection can also occur in the stigma. Intercellular hyphal growth follows a route similar to pollen tubes, through the transmission tissue of the stigma and style and then between the outer integuments of the ovule and the inner epidermis of the ovary wall. Further investigation showed that although ovary tissue is eventually replaced by mycelium of the ergot fungus, the first 4 or 5 days of colonization of the ovary are remarkable for the surprisingly delicate effects on the host tissues. Production of pectolytic enzymes and β-glucanase appears to enable the fungus to grow between cells without eliciting a lethal response in adjacent tissues, and vascular tissue at the base of an ovary remains intact despite the close proximity of hyphae. Dickerson *et al.* (1978) have suggested that the glucanases produced by *C. purpurea* prevent the formation of callose plugs in sieve tubes, even in the later stages of colonization of the ovary, and thus maintain a free flow of sap from phloem tissue. Invasion of the ovary eventually leads to its total replacement by hyphae as an ergot sclerotium is formed (Luttrell, 1980; Shaw & Mantle, 1980). In contrast to the dark, pigmented sclerotium, there is also an intermediate stage of white, hyphal growth and conidiospore production as the fungus permeates the ovary tissue. Within 6–10 days after infection, conidiospores are carried to the outside of the infected floret as a suspension in an exudate of vascular sap. This 'honeydew' may contain as many as 10^8 spores/ml and is therefore a potent source of inoculum for a secondary phase of infection, particularly where there is an opportunity for transfer of spores between early flowering grasses and cereals (Mantle *et al.*, 1977). At harvest, some ergots fall to the ground but many are gathered with the seed and therefore become important as contaminants. Ergots in the soil germinate in the spring and early summer to produce perithecia which eject ascospores into the air thus initiating a new cycle of infection (Mantle & Shaw, 1976).

There are many similarities between the life-cycles of *C. purpurea* and *Gloeotinia temulenta* which commonly causes blind seed disease in seed crops of ryegrass and other commercial grasses (Couch, 1973). *G. temulenta* can establish disease in various forms, depending on the timing of infection in relation to floral development. Exposed, receptive stigmas are susceptible to early infection although this generally leads to destruction of the embryo and cessation of seed development. Later infection, after the embryo and endosperm have differentiated, leads to non-viable 'blind' seeds which when sown with healthy seeds can remain intact in the soil until early summer when fungal mycelium in the seed gives rise to apothecia. Ascospores, discharged into the air around flowering grasses, complete the cycle of infection (Wright, 1967).

Ustilago nuda also infects wheat and barley at flowering but, in contrast to ergot and blind seed diseases, the first phase of colonization results in insignificant damage to the ovary or to the embryo of a developing seed. Little appears to have been added to our knowledge of infection of an ovary by *U. nuda* since the work of Batts (1955) who concluded that penetration occurs through the ovary wall and not the stigma or style as had been previously accepted. However, in the absence of more recent investigations using modern microscopy techniques, it is perhaps unwise to conclude that penetration via stigma and style never occurs. Subsequent colonization of the embryo is a fairly slow process, taking about 30 days. Batts (1955) observed intracellular hyphae passing consecutively through a number of cells of the pericarp but, even where hyphae passed directly through cytoplasm, encapsulation and sheathing appeared to ensure that there was minimal damage to the host cell. Growth of hyphae is most prominent in the scutellum but more importantly the fungus invades the growing point of the embryo where the mycelium becomes dormant until the seed germinates. Infected and healthy seeds are indistinguishable, except by extraction and microscopic examination of the embryos as in the standard embryo test (Popp, 1959). From the seedling stage to the start of heading there are no obvious outward signs of infection although, in addition to the crucial infection of the stem apex, there is also limited and non-destructive intercellular growth of the fungus in nodal regions (Batts & Jeater, 1958). Colonization of the inflorescence occurs at a very early stage in its development. Replacement of floral tissues by teliospores can be seen by the time the whole barley ear is about 1 cm long. The life cycle of the pathogen is completed as entirely smutted ears emerge from the leaf sheaths of infected tillers and teliospores are dispersed to healthy plants as flowering commences.

Seedling infection by smuts

Teliospores of *Ustilago avenae*, the cause of loose smut of oats, are dispersed at flowering time but unlike *U. nuda* the hyphae from germinating spores invade only the glumes and pericarp of the seed before harvest. Dormant mycelium becomes active when seeds germinate. Hyphae then penetrate the developing seedling. Subsequent colonization of the host is then similar to that by *U. nuda* (Mills, 1966).

Bunt, or covered smut of wheat, caused by *Tilletia caries*, does not completely destroy the inflorescence of its host. Replacement of host tissues is restricted to the ovary as a 'bunt ball' containing a mass of teliospores is formed. In a seed crop, healthy seeds are therefore only contaminated by dry teliospores as bunt balls break during harvesting. Germination of spores at seed germination leads to penetration of seedlings. The early stages of infection are uncertain because the very thin hyphae are obscure in tissue sections. Subsequently mycelium reaches the stem apex and invades the developing inflorescence.

Factors affecting the incidence of ergot and loose smut

Weather conditions

Cool, damp weather extends the period of flowering, the only time when the host is susceptible, and is favourable for infection. Although the dry spores of *U. nuda* are wind-dispersed, high humidity encourages spore germination and infection, whereas light rain assists in depositing spores on barley heads and splashing them into open florets (Malik & Batts, 1960). Ergot is perhaps even more dependent on damp conditions. High moisture content of soil and in the air around susceptible plants encourages the formation of fruiting bodies from germinating sclerotia and promotes ascospore discharge (Mantle & Shaw, 1976). In addition, infected florets exude more conidiospore-laden honeydew, thereby increasing the potential for secondary infection. After a few days of hot, dry conditions, the flowering period of the host is shortened and both honeydew and exposed fruiting bodies may dry out and cease to be important sources of inocula. This point is emphasised because it is possible that a cultivar may appear to show field resistance to ergot mostly, or even entirely, by virtue of its particular flowering period having coincided with a few days of hot, dry weather.

Open flowering habit

A second factor is the role of disease escape associated with the degree and duration of gaping of florets at anthesis. Quite apart from physiological aspects of resistance which are not reviewed here, some barley cultivars exhibit field resistance to loose smut by virtue of their closed-flowering habit (Pedersen, 1960). This feature was originally investigated by Malik & Batts (1960) by allowing drops of coloured water to fall on to ears of barley cultivars at anthesis. Staining of style or ovary tissues indicated the extent to which droplets had entered the open florets and was found to correlate with observed field resistance of the cultivars. Since the development of Fl hybrid wheat and barley, which require male-sterile plants with gaping florets for seed multiplication, there has been renewed interest in the open flowering characteristic in relation to the incidence of both loose smut and ergot (Done, 1973).

Pollination

The effects of pollination and fertilization on susceptibility of host plants to infection by ergot and loose smut appear to be particularly important in self-pollinating cereals like wheat and barley where anthesis usually precedes flower opening. Several studies have demonstrated that some degree of physiological resistance to ergot in cereals is apparent soon after pollination although the mechanism remains unclear. One report (Darlington & Mathre, 1976) suggests that resistance is established within 30 min, whereas other investigations have shown a less rapid decline of susceptibility to ergot during the first 48 h after pollination (Puranik & Mathre, 1971; Watkins & Littlefield, 1976). On the other hand, it has been shown that unfertilized, male sterile cultivars can remain susceptible to ergot for up to 10 days (Puranik & Mathre, 1971; Shaw, 1975). Mathre (1978) has suggested that the mechanism of pollen-induced resistance may be related to hormonal changes in fertilized tissue affecting the penetration phase of infection.

However, if hyphal growth of *C. purpurea* through the stigma and style is preceded by a pollen tube, collapse of the transmission tissue may possibly hinder the establishment of the pathogen. A similar deduction has been made by Thakur & Williams (1980) for ergot in pearl millet caused by *Claviceps fusiformis*.

In barley, most pollen is shed within the spikelet before the florets open and infection by loose smut appears to be less critically affected by pollination. Successful inoculation of barley spikelets has been claimed even at late stages of anthesis when fertilized ovaries are beginning to differentiate into seeds (Malik & Batts, 1960). The possibility that penetration by *U. nuda* occurs through the ovary wall and not via the stigma or style may account for the lack of rapid induction of resistance to this pathogen in response to pollen tube penetration. Nevertheless, the period of maximum susceptibility in the field is generally considered to commence as flowers gape to expose fertilized and unfertilized ovaries.

Host damage in relation to yield loss

Reflection on the type and extent of damage to flowers, seeds and fruits inflicted by pathogens could leave the impression that such pathogens are very destructive. A blighted flower is a lost fruit, replacement of ovary tissue by hyphae of *C. purpurea* is a lost seed, and systemic invasion of a cereal plant by one of the smut pathogens may cause total loss of yield. Yet in the UK many of the diseases described in this paper are currently regarded as relatively unimportant so far as national crop yields are concerned. This is because in agriculturally developed countries the incidence of most reproductive diseases caused by seed-borne pathogens has been regulated by seed certification schemes (Hewett, 1979) and there has been extensive use of chemical seed-treatments. Common bunt of wheat was once regarded as one of the most important diseases of cereals (Large, 1940) but it has been brought under control in agriculturally developed countries through the effective use of chemical seed-treatment. However it still causes serious losses in under-developed countries. In North Africa and Central Asia, where it is estimated that less than 40% of seed is treated, recent estimates indicate a 5–7% annual loss of wheat yield due to common bunt, which still ranks as second in importance after rust diseases (Hoffmann, 1982).

Loose smut disease can also cause serious crop losses in some years if inoculum of *U. nuda* is allowed to increase unchecked from one generation of seed to the next. Doling (1966) estimated that within an untreated and susceptible crop, loose smut can increase 2–20 times during one growing season, although clearly this would depend on the weather conditions at anthesis. In 1981 the Official Seed Testing Station (Cambridge, England) reported that the amount of infected seed of winter barley cultivars in the UK was potentially serious; about 20% of samples of seed subjected to an embryo test had more than 0.3% infection (Anon., 1981). As much as 5.2% infection was found in one sample. Yield loss is generally calculated on the basis that the percentage smutted ears in a crop is directly proportional to the percentage yield loss (Demby, 1971). However, in the example given above the actual number of smutted ears in subsequent crops would have been considerably reduced by selection of the more healthy seed stocks and by the use of systemic fungicides as seed treatments.

Direct loss of grain yield due to ergot disease is principally the result of total replacement of ovary tissues by a sclerotium in an infected floret. In addition, there can be less obvious damage in the form of aborted ovaries and damaged grains resulting from unsuccessful attempts to infect (Shaw, 1975). There may also be a reduction in grain yield due to the diversion of nutrients from healthy to infected florets. Ergot honeydew can have 10–50 times lower water potential than phloem sap, thus creating a substantial sink for nutrients (Mower & Hancock, 1975). Nevertheless, it is difficult to imagine circumstances in the UK where total yield losses due to ergot in a normal, fully fertile cereal crop would exceed, or even approach, 1%.

Consequential losses in the seed and grain trades

Evaluation of yield loss due to reproductive diseases caused by seed-borne pathogens may lead to unwarranted complacency if the broader implications of post-harvest losses and costs incurred by seed and grain traders are not also considered. In the UK the main strategy for control of loose smut in susceptible cultivars has been to limit seed-borne multiplication of this pathogen from one season to the next. According to UK standards, certified cereal seed should not have greater than 0.2% loose smut to satisfy the Higher Voluntary Standard, or 0.5% to satisfy the minimum EEC standard (Anon., 1980). The seed merchant, who is responsible for maintaining this standard, can estimate the likely degree of loose smut infection in a seed crop during crop inspection and by observing control plots grown from a sample of the previous generation of seed. Where unacceptable levels of loose smut are indicated, an embryo test (currently costing about £1 per tonne of seed) may be carried out to determine the percentage infection. If the statutory or voluntary standards are exceeded, the merchant may either reject the grain for use as seed or apply an approved systemic seed dressing. The latter may lead to at least a 20% increase in the selling price of cleaned and treated seed compared with seed treated with a non-systemic fungicide that is only effective against pathogens in the seed testa. Alternatively, rejection of the crop for use as seed would currently reduce its commercial value by 8–15% depending on the generation of seed grown.

A similar analysis of post-harvest loss and costs due to ergot contamination of both seed and milling cereals reveals an even more complex situation. First, consider ergot sclerotia in seed grain which are potentially sources of inocula for subsequent crops if they are sown with the seed. UK seed standards ensure that only small quantities of ergot are found in certified seed. To satisfy the Higher Voluntary Standard not more than one piece of sclerotium should be found in a 1 kg sample (or three pieces in a 0.5 kg sample to satisfy the minimum EEC standard) (Anon., 1980). In practice, many of the sclerotia in cereal seed can be removed by sophisticated seed cleaning machinery as long as the pieces are sufficiently large but this can be expensive since the finer machine settings which remove smaller pieces of sclerotium also reject more grain. It is difficult to estimate the increased screening loss of grain, but when it is necessary to remove small pieces of ergot sclerotia from wheat, losses may be 2–4%. However, this must be considered a variable factor depending on the degree of contamination and on the size and shape of both sclerotia and grain.

Ergot in grain for milling and animal feed presents quite a different hazard, in this case to the consumer, because sclerotia generally contain toxic ergot alkaloids. Since it is standard practice in the UK for millers to arrange contracts that require cleaned grain to be free from ergot, the chances of toxic quantities of ergot occurring in home-produced cereal products for human consumption are remote. One consequence of the rigorous standards and the difficulty of cleaning contaminated grain to an ergot-free standard is that the frequency of rejection of ergot-contaminated grain crops offered for milling is possibly greater than for similarly contaminated grain offered as seed. The records of one grain trading company, in an area where ergot was a problem in 1981, show that about 12% of the wheat and barley presented for milling and compounding was contaminated (Anon., 1982). To a farmer growing milling quality wheat, rejection could mean the loss of a milling premium of 2–10%, depending on the type of milling and on the current supply and demand. In addition, the merchant would probably face a substantial increase in transport costs as the contaminated grain has to be diverted to an alternative market. As with loose smut of cereals, it is clear that the potential post-harvest cost of contamination of grain can be disproportionately large by comparison with the apparently small yield loss in a crop where there is a relatively small percentage of infected plants.

Acknowledgement

I would like to thank Dr P.G. Mantle of Imperial College, London, for his constructive criticism and discussion of this chapter.

References

Anon. (1980) The Cereal Seeds Regulations, *Statutory Instruments*, No. **900**. Her Majesty's Stationery Office, London.

Anon. (1981) *Report of the National Institute of Agricultural Botany*, No. **62**, p. 90.

Anon. (1982) Ergot: how much is dangerous? *British Farmer and Stockbreeder*, 18th September, p. 23.

Batts C.C.V. (1955) Infection of wheat by loose smut, *Ustilago tritici* (Pers.) Rostr. *Nature, (London)* **175**, 467–8.

Batts C.C.V. & Jeater A. (1958) The development of loose smut (*Ustilago tritici*) in susceptible varieties of wheat, and some observations on field infection. *Transactions of the British Mycological Society* **41**, 115–25.

Byrde R.J.W. & Willetts H.J. (1977) *The Brown Rot Fungi of Fruit*. Pergamon Press, Oxford.

Campbell W.P. (1958) Infection of barley by *Claviceps purpurea*. *Canadian Journal of Botany* **36**, 615–9.

Chou M.C. & Preece T.F. (1968) The effect of pollen grains on infections caused by *Botrytis cinera* Fr. *Annals of Applied Biology* **62**, 11–22.

Ciccarone A. (1960) Reproduction is affected. In: *Plant Pathology* Vol. I (Ed. by J.G. Horsfall & A.E. Dimond), pp. 249–76. Academic Press, New York.

Couch H.B. (1973) *Diseases of Turfgrasses*. Robert E. Krieger Publishing Company, New York.

Darlington L.C. & Mathre D.E. (1976) Resistance of male sterile wheat to ergot is related to pollination and host genotype. *Crop Science* **16**, 728–30.

Demby W.M. (1971) In: *Crop loss assessment methods*: FAO manual on the evaluation and prevention of losses by pests, diseases and weeds. (Ed. by L. Chiarappa), p. 48. Commonwealth Agricultural Bureaux, Farnham Royal.

Dickerson A.G., Mantle P.G., Nisbet L.J. & Shaw B.I. (1978) A role for β-glucanases in the parasitism of cereals by *Claviceps purpurea*. *Physiological Plant Pathology* **12**, 55–62.
Doling D.A. (1966) Loose smut in wheat and barley. *Agriculture*, London **73**, 523–7.
Done A.C. (1973) Implications for the introduction of hybrid cereals on disease patterns. *Annals of Applied Biology* **75**, 144–9.
Hewett P.D. (1978) Epidemiology of seed-borne disease. In: *Plant Disease Epidemiology* (Ed. by P.R. Scott & A. Bainbridge), pp. 167–76. Blackwell Scientific Publications, Oxford.
Hewett P.D. (1979) Regulating seed-borne disease by certification. In: *Plant Health* (Ed. by D.L. Ebbels & J.E. King), pp. 163–73. Blackwell Scientific Publications, Oxford.
Hoffmann J.A. (1982) Bunt of wheat. *Plant Disease* **66**, 979–86.
Large E.C. (1940) *The Advance of the Fungi*. Jonathan Cape, London.
Luttrell E.S. (1980) Host-parasite relationships and development of the ergot sclerotium in *Claviceps purpurea*. *Canadian Journal of Botany* **58**, 942–58.
McClellan W.D. & Hewitt W.B. (1973) Early Botrytis rot of grapes: time of infection and latency of *Botrytis cinerea* Pers. in *Vitis vinifera* L. *Phytopathology* **63**, 1151–7.
Malik M.M.S. & Batts C.C.V. (1960) The determination of the reaction of barley varieties to loose smut. *Annals of Applied Biology* **48**, 39–50.
Mantle P.G. & Shaw S. (1976) Role of ascospore production by *Claviceps purpurea* in aetiology of ergot disease in male sterile wheat. *Transactions of the British Mycological Society* **67**, 17–22.
Mantle P.G., Shaw S. & Doling D.A. (1977) Role of weed grasses in the etiology of ergot disease in wheat. *Annals of Applied Biology* **86**, 339–51.
Mathre D.E. (1978) Disrupted reproduction. In: *Plant Disease*. Vol. III (Ed. by J.G. Horsfall & E.B. Cowling), pp. 257–78. Academic Press, New York.
Mills J.T. (1966) The development of loose smut (*Ustilago avenae*) in the oat plant with observations on spore formation. *Transactions of the British Mycological Society* **49**, 651–63.
Mower R.L. & Hancock J.G. (1975) Mechanism of honeydew formation by *Claviceps purpurea*. *Canadian Journal of Botany* **53**, 2826–34.
Neergaard P. (1977) *Seed Pathology*. Macmillan, London.
Noble M. (1971) Seed pathology. In: *Diseases of Crop Plants* (Ed. by J.H. Western), pp. 21–36. Macmillan, London.
Ogawa J.M. & English H. (1960) Blossom blight and green fruit rot of almond, apricot and plum caused by *Botrytis cinerea*. *Plant Disease Reporter* **44**, 265–8.
Pedersen P.N. (1960) Methods of testing the pseudoresistance of barley to infection by loose smut *Ustilago nuda* (Jens.) Rostr. *Acta Agriculture Scandanavica* **10**, 312–32.
Popp W. (1959) A new approach to the embryo test for predicting loose smut of wheat in adult plants. *Phytopathology* **49**, 75–7.
Puranik S.B. & Mathre D.E. (1971) Biology and control of ergot on male sterile wheat and barley. *Phytopathology* **61**, 1075–80.
Rosen H.R. (1936) Mode of penetration and of progressive invasion of fire blight bacteria into apple and pear blossoms. *Arkansas Agricultural Experimental Station Bulletin* **331**, 1–68.
Schroth M.N., Thomson S.V., Hildebrand D.C. & Moller W.J. (1974) Epidemiology and control of fire blight. *Annual Review of Phytopathology* **12**, 389–412.
Shaw B.I. & Mantle P.G. (1980) Host infection by *Claviceps purpurea*. *Transactions of the British Mycological Society* **75**, 77–90.
Shaw S. (1975) Aetiology of ergot disease of male-sterile wheat. Ph.D Thesis, University of London.
Thakur R.P. & Williams R.J. (1980) Pollination effects on pearl millet ergot. *Phytopathology* **70,** 80–4.
Watkins J.E. & Littlefield L.J. (1976) Relationship of anthesis in Waldron wheat to infection by *Claviceps purpurea*. *Transactions of the British Mycological Society* **66**, 362–3.
Weaver L.O. (1950) Effect of temperature and relative humidity on occurrence of blossom blight of stone fruits. *Phytopathology* **40**, 1136–53.
Wright C.E. (1967) Blind seed disease of ryegrass. *Euphytica* **16**, 122–31.

14 Damage, symptoms and crop loss caused by vascular pathogens

P.W. TALBOYS
East Malling Research Station, Maidstone,
Kent ME19 6BJ, UK

Introduction

The characteristic and defining feature of vascular wilt diseases is that, throughout most of the period of pathogenesis, the causal organism is constrained within the non-living elements of the xylem. It is only when the constraint is removed, as the host approaches natural or pathogen-induced senescence, that the pathogen colonizes phloem, cortex and pith, as well as the xylem parenchyma and medullary ray tissues. The classical vascular wilt pathogens include *Verticillium albo-atrum*, *V. dahliae* and *V. tricorpus*, and the various pathotypes of *Fusarium oxysporum*. In Dutch elm disease (*Ceratocystis ulmi*) and oak wilt (*Ceratocystis fagacearum*) there is earlier but limited invasion of living xylem elements but the general nature of pathogenesis in these diseases so closely resembles that of the 'true' vascular wilts that it would be unrealistic to exclude them. The group also includes diseases caused by bacterial pathogens, e.g. *Pseudomonas solanacearum*.

In vascular wilts caused by soil-borne fungi there is a considerable interval between the initiation of infection and the onset of external symptoms. Artificial inoculation entailing introduction of large quantities of the pathogen directly into the vascular system may reduce the interval to a few days but under field conditions there is commonly a delay of several weeks. During this period the pathogen penetrates to the vascular system at a number of points determined by the concentration of infective units in the soil and by the resistance of extravascular tissues at the sites of contact between roots and inoculum; initial penetration is followed by invasion of vessels by hyphal growth and by transport of conidia in the transpiration stream. Some minimal level of vascular invasion is a prerequisite for development of the wilt syndrome.

This chapter is concerned with the consequences of the presence and metabolic activities of these specialized pathogens within the xylem of the host plant. The *nature* of their activities (enzyme and toxin production, hormone production or activation, effects on respiration and photosynthesis) and the mechanisms leading to damage, are discussed elsewhere in this volume and will not be considered in detail here; their manifestations are expressed as structural and functional damage to xylem elements and associated extravascular tissues, and as the external

Wood R.K.S. & Jellis G.J. (1984) *Plant Diseases: infection, damage and loss.* Blackwell Scientific Publications, Oxford.

symptoms characteristic of this group of diseases. Most of the features to be described are common to many different diseases but the examples will be taken predominantly from the verticillium wilts amongst which almost all of these elements are represented.

Structural and functional damage

Transfer of water from the soil to the atmosphere via the plant is largely a physical process determined by the difference in water potential between the absorptive surfaces of the root and the dispersive surfaces in the leaf and by the conductance (permeability) of the pathway between them. The lumina of the non-living xylem elements and the intermicellar spaces in primary cell walls form a continuum, the apoplast. This provides for the free movement of water and solutes through the plant without the direct intervention of living cells except at the endodermis in the root and at the stomata in the leaf; the latter constitute a control system operating through feedback mechanisms to modulate water loss. Damaging water deficits in the aerial parts of the plant can result from various forms of occulsion in the xylem vessels. Such deficits may also occur, less obviously, if occlusion of the intermicellar spaces, or accumulations of hydrophobic molecules (e.g. oxidized phenols) on the inner surfaces of the terminal tracheid elements, prevents the transfer of water from the xylem to the bundle sheath cells and thence to the leaf mesophyll, palisade and epidermal tissues. Thus, occlusions at various locations in the apoplast may all contribute to the leaf water deficits that Hall & MacHardy (1981) regard as 'the central fact of water relations in hadromycosis'.

Vascular occlusion

By definition vascular pathogens occupy space within the vascular system (Figure 14.1) but the effect of fungal mycelium on water movement is generally considered to be inadequate to account for the observed reductions in flow. However, bacterial pathogens enmeshed in their mucilaginous matrix appear to play a directly occlusive role in pathogenesis. The nature and mode of action of other forms of vascular occlusions have been discussed from different points of view by Talboys (1968) who considered their role in pathogenesis, and Beckman & Talboys (1981) who dealt with their apparent function as components of disease resistance. These vascular occlusions include: gels that appear to result from modification of primary cell walls and middle lamella material of pit membranes, possibly through the intervention of fungal pectic enzymes (Cooper, 1981); swelling of perforation plates and vessel end walls (Beckman, 1969a, 1969b); gums apparently exuded from xylem parenchyma cells (Figure 14.2), and usually impregnated with oxidized phenols (Talboys, 1968); 'smooth', 'bubbly' and 'fibrous' coatings on the walls of xylem elements (Robb *et al.*, 1979a 1979b, 1982), 'concentric accretions' (Figure 14.3) of uncertain origin (Talboys, 1968); fungal polysaccharides that may occlude the inter-

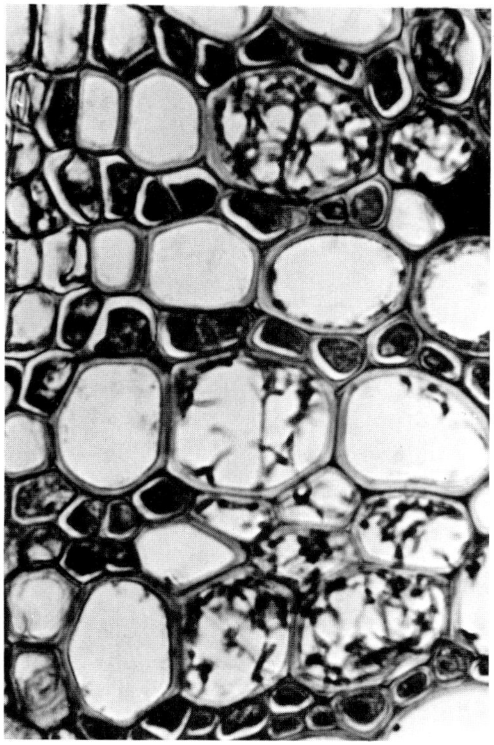

Figure 14.1. Intensive colonization of vessels in the petiole of a hop leaf infected with a virulent strain of *Verticillium albo-atrum*.

micellar spaces of primary walls in pit membranes; and the intrusive balloon-like tyloses (Figure 14.4), which are formed by distension of pit membranes into the vessel from adjacent xylem parenchyma cells, and which undergo active growth, with the formation of new cell wall material which sometimes includes reticulate secondary thickening. Prominent nuclei and masses of starch grains suggest intensive metabolic activity in tyloses (Talboys, 1958). Gels, gums and tyloses can be induced by various foreign organisms including non-pathogens which have been introduced into the vascular system (VanderMolen *et al.*, 1977). Many plants also react to wounding by forming either gums or tyloses in damaged vessels and it appears that these are generalized responses to any form of injury; usually one or other of the responses is characteristic for a given plant species.

Vascular discoloration

Vascular invasion is usually accompanied by the development of brown or greenish-black discoloration in the xylem attributable to the oxidation of phenolic compounds.

Mace (1963) reported the occurrence, in banana roots, of isolated phenol-rich

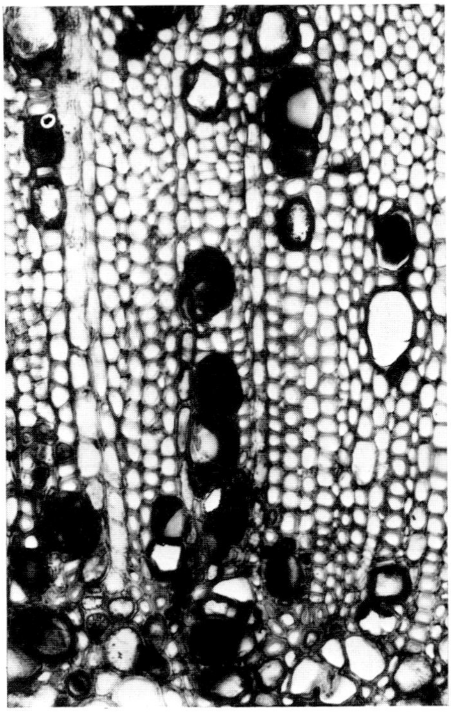

Figure 14.2. Vessels occluded by dark gum deposits in Brompton plum stem infected by *Verticillium dahliae*.

cells with, apparently, separately compartmented peroxidases and polyphenoloxidases. These cells are readily damaged by pathogens and other agencies and their contents then diffuse into adjacent tissues, the phenols oxidizing to dark-coloured compounds. Beckman *et al.* (1974) and Beckman & Talboys (1981) suggested that the oxidizing phenols supplement gelation and tylosis as a defensive response. In many plants the capacity for a rapid browning reaction in response to vascular invasion appears to be general in parenchymatous cells associated with the xylem.

Damage to leaf tissue

Most of the water transpired from leaves passes through the terminal tracheids of the vascular system and is transferred to the mesophyll, palisade and epidermal cells via the bundle sheath cell walls. It follows that these cells would be the first recipients of potentially toxic molecules translocated from an infected vascular system; they would also be under immediate stress if water transfer from terminal tracheids were prevented. The structure of bundle sheath cells varies considerably between plant species but they often resemble endodermal cells in having a substantial content of phenolic compounds. In some diseases, such as hop verticillium wilt, necrosis resulting from vascular pathogenesis is first apparent in the bundle

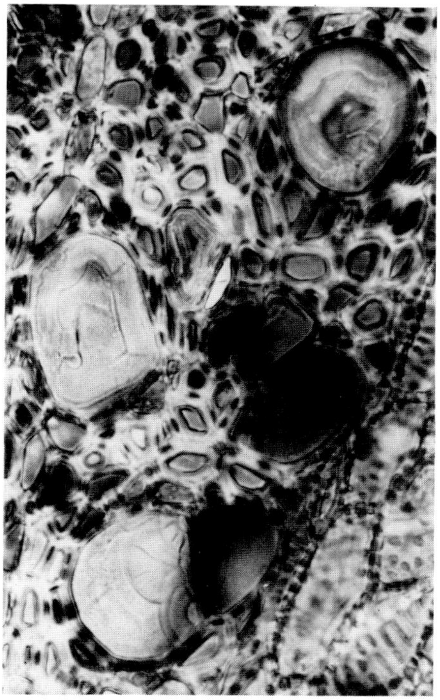

Figure 14.3. Concentric and hemispherical accretions and brown gum deposits in vessels of *V. dahliae*-infected raspberry root.

sheath cells, which show typical phenol oxidation reactions (Figure 14.5). The mesophyll and palisade tissues subsequently collapse and become desiccated, suggesting that the browning reaction may be a consequence of, or may contribute to, the failure of water transport from the xylem elements. Death of mesophyll and palisade tissues is generally preceded by degeneration of chloroplasts and loss of green coloration.

The vascular wilt syndrome

The progressive advance of a soil-borne vascular pathogen through catenary sequences of vessels in the xylem generates in widely differing plant species a consistent pattern of disease development. The disease typically affects the leaves on the main axis in acropetal succession. This is most clearly apparent in tall herbaceous plants but the same sequence occurs in plants with a very short axis, such as the strawberry, in which the oldest (outermost) leaves are the first to be affected; infection of the youngest leaves may be greatly delayed, and in some cultivars the plant may even grow away from the disease. During the growth of a plant successive shells of xylem are being laid down. The xylem supplying the youngest leaves is likely to be continuous with that of the youngest roots and may not necessarily be invaded by radial growth from older and earlier-infected xylem.

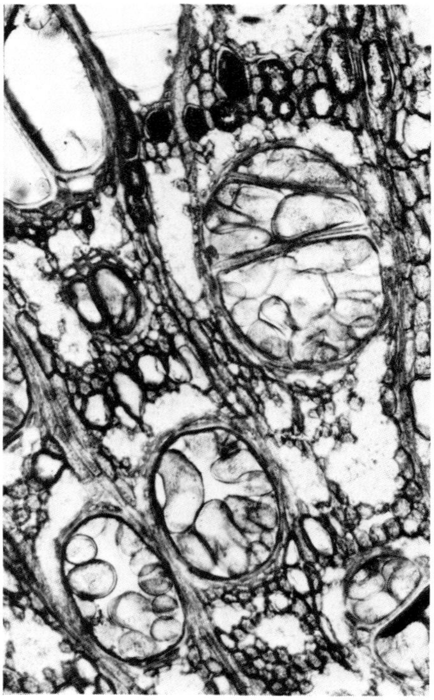

Figure 14.4. Tyloses occluding the vessels in a root of a moderately wilt-resistant cultivar of hop infected by a virulent strain of *V. albo-atrum*.

The extent of lateral or radial spread through xylem vessel and fibre walls is limited by the distribution of sheets of xylem parenchyma.

A second common feature of vascular wilt diseases is the one-sided development of symptoms associated with limitation of the pathogen to a small sector of the xylem. The extent of symptoms on individual leaves is closely related to the distribution of infected vessels in the leaf trace and the patterns can be simulated by injecting dyes into a few vessels. One-sided development of symptoms usually occurs only when there is moderate resistance of the pathogen and is common in diseases of trees and shrubs caused by *V. dahliae*, where symptoms often occur only in one branch and vascular discoloration can be traced back to infection of a single small root.

The symptoms attributable to vascular pathogens can include: epinasty; diurnal wilting, usually leading to irreversible wilting; chlorosis; vein clearing; progressive foliar desiccation without wilting; foliar necrosis; defoliation; growth retardation; stem xylem hyperplasia; adventitious rooting; and stem and root necrosis. Various combinations of these symptoms occur and differences may be associated with the virulence of the pathogen and the resistance of the host; with the growth form of the host, viz. whether it has an annual shoot system or is a woody perennial (Talboys, 1968); and with environmental factors.

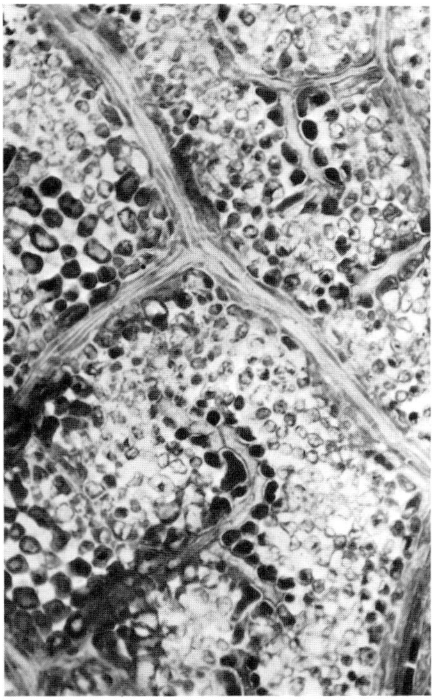

Figure 14.5. Terminal tracheids surrounded by necrotic bundle sheath cells in a hop leaf during development of acute (tiger-stripe) symptoms caused by a virulent strain of *V. albo-atrum*.

Epinasty

This symptom, a downward bending of the leaf, results from differential cell expansion of abaxial and adaxial tissues in the petiole and lamina. It is often the first indication of a wilt disease. In many plants the lamina of the epinastic leaf becomes orientated vertically, parallel with the stem, giving a superficial appearance of wilting. However, the lamina and petiole are turgid, not flaccid as in a wilting leaf. Epinasty is strikingly developed in verticillium wilt of tomato, but it also occurs in this host in response to pathogens that cause stem lesions such as *Botrytis cinerea*. It can be induced by treatment with ethylene, and may be a response to ethylene formed in pathogen-infected tissues either as a product of the pathogen itself or as a consequence of damage to host cells. Alternatively, epinasty may result from pathogen-induced hyperauxiny, although the effects of IAA also may be mediated by ethylene. The complex interrelations and effects of ethylene and other hormones have been reviewed by Pegg (1981).

Wilting

Flaccidity of leaves and shoots is a frequent but not invariable symptom of diseases caused by vascular pathogens; it does not normally occur, for example, in verticillium wilts of hop, sunflower and raspberry. There has been much controversy about the cause of wilting and particularly about the relative importance of toxins and vessel blockage in symptom development (Talboys, 1968; Duniway, 1976; Ayres, 1978), but Hall & MacHardy (1981) concluded that wilting is invariably attributable to some form of occlusion in the water transport system, although the locations of such occlusions can vary and may be characteristic for each disease.

Wilting occurs when the water potential generated by transpiration is more strongly negative than the solute potential of the leaf tissues, so that water is withdrawn from the cells and they lose turgor. This condition develops when water lost in transpiration is not replaced from the soil sufficiently rapidly, either because the matric potential of the soil is low or because the transport of water through the plant is retarded by occlusions in the vascular system, or because of a combination of both factors. Hence, under warm dry conditions a moderate degree of occlusion may be sufficient to induce wilting during the day but does not prevent recovery at night when transpiration is reduced; irrigation may prevent wilting during the day. Diurnal wilting is often an early symptom of vascular disease; occasionally, as in some moderately wilt-resistant cultivars of strawberry, it is the only symptom that occurs in some seasons. More commonly, with increasing vascular occlusion, diurnal wilting progresses to irreversible wilting, even when water is freely available in the soil.

In some vascular diseases there is an apparent anomaly; leaves detached from an infected plant wilt more slowly than comparable leaves from a non-infected plant. This difference is possibly attributable to drought hardening of the diseased plant; water stress leads to accumulation of abscisic acid which induces closure of the stomata and a consequent reduction in the rate of transpiration (Milborrow, 1981). Large increases in the abscisic acid content of leaves of *Verticillium*-infected cotton plants were reported by Wiese & DeVay (1970).

Discoloration and chlorosis

Chlorosis, the loss of chlorophyll associated with degeneration of chloroplasts, is a more consistent feature than wilting in the syndrome of vascular wilt diseases, although in some hosts the characteristic yellowing is obscured by the development of red or purple anthocyanin pigments. It may precede, accompany or follow wilting, and it also occurs in non-wilting syndromes.

The pattern and rate of development of chlorosis is variable in a given disease, and differs between diseases. Its first manifestation is often a subtle change in tint from the bluish-green of a healthy plant to a yellowish-green. Thereafter it may progress through an overall mottling or blotching to a uniform yellowing of the leaf, or it may be initially localized in wedge-shaped sectors bounded by main or minor veins, or in marginal and interveinal areas. In mild or chronic infections chlorosis sometimes remains localized but in acute disease whole leaves are generally affected, and eventually the whole plant develops some degree of chlorosis.

Chlorosis commonly leads to necrosis and abscission but abscission may take place before the onset of necrosis. This is frequent in woody plants, where relatively mild yellow mottling may be followed by rapid defoliation of whole branches.

Chlorosis characteristically affects the leaves in acropetal succession along the stem, although in trees and shrubs this tends to be less obvious than in herbaceous plants. One difference between herbaceous and woody plants is that the leaves from overwintered buds on woody plants are all of comparable physiological age, whereas leaves on herbaceous plants (and also on the current year's growth of woody plants) differ greatly in their physiological condition; the first-formed may be approaching senescence while new leaves are still being formed. It is not clear to what extent the physiological state of the leaf, as distinct from its position on the axis, determines its sensitivity or liability to the processes of pathogenesis in vascular wilt diseases. Leaf age is often critical for the establishment of infection by airborne pathogens landing on the outside of the leaf; it would be reasonable to suppose that the response to internal infection via the vascular system might also be different in young and old leaves.

Vein clearing

A translucent appearance of the veins in the leaf lamina has been reported in fusarium wilts of tomato, cotton and *Phaseolus* beans. The appearance suggests a toxic action closely limited to the tissues adjacent to vascular elements; fusaric acid under slightly acid conditions tends to remain in close proximity to the xylem and has been proposed as the causal agent of the vein-clearing symptom, which does not seem to occur in wilt disease caused by *Verticillium* spp. or *Ceratocystis* spp.

Progressive foliar desiccation without wilting

In those vascular diseases in which there is no obvious wilting there is nevertheless progressive desiccation of the leaf, sometimes with a strongly defined pattern of chlorosis and necrosis, as in the syndrome of severe hop verticillium wilt and raspberry verticillium wilt. Marginal and interveinal chlorosis lead rapidly to necrosis, which is generally initiated at the leaf margins and in the interveinal areas of the leaf midway between the secondary veins. The veins remain green until most of the leaf is necrotic but the rapidly diminishing area between the veins and the spreading necrotic zones becomes intensely chlorotic, giving a striking yellow and black striped pattern to which the term 'tiger striping' has been applied (Figure 14.6).

In hop wilt the development of this pattern is accompanied by a well defined sequence of changes in stomatal behaviour. When the leaves first change from a blue-green to a yellow-green tint the stomata close and apparently remain closed as long as the underlying tissues remain green. In parts of the leaf that become strongly chlorotic the stomata open and as necrosis supervenes the guard cells

Figure 14.6. Acute (tiger-stripe) symptoms of verticillium wilt in the hop.

collapse, leaving the aperture more or less open. In leaves at an advanced stage of desiccation, with only narrow bands of chlorotic tissue between the green veins and the necrotic areas, there is often a steep gradient from closed to open to collapsed stomata within a band *c.* 2 mm wide.

If a hop plant showing tiger stripe symptoms is severed at the base and allowed to take up a mobile dye such as acid fuchsin or decolorized basic fuchsin (Talboys, 1958) the dye will slowly move through the vascular system into the leaves as far as the boundary between chlorotic and necrotic tissues, despite the presence of mycelium in many vessels in the stem and petioles. Clearly there is sufficient movement of water to maintain turgor in those parts of the leaf to which it has access, and death of the interveinal areas does not seem to be attributable to any systemic occlusion of the vascular system. Nevertheless, mention has already been made of the browning reaction in the cells of the bundle sheath surrounding the terminal tracheids, followed by collapse of chlorotic mesophyll cells outside the bundle sheath. It is suggested that, if occlusion is the primary cause of the tiger-stripe wilt syndrome it must take place at the interface between the terminal tracheids and the bundle sheath cells, i.e. in the non-lignified areas of the tracheid walls or the sheathing cells. The nature of any such occluding materials is open to question but there are several possibilities: polysaccharide secreted by the pathogen; galacturonide and other residues resulting from the action of fungal enzymes on host cell walls; products of fungal lysis induced by host enzymes; oxidized polyphenols released from bundle sheath cells as a result of damage caused by toxic metabolites of the pathogen; or a combination of two or more such factors.

It is paradoxical that the onset of chlorosis and necrosis occurs in the non-infected parts of the leaf whereas the main veins remain green to a late stage although internally the xylem strand contains many vessels colonized by the pathogen. Possibly the condition is comparable with the 'green island' effect caused by some biotrophic pathogens. Alternatively it may relate to the relatively dense packing of the cells and the absence of stomata along the veins, so that relatively little water (carrying fungal metabolites) moves outwards from the xylem along the main veins compared with that in the interveinal areas where transpiration is greatest even with stomata mostly closed.

The patterns of chlorosis and necrosis in less severe (mild) wilt syndromes tend to reflect a more limited extent and intensity of vascular colonization than that characterizing acute syndromes, and different host responses that may be the cause or the consequence of different levels of infection. Thus in mild hop wilt (Figure 14.7) the small sectors of chlorotic/necrotic tissues delimited by veins are associated with vascular occulsion by tyloses in a few vessels in petiolar vascular bundles, and with amounts of mycelium that are often so small that they are difficult to detect microscopically. Necrosis of the extravascular tissues in the leaf tends to be initiated near infected vessels, not remote from them as in the acute syndrome: the mechanisms of symptom induction appear to be different in the two forms of the disease, direct occlusion of vessels playing a larger role in pathogenesis in the mild

Figure 14.7. Sectorial chlorosis and necrosis in the mild syndrome of hop verticillium wilt.

disease but perhaps also contributing to the limitation of colonization in that syndrome (Beckman & Talboys, 1981).

Foliar necrosis

Irrespective of the mechanisms of pathogenesis in vascular wilt diseases, whether occlusive or toxigenic or enzymic, or a combination of these, the eventual outcome is usually some degree of foliar necrosis, mostly following chlorosis. This represents an absolute loss of photosynthetic capacity and can obviously be expected to have a direct impact on the growth and yield of the plant, although some reduction in photosynthetic area may be compensated for by increased photosynthetic efficiency in areas remaining green.

Defoliation

In most dicotyledonous plants damage to leaves commonly results in abscission, provided that the tissues of the abscission zone in the petioles are not killed before they undergo the physiological changes that precede fracture at the separation layer. Wilt diseases generally cause defoliation, the leaves falling in acropetal succession at a characteristic stage of chlorosis or necrosis that varies from one disease to another. In plants that lack an abscission mechanism, or if the progress of the disease is so rapid as to kill the tissues in the abscission region prematurely, the dead leaves hang *in situ* until they break away or disintegrate.

Root and stem necrosis

Damage to the root system is generally not a feature of the early stages of pathogenesis in soil-borne vascular wilts, and indeed it is frequently difficult to identify visually which roots are infected. However, severely diseased plants generally have a much reduced root system and there may be considerable necrosis, but it is not clear to what extent this necrosis is directly attributable to the pathogen, or whether premature senescence resulting from damage to the aerial parts renders the roots susceptible to invasion by soil micro-organisms. Root necrosis is not a useful diagnostic feature.

Stem necrosis is a frequent but not invariable feature of wilt diseases of herbaceous plants; it is less common in woody plants except in outstandingly susceptible hosts, e.g. some species of *Acer*, because infected wood tends to be buried by later-developing xylem and is effectively sealed off. In herbaceous plants death of the stem generally occurs after the abscission of the leaves. It appears that pathogen-induced or normal senescence permits rapid invasion and killing of extravascular tissues by the pathogen but in some wilt diseases death and desiccation of the stem are catastrophically rapid and complete, suggesting a systemically toxic action rather than colonization, perhaps attributable to autocatalytic ethylene production in the host tissues. In mild syndromes even defoliated stems may remain alive for several weeks.

Growth retardation

When vascular wilt diseases are so severe that the host plant dies rapidly, other effects on growth are not usually a notable feature. Less severe disease often results in shortened stems and reduced leaf areas (Selman & Pegg, 1957; Harrison & Isaac, 1968). Severe dwarfing of the plant is a frequent feature of strawberry verticillium wilt (Figure 14.8).

Verticillium wilts in woody plants sometimes cause localized reductions in the radial growth of the cambium leading to longitudinal grooving in infected sectors of the stem. Grooves arise above the points at which roots containing *Verticillium* enter the stem and may be continuous into some of the lateral branches, running parallel with infected strands of vessels in the xylem.

Effects on plant growth may be associated with changes in the content of growth-regulating substances in host tissues but the mechanisms are not fully elucidated: water stress, loss of photosynthetic capacity, the hormone-producing or inducing potential of the pathogens and the direct effects of fungal toxins, may all be interactive factors underlying growth abnormalities in the diseased plant.

Stem xylem hyperplasia

In some wilt diseases a non-lethal infection can induce renewed or prolonged cambial activity, leading to considerable xylem hyperplasia with a consequently

Figure 14.8. Strawberry wilt (*Verticillium dahliae*) in the cultivar Rearguard showing the severe stunting that frequently accompanies necrosis of the older leaves in heavily infected plants.

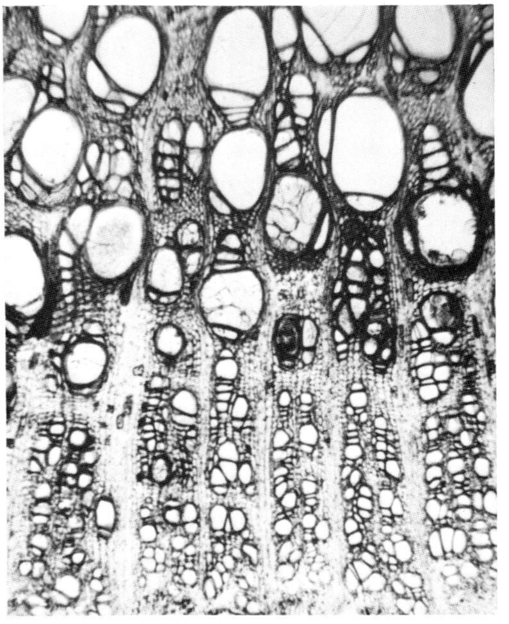

Figure 14.9. Normal (above) and hyperplastic xylem (below) in the lower stem of a hop plant that showed mild symptoms of wilt (*Verticillium albo-atrum*).

increased stem diameter and enhanced periderm formation. This is strikingly developed in the mild syndrome of hop verticillium wilt (Talboys, 1958) but has also been reported in oak wilt (Schoeneweiss, 1959). Xylem hyperplasia in the hop (Figure 14.9) is always associated with limited vascular infection and extensive tylosis, and appears to represent a repair mechanism by which a continually renewed outer shell of xylem compensates for loss of function in occluded primary and secondary xylem which would otherwise lead to death of the plant. One of the effects noted by Pegg (1976) in ethylene-treated tomato plants infected with *V. albo-atrum* was an increase in vessel differentiation to form a band of fungus-free secondary xylem, which may have contributed to the enhanced resistance that was associated with the ethylene treatment.

Adventitious rooting

Adventitious rooting from the lower parts of stems is a further positive growth response to vascular infections in some hosts and, like xylem hyperplasia, appears to be a repair mechanism. It occurs in tomato in response to vascular wilt pathogens but also to some non-vascular pathogens.

Damage assessment

Plants with an acute vascular wilt disease become defective in water transport,

15 Galls caused by fungi and bacteria

G.R. DIXON
School of Agriculture,
Aberdeen AB9 1UD, UK

Introduction

Terms which are used to describe manifestations of disorganized host growth following pathogen invasion include blister, enation, gall, tumour and witches' broom. Characteristically a gall or tumour is a mass of host tissue composed usually of thin-walled cells with few, if any, intercellular spaces. Thus gall-inducing pathogens damage their hosts by causing unregulated cell division and development which disrupts the anatomical structure of a range of host organs. This disruption is caused by the processes of hypertrophy (excessive growth due to enlargement of individual cells) and hyperplasia (excessive growth due to increased division of cells). Host symptoms and damage resulting from both processes are found to a greater or lesser extent with most gall-inducing pathogens. A classification of galling into two forms was attempted by Butler & Jones (1961) for fungi and bacteria and by Bos (1970) for viruses. These authors differentiate between 'histoid' galls, where there is marked tissue swelling, and 'organoid' galls, where there is marked morphological change but limited anatomical effects as, for example, the changes to symmetry and habit brought about by some pathogens causing downy mildew or rust diseases. A more limited definition is given by Anon. (1973) as 'a localized proliferation of plant tissue producing a swelling or outgrowth commonly having a characteristic shape unlike any organ of the normal plant'. This appears to cover only histoid galls and as other descriptions of damage are available, i.e. blisters, enations and witches' brooms, it is possibly more useful. Inclusion of 'localized' in this definition may be more limiting than it needs to be since systemic spread and subsequent gall formation are features of some diseases, such as that caused by clover wound tumour virus. Bos (1970) further distinguishes between primary and secondary malformations (galls), treating primary malformations as an initial symptom and secondary malformations as those taking place only after other (possibly physiological) events. There may be some merit in this distinction in relation, for example, to the cycles of events with *Plasmodiophora brassicae* (clubroot) and *Synchytrium endobioticum* (potato wart).

Characteristics of some gall-inducing pathogens

Features of a range of gall-inducing pathogens and the symptoms which they induce

Wood R.K.S. & Jellis G.J. (1984) *Plant Diseases: infection, damage and loss.* Blackwell Scientific Publications, Oxford.

are summarized in Table 15.1. A noticeable characteristic of these pathogens is the diverse and disconnected range of fungi and bacteria which incite galling symptoms. Other pathogens, such as some downy mildew, will cause analogous symptoms but without gross developments of tumourous tissue. One of the few cases where this has been investigated in detail is the work reported by Ingram (1982) for sugar beet leaves infected by *Peronospora farinosa*. Infection caused stunted leaf growth, i.e. a decrease in leaf area, length, fresh weight and dry weight accompanied by general chlorosis. A marked increase in the thickness of infected leaves resulted from an increase in the size and number of cells in the mesophyll and some parts of the vascular region. There was a complete disappearance of intercellular spaces and loss of distinction between the palisade and spongy mesophyll tissues. Comparisons between the growth of infected and non-infected leaves indicated that the degree of morphological disturbance in infected leaves was correlated with the stage of development of the lamina at the time of infection.

Few viruses are specifically associated with galling symptoms although enations often result from virus–host interactions. Enations, however, do not alter the form of a host organ to which they are attached and are of limited size. Clover wound tumour virus is the most persistently quoted virus to be associated with galling or tumourous symptoms. It has been reviewed by Black (1965). In systemically infected plants external tumours develop on leaves and stems near inoculation wounds. In infected roots galls appear spontaneously, forming close to cells in the pericycle which have been wounded when developing lateral roots break through the cortex. This virus is also responsible for many small internal tumours in the phloem of leaf, stem and root. Tumours are initiated by tangential division in cells of the pericycle opposite the primary phloem. Abnormal cell multiplication and not cell enlargement is responsible for subsequent tumour development. Growth of the tumour continues by cell division in the meristematic tissue between xylem and phloem and between xylem extensions. Wound tumour virus has not been found infecting plants in nature and was only discovered by accident in nymphs of the leaf hopper *Agalliopsis novella* (Smith, 1972).

Some micro-organisms are associated with gall formation but not with pathogenesis, the most obvious example being the nodules resulting from root invasion of legumes by *Rhizobium* spp. (recently reviewed by Bergersen, 1982). A well established population of these bacteria is a prerequisite for the prompt nodulation of legume seedlings. Population establishment is influenced by such factors as soil pH, water content, divalent cations and the nature of organic nutrients provided by the presence of other micro-organisms. Susceptibility to infection varies widely between genera and species of legume. Invasion takes place through root hairs at the zone of elongation of the root tip or via root wounds. In advance of infection, usually near to a protoxylem point, inner cortical cells commence division and become orientated such that they ultimately form the nodule meristem. Nodules contain centrally a relatively large volume of tissue composed of enlarged parenchyma cells the cytoplasm of which is packed with bacterial cells. Each

Table 15.1. Features of some gall-inducing pathogens and their hosts.

Pathogen	Host Range	Host Characters
A. Soil-borne types		
Agrobacterium tumefaciens (crown gall)	331 genera in 93 dicotyledonous families	Once tumour initiation has begun abnormal growth is autonomous and gall development may proceed in the absence of bacterial cells
Plasmodiophora brassicae (clubroot)	The primary stage of root hair infection can occur in several host families but the secondary stage of gall formation is confined to members of the Cruciferae	The extent and form of galling varies with host species, age, size and form of host roots, host growth rate and soil conditions such as pH, moisture, temperature and fertility
Synchytrium endobioticum (potato wart)	Natural infection is restricted to the potato although other plants have been infected artificially	A primary asexual phase follows zoospore invasion and gives rise to epidermal rosettes of gall tissue In the secondary asexual phase infected cells become more deeply embedded in host tissue due to hyperplasia
Urophlyctis alfalfae (crown wart)	*Medicago sativa* (lucerne) in the UK but also *M. falcata* in USA	Galls develop as swellings of adventitious buds at or below the stem base. Scale leaves and stipules may also be deformed
B. Aerial types		
Taphrina deformans (peach leaf curl)	Members of the Rosaceae	Leaf blistering results from excessive multiplication of the palisade mesophyll cells which become shapeless and there is a loss of intercellular spaces
Exobasidium spp. (leaf and stem galls)	Ericaceae Empetraceae Theaceae Lauraceae	Host leaf and stem cells expand with the loss of intercellular spaces
Cronartium spp. (blister rusts)	Aecial stage on *Pinus* spp. and telial stage on herbaceous perennials, shrubs and deciduous trees	The aecial stage incites spindle-shaped galls on branches and stems of *Pinus* spp. due to tissue hyperplasia
Gymnosporangium spp. (leaf and stem galls)	Aecial stage on rosaceous trees and shrubs particularly *Malus* spp. and telial stage on *Juniperus* spp.	The telial stage incites galls on *Juniperus* foliage; galls are composed of disrupted host tissue which is permeated by mycelium

nodule is surrounded by a spongy cortex divided into inner and outer layers by an endodermis. In nodules with indeterminate growth the endodermis is open at the undifferentiated tip of the nodule whereas in determinate types the endodermis is completely closed and the cell walls are thickened forming a prominent scleroid layer. Beneath the endodermis lie vascular traces which pass around a central tissue to join at the nodule base and enter the host vascular system (Figure 15.1). On some legumes nodules are ephemeral, lasting only a few weeks before being replaced by new ones on freshly grown roots. Perennial nodules are found on other hosts where

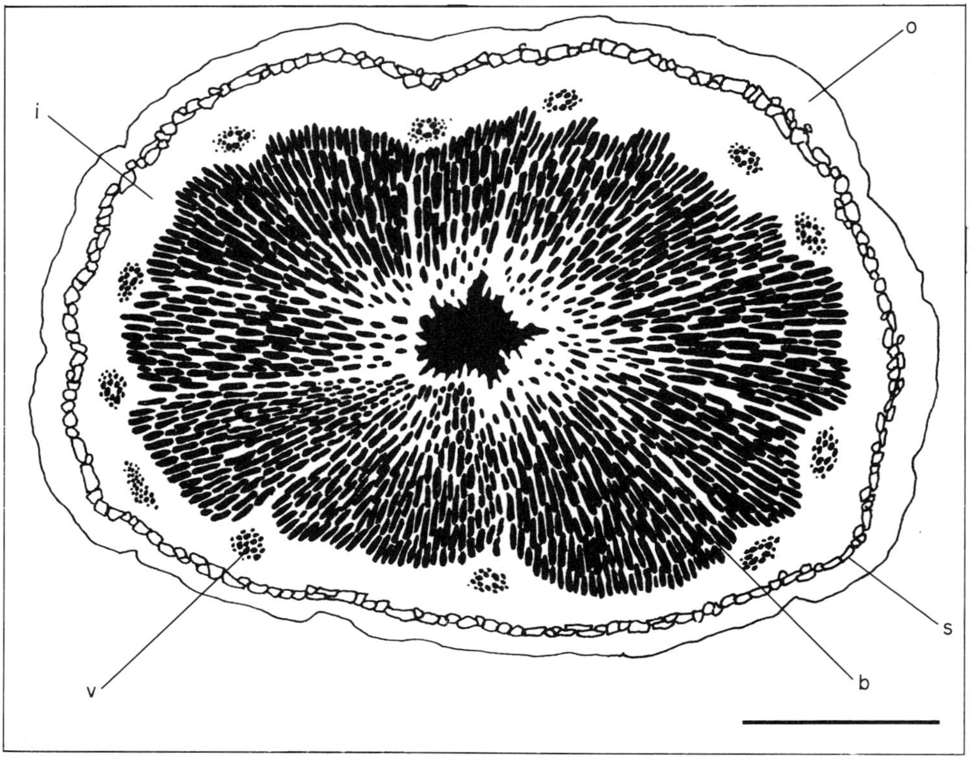

Figure 15.1. Section of a *Rhizobium* nodule, b = bacteroid containing central tissue, 0 = outer cortex, i = inner cortex, v = vascular tissue, s = scleroid layer, bar = 1 mm.
Redrawn from Bergersen (1982)

the nodules develop new tissue which arises from the old as growth resumes in spring.

Mechanisms of gall formation

For many pathogens knowledge of the mechanisms of gall formation is limited. The following account contrasts those mechanisms elucidated so far for two organisms, *Agrobacterium tumefaciens* (crown gall) which affects a wide host range and where once gall initiation has begun subsequent disorganized growth appears to be autonomous and *Plasmodiophora brassicae* (clubroot) which causes galling in a restricted range of plants and where it seems that galling occurs only in the presence of the pathogen.

Agrobacterium tumefaciens, crown gall

A causative relationship between crown gall disease of dicotyledonous plants and the Gram-negative soil bacterium *A. tumefaciens* was established by Smith &

Townsend (1907). Since then at least five hypotheses have been advanced to describe the tumour-inducing principle (TIP) (Dixon, 1981). Recent evidence strongly suggests that large DNA molecules (plasmids) which are present in *A. tumefaciens* cells confer on them tumourigenic and virulence properties. Tumour development is associated with the presence of two opines in the host cell; these are octopine (arginine condensed with pyruvic acid) and nopaline (arginine condensed with ketoglutaric acid). Once host cells have become tumourous they may be grown in the absence of bacteria and bacteria-like derivatives and they continue to synthesize octopine and nopaline. Tumour-inducing (T_i) plasmids have been identified in *A. tumefaciens* cells and it has been demonstrated that tumourigenicity, opine catabolism and synthesis are directly related to one another and to particular plasmids (Bomhoff *et al.*, 1976). The simplest explanation is that during the transformation of healthy host plant cells to a tumourous condition, specific genetic material is transferred from the bacterium to the plant cell. Consequently the induction of crown gall tumours results in the expression of a new genetic function whose phenotype is T_i plasmid specific rather than host specific. There is a transfer of genetic information from *A. tumefaciens* to the plant cell and the source of this information is the T_i plasmid (Hepburn, 1982). Portions of DNA from the plasmid pT_i B6–806 have been identified in the cells of plant tumours resulting from *A. tumefaciens* infection (Chilton *et al.*, 1977). More recently it has been demonstrated that two portions of plasmid DNA are transferred to plant cells either collectively or separately (Thomashow *et al.*, 1980). These authors suggest that plasmid DNA is transferred during the infection process (T-DNA) and is integrated into host-cell chromosomal DNA. The use of T_i plasmids as vectors to permit the insertion of foreign genes into host plant cells is reported by Herrera-Estrella *et al.* (1983) and Bevan *et al.* (1983). Advances in this topic are reviewed by Kahl & Schell (1982). It is well established that invasion by *A. tumefaciens* requires damaged or wounded plant cells (Lippincott & Lippincott, 1969). Subsequently bacterial cells become bound to the host plant cell wall (Smith & Hindley, 1978). The binding process is thought to be related to the presence of plasmids in *A. tumefaciens* cells because cells free from plasmids cannot bind to the plant cell wall. Binding takes place between a lipopolysaccharide component of the bacterial cell wall and one or several components of the host cell wall which are only exposed in young cells. Following binding, the T_i plasmid enters the plant cell nucleus where it can be identified for up to 9 days (Nuti *et al.*, 1980). It appears that the bacterium has only a passive role in these events because *Petunia* protoplasts have been transformed to a tumourous condition using only purified T_i plasmid (Davey *et al.*, 1980). A causal relationship between plasmids present in *A. tumefaciens* and gall formation in host plants would not appear to be an isolated phenomenon. The ability to nodulate peas has been transferred from a strain of *Rhizobium leguminosarum* with nodulating ability to a non-nodulating strain of *R. leguminosarum* and to three other species of *Rhizobium* which normally nodulate legumes other than peas, via the transfer of plasmids (Johnston *et al.*, 1978).

Plasmodiophora brassicae, clubroot

Plasmodiophora brassicae was identified as the causal agent of clubroot of crucifers by Woronin (1878). Knowledge of the mechanisms of gall formation in clubroot disease is scanty compared with that for crown gall. A reason for this is the complex life cycle of *P. brassicae*; there is a primary root hair stage which may occur in hosts other than crucifers and a secondary gall-forming stage affecting cortical and vascular tissues of only crucifers. The life cycle has not been studied in a sequential manner throughout the primary and secondary stages. Penetration and invasion by primary zoospores derived from resting spores have been described in detail by Williams *et al.* (1973). Roots of cabbage (cv. Jersey Queen) seedlings were immersed in a suspension of primary zoospores for 0.5–1 h and then maintained in nutrient solution free from *P. brassicae*. Zoospores collided with the root hairs several times before eventually becoming attached. The point at which the zoospore became attached was not random but was on the side of the spore opposite to the flagellar insertion. There followed an apparently quiescent period during which a thick cell wall formed around the zoospore (encystment). Fifteen minutes prior to penetration, a vacuole within the cyst enlarged to fill half the cyst with lipid bodies closely adpressed to the vacuole. At the start of penetration a sock-like structure (rohr) evaginated from the spore and became pressed against the outer wall of the host, forming a bulb-shaped adhessorium. A needle-shaped structure (stachel) passed down the rohr, through the adhessorium and through the root hair wall. Immediately a stream of cytoplasm was transferred from the pathogen spore to the host cell. Within 4 h the nucleolus began to increase in size, by 6 h it had doubled and then continued to enlarge for 24 h. This was accompanied by increased nuclear RNA and protein (Bhattacharya & Williams, 1971). Although nuclear total protein content increased considerably the level of histones was slightly decreased. Within the histones the lysine-rich fraction decreased by 25% whereas the arginine-rich fraction remained unaltered. Regardless of age at the time of infection or the number of primary parasite plasmodia within a root hair cell, the nucleolus did not swell beyond 1.5–3.0 μm in diameter. No detectable changes in the levels of host DNA occurred as a result of infection. Accompanying these chemical changes were alterations to root hair cell morphology. Instead of normal linear root hair extension, the tip became enlarged and club-shaped or branched. Swelling of most root hair cell tips was detected within 4–6 h after invasion. Distortion took place to a greater extent in younger host cells.

Whether or not the pathogen leaves the host cells following these primary stages and then re-enters to invade the cortical and stelar tissue is a matter of debate (Aist & Williams, 1971; Dekhuijzen, 1979). Some information is available, however, on the events occurring within host cells following the onset of galling; nuclear RNA, DNA and histone levels have been positively correlated with the number of parasite nuclei within each cell (Williams, 1966). In cells containing large plasmodia the nucleolar volumes were increased 30-fold and the DNA contents 16-fold compared with non-infectd cells. It has also been established that changes in cytokinin, auxin and amino nitrogen occur in galled tissue (Katsura *et al.*, 1966; Matsubara & Nakahira, 1968; Ingram, 1969a, 1969b; Kavanagh *et al.*, 1969; Reddy & Williams, 1970; Dekhuijzen & Overeem, 1971; Chiang & Nip, 1971, 1973). These changes have been implicated in the hypertrophy and hyperplasia of roots infected by *P. brassicae.* Whether the pathogen is a source of these growth regulators or stimulates the host to synthesize abnormal amounts of them is unknown (Ingram, 1969b). The latter would seem to be more likely especially as Butcher *et al.*, (1974) found that there was decompartmentalization of indole glucosinolate, glucobrassicin and the enzymes glucosinolase and nitrilase in clubbed tissue, leading to an accumulation of abnormally high levels of indole-3-acetic acid.

Discussion

The morphology and anatomy of galls are well understood for both pathogenic and symbiotic associations. Species of micro-organisms which incite these effects are widely distributed among the fungi and bacteria but with little apparent relationship between the particular types involved. It is now realized that pathogens such as the downy mildews may incite anatomical and physiological effects akin to those caused by gall-inducing organisms. An explanation of the mechanism of gall formation in plants affected by *A. tumefaciens* and *Rhizobium* spp. now appears possible. There is a transfer of genetic information in plasmid form from the bacteria to cells of the higher plant hosts and once this information has been incorporated into the host nucleus it may provide a trigger which destabilizes host cell metabolism. In consequence, once gall formation is initiated further unregulated growth is autonomous. As yet, however, there is no indication of why nodules in the *Rhizobium*–legume association are limited in size and effect, whereas those caused by *A. tumefaciens* are not. In contrast to these gall-forming bacteria, information about the events leading to galling incited by *P. brassicae* is far more scanty. A prime factor here is the more extended and imperfectly known life cycle. Various growth regulatory compounds, which are recognized to have some rôle in symptom development in a wide range of host–parasite combinations, have been implicated in gall formation. There may be differences in the physiological and genetic abilities of primary and secondary zoospores. Certainly the former are less specific in their host requirements than are the latter. Some changes in host nuclear material have been associated with both types of zoospore; it may be informative to note that so far primary zoospores have been shown to affect only host RNA levels whereas secondary zoospores affect both RNA and DNA levels.

References

Aist J.R. & Williams P.H. (1971) The cytology and kinetics of cabbage root hair penetration by *Plasmodiophora brassicae. Canadian Journal of Botany* **49**, 2023–34.

Anon. (1973) A guide to the use of terms in plant pathology. *Phytopathological Papers* No. **17**. Commonwealth Mycological Institute, Kew.

Bergersen F.J. (1982) *Root Nodules of Legumes: Structure and Functions*. John Wiley, Chichester.

Bevan M.W., Flavell R.B. & Chilton M-D. (1983) A chimaeric antibiotic resistance gene as a selectable marker for plant cell transformation. *Nature* (London) **304**, 184–7.

Bhattacharya P.K. & Williams P.H. (1971) Microfluorimetric quantitation of nuclear proteins and nucleic acids in cabbage root hair cells infected by *Plasmodiophora brassicae. Physiological Plant Pathology* **1**, 167–75.

Black L.M. (1965) Symptoms of virus diseases in plants. In: *Handbüch der Pflanzen Physiologie* (Ed. by W. Ruhland), pp. 236–66. Springer-Verlag, Berlin.

Bomhoff G., Klapwijk P.M., Kester H.C.M., Schilperoort R.A., Hernalisteens J.P. & Schell J. (1976) Octopine and nopaline synthesis and breakdown genetically controlled by a plasmid of *Agrobacterium tumefaciens. Molecular and General Genetics* **145**, 177–81.

Bos L. (1970) *Symptoms of Virus Diseases in Plants*. PUDOC, Wageningen.

Butcher D.M., El-Tigani S. & Ingram D.S. (1974) The role of indole glucosinolates in the clubroot disease of the Cruciferae. *Physiological Plant Pathology* **4**, 127–40.

Butler E.J. & Jones S.G. (1961) *Plant Pathology*. Macmillan, London.

Chiang M.S. & Nip W.K. (1971) Free amino acid content in root tissue of clubroot resistant and susceptible cabbages. *Canadian Journal of Plant Science* **51**, 66–7.

Chiang M.S. & Nip W.K. (1973) Free amino acid content in leaf and root tissues of clubroot-resistant and clubroot-susceptible cabbages. *Euphytica* **22**, 393–8.

Chilton M-D., Durmmond M.H., Merlo D.J., Sciaky D., Montoya A.L., Gordon M.P. & Nester E.W. (1977) Stable incorporation of plasmid DNA into higher plant cells : the molecular basis of crown gall tumourigenesis. *Cell* **11**, 263–71.

Davey M.R., Cocking E.C., Freeman J., Pearce N. & Tudor I. (1980) Transformation of *Petunia* protoplasts by isolated *Agrobacterium* plasmids. *Plant Science Letters* **18**, 307–13.

Dekhuijzen H.M. (1979) Electron microscopic studies on the root hairs and cortex of a susceptible and a resistant variety of *Brassica campestris* infected with *Plasmodiophora brassicae. Netherlands Journal of Plant Pathology* **83**, 1–17.

Dekhuijzen H.M. & Overeem J.C. (1971) The role of cytokinins in clubroot formation. *Physiological Plant Pathology* **1**, 151–62.

Dixon G.R. (1981) *Vegetable Crop Diseases*. Macmillan, London.

Hepburn A.G. (1982) The biology of the crown gall — a plant tumour induced by *Agrobacterium tumefaciens*. In: *Bacteria and Plants* (Ed. by M. Rhodes-Roberts & F.A. Skinner) pp. 101–3. Academic Press, New York.

Herrera-Estrella L., Depicker A., van Montagu M. & Schell J. (1983) Expression of chimaeric genes transferred into plant cells using a T_i-plasmid-derived vector. *Nature* (London) **303**, 209–13.

Ingram D.S. (1969a) Growth of *Plasmodiophora brassicae* in host callus. *Journal of General Microbiology* **55**, 9–18.

Ingram D.S. (1969b) Abnormal growth of tissues infected with *Plasmodiophora brassicae. Journal of General Microbiology* **56**, 55–67.

Ingram D.S. (1982) The biochemistry of host-parasite interaction. In: *The Downy Mildews* (Ed. by D.M. Spencer). pp. 143–64. Academic Press, New York.

Johnston A.W.B., Beynon J.L., Buchanan-Woolaston A.V., Setchell S.M., Hirsch P.R. & Beringer J.E. (1978) High frequency transfer of nodulating ability between strains and species of *Rhizobium. Nature* (London) **276**, 634–6.

Kahl G. & Schell J.S. (Editors) (1982) *Molecular Biology of Plant Tumours*. Academic Press, New York.

Katsura K., Egawa H., Toki T. & Ishii S. (1966) On the plant hormones in the healthy and *Plasmodiophora* infected roots of *Brassica rapa* var. Neosuguki Kitam. *Annals Phytopathological Society* (Japan), **32**, 123–9.

Kavanagh J.A., Reddy M.N. & Williams P.H. (1969) Growth regulators in clubroot of cabbage. *Phytopathology* **59**, 1035.

Lippincott B.B. & Lippincott J.A. (1969) Bacterial attachment to a specific wound site as an essential stage in tumour induction by *Agrobacterium tumefaciens. Journal of Bacteriology* **97**, 620–8.

Matsubara S. & Nakahira R. (1968) Cytokinin activity in an extract from the gall of *Plasmodiophora*-infected root of *Brassica rapa* L. *Botanical Magazine* (Tokyo) **80**, 373–4.

Nuti M.P., Ledeboer A.M., Durante M., Nuti-Ronchi V. & Schilperoort R.A. (1980) Detection of T_i-plasmid sequences in infected tissues by *in situ* hybridization. *Plant Science Letters* **18**, 1–6.

Reddy M.N. & Williams P.H. (1970) Cytokinin activity in *Plasmodiophora brassicae*-infected cabbage tissue cultures. *Phytopathology* **60**, 1463–5.

Smith K.M. (1972) *A Textbook of Plant Virus Diseases*. Longman, London.

Smith E.F. & Townsend C.O. (1907) A plant-tumour of bacterial origin. *Science* (New York) **25**, 671–3.

Smith V.A. & Hindley J. (1978) Effect of agrocin 84 on attachment of *Agrobacterium tumefaciens* to cultured tobacco cells. *Nature* (London) **276**, 498–500.

Thomashow M.F., Nutter R., Montoya A.L., Gordon M.P. & Nester E.W. (1980) Integration and organisation of T_i plasmid sequences in crown gall tumours. *Cell* **19**, 729–39.

Williams P.H. (1966) A cytochemical study of hypertrophy in clubroot of cabbage. *Phytopathology* **56**, 521–4.

Williams P.H., Aist J.R. & Bhattacharya P.K. (1973) Host-parasite relations in cabbage clubroot. In: *Fungal Pathogenicity and the Plant's Response* (Ed. by R.J.W. Bryde & C.V. Cutting), pp. 141–58. Academic Press, New York.

Woronin M. (1878) *Plasmodiophora brassicae*. Ürheber der Kohlpflanzen-Hernie. *Jahrbüch wissenschaftliche Botanik* (Berlin) **11**, 548–74.

16 Infection by viruses and subsequent host damage

A.J.H. CARR
Welsh Plant Breeding Station, Aberystwyth,
Dyfed, SY23 3EB, UK

Introduction

Some of the most damaging plant diseases are caused by viruses and by mycoplasma-like (MLOs) and rickettsia-like (RLOs) organisms which, in vector and epidemiological characteristics, behave like viruses. Examples which come readily to mind include swollen shoot of cacao, sugar cane mosaic, curly top of sugar beet and the complex of three viruses (beet yellows, beet mild yellowing and beet western yellows) causing sugar beet yellows, all of which can have devastating effects on their respective hosts.

Losses of sugar beet caused by virus yellows have been put at 25%, a loss of some £14m to the UK economy (Heathcote, 1978). According to Hull (1961) there is a loss of 3% for every week a plant shows symptoms. Surveys carried out by the Agricultural Development and Advisory Service (ADAS) estimate that 60–70% of cereals are infected with barley yellow dwarf virus with an average loss in grain yield of 5–10%. This represents an annual loss of £60m–120m. Because viruses are systemic, vegetatively propagated crops such as potatoes are at particular risk (Jellis & Boulton, this volume), as are crops infected with seed and/or pollen-borne viruses (fortunately rare) such as bean mosaic and barley stripe mosaic (both seed and pollen-borne) and lettuce mosaic (seed-borne).

Despite the above examples it has generally not proved possible with plant viruses to predict accurately from symptom intensity and infection level the yield losses to be expected, as has been achieved for many fungal diseases. The intention here is to explore the reasons for this failure and to indicate lines of approach that are helping, or may help, to realize this objective.

Symtomatology

Plant virus symptoms are rarely diagnostic and may at all times in some hosts or for part of the time in others not be visible, yet the virus may still cause marked yield losses. Sometimes, as with barley yellow dwarf virus (BYDV) in ryegrass, the extent of damage bears little relationship to the intensity of visible symptoms. Wilkins & Catherall (1977) have shown that this is because the virus stimulates up-

Wood R.K.S. & G.J. Jellis (1984) *Plant Diseases: infection, damage and loss.* Blackwell Scientific Publications, Oxford.

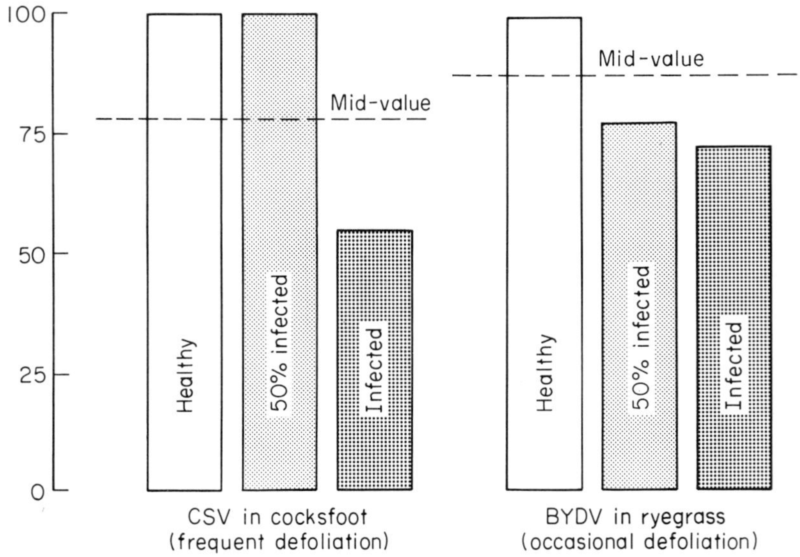

Figure 16.1. Yield of one-year-old simulated swards as a percentage of healthy plot yield. (From Catherall, 1966)

ward growth in spring but suppresses it in autumn, leading to a differential growth rhythm. By contrast, the same workers (pers. comm.) have shown a virtually direct relationship between the level of ryegrass mosaic virus (RMV) infection and yield loss.

Virus symptoms are diverse in nature but broadly may be classified into three kinds: mosaics, mottles, ringspots, streaks and the like; discolouration (yellowing or bronzing) of entire leaves, tillers or plants; and gross changes in morphology such as swollen stems, fasciation, tumours, reduced or excessive tillering or the production of juvenile tillers, and dwarfing and rosetting. Changed morphology may affect not only the productivity of the individual plant but also its competitive ability with its healthy neighbours. Nowhere is this more evident than under the competitive conditions in a grass sward. Catherall (1966) compared the effects of two grass viruses, cocksfoot streak (CSV) and BYDV, on their respective hosts. Both viruses caused a considerable reduction in yield in an 'all-infected' situation, yet only BYDV caused a significant reduction in overall yield in simulated swards that were 50% infected (Figure 16.1). The most likely explanation for this (Figure 16.2) is that infection with CSV causes a reduction in tillering but not in height so that there was complete compensation by healthy neighbouring plants which were competing for available lateral space; infection with BYDV, however, leads to dwarfing accompanied by increased tillering so that healthy plants are unable to compete with infected and so compensate for yield loss.

When a major disease is caused not by one but by a combination of viruses, the effects of the combination may exceed the sum of the individual constituents. A case in point is corn lethal necrosis, a combination of maize chlorotic mottle virus and either maize dwarf mosaic or wheat streak mosaic viruses. Losses of 50–90%

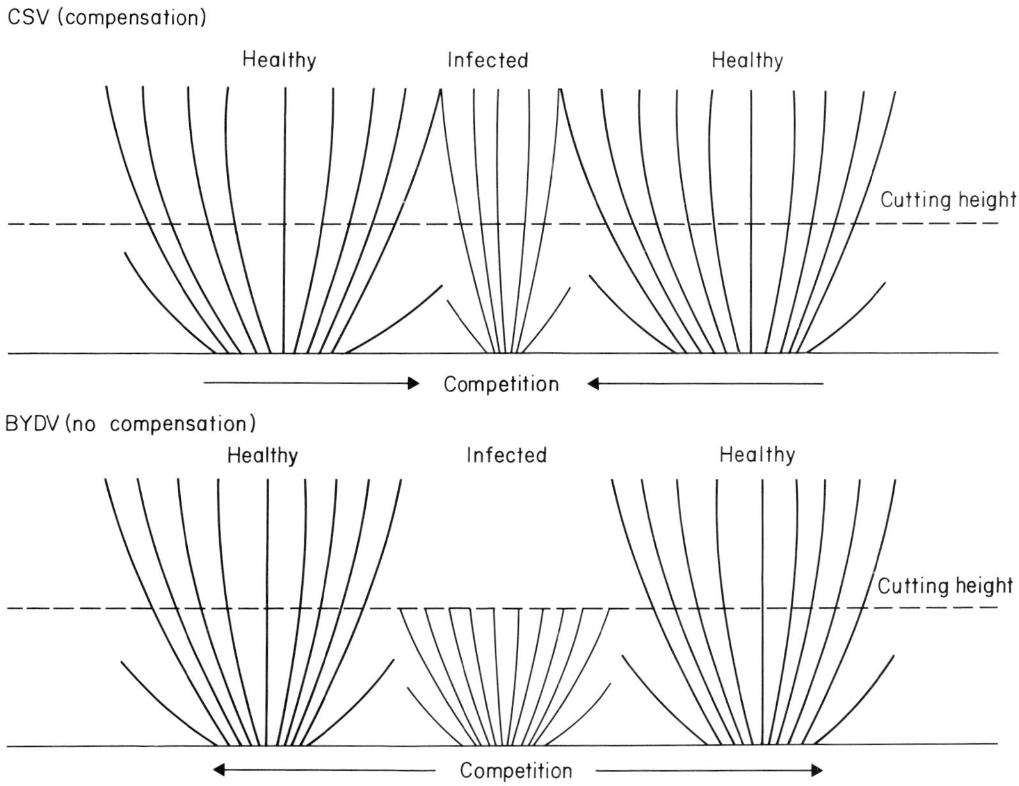

Figure 16.2. Diagrammatic representation of sward behaviour following infection with viruses having different effects on plant habit.

caused by this complex have been recorded in Kansas, USA (Uyemoto *et al.*, 1981). In some instances, a 'helper' is required for successful infection. Aphids can transmit carrot mottle virus, one component of carrot motley dwarf disease, only from plants having a mixed infection with the helper, carrot red leaf virus.

Influence of vectors

Predicting the likely effects which viruses may have on crop yields requires an understanding of the population dynamics of the vectors which transmit the vast majority of them. Persistent viruses such as BYDV are normally brought in from outside the crop; non-persistent viruses are more readily transmitted within a crop. This, in turn, influences the distribution over the field. The former often occur in patches whereas the latter are generally more uniformly distributed throughout the crop.

Autumn-sown cereals are generally at greater risk from BYDV than are spring-sown, because the susceptible growth stage occurs at a time when the appropriate aphid vectors are at a peak and many are viruliferous, having acquired the virus from the perennial grasses. Similarly, the eryophyid mites that transmit RMV in the

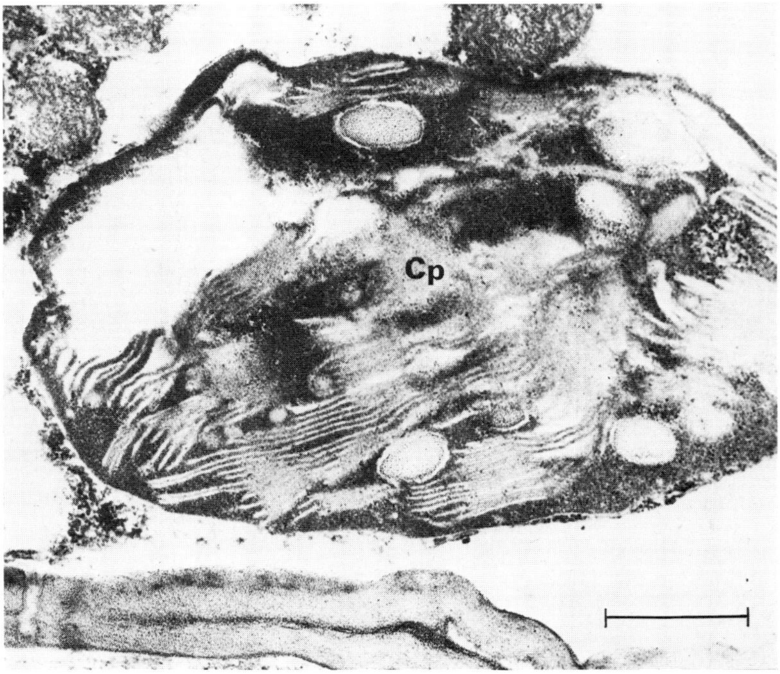

Figure 16.3. Section of cell infected with agrostic streak virus showing degeneration of the chloroplast (Cp). Bar = 1 μm. (Electron micrograph by J.A. Chamberlain)

non-persistent manner have their dispersal between June and August (A'Brook, 1975); hence autumn-sown ryegrasses are more likely to escape infection than are those sown in spring. It has been reported by ADAS that with autumn sowing only about 5% of plants were infected, yielding 308 kg seed/ha, whereas 75% of spring-sown plants were infected and in consequence the seed yield was only 196 kg/ha.

Soil-borne vectors, largely nematodes and lower fungi, present special problems in assessing the effects of the plant viruses which they transmit, mainly because the time of year at which they become viruliferous is critical. Spring-sown oats rarely become infected with oat mosaic virus, even on land severely infested with the fungal vector, *Polymyxa graminis*, whereas autumn-sown crops on similar land have been reported to have less than 5% of tillers reaching normal height and grain yield reduced by up to 80%. This was undoubtedly related to the time of release of the motile, virus-carrying zoospores (Catherall & Hayes, 1970).

Effects of management

The effects viruses have on yield are undoubtedly influenced by management, especially fertilizer application. A'Brook & Heard (1975) found that the effects of RMV became more pronounced at higher levels of nitrogen application. Four hundred kg N_2/ha increased dry matter yield of healthy swards by 13 630 kg/ha,

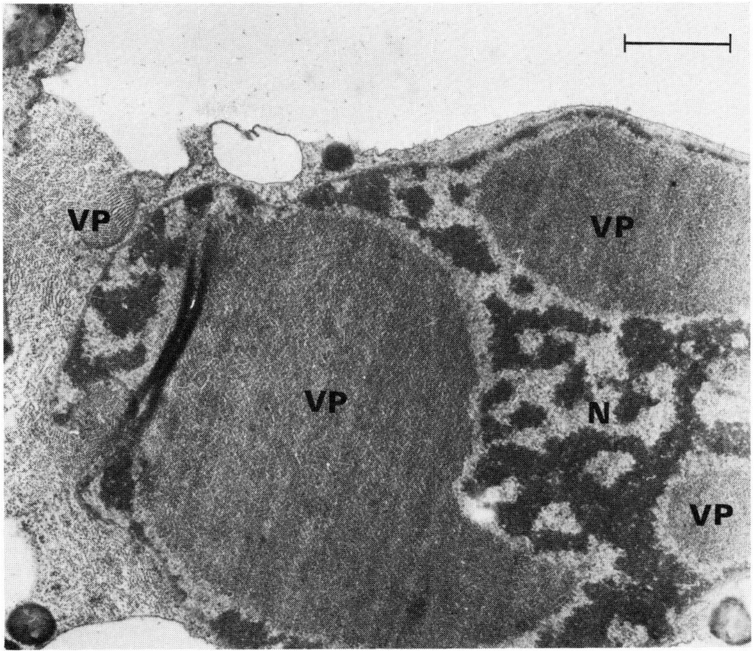

Figure 16.4. Section of oat cell infected with cynosurus mottle virus showing degenerate nucleus (N) containing several aggregates of virus particles (VP). Bar = 1 μm.
(Electron micrograph by J.A. Chamberlain)

whereas for swards in which 80% of ryegrass plants were infected the increase was only 9 720 kg/ha. Harvesting, by cutting or grazing, will spread sap-transmissable viruses; in grassland swards it may also alter sward balance by influencing the degree of competition and compensation between healthy and infected plants referred to earlier.

Effects on the host

The effects viruses have in the field depend on the effects they have on individual host plant metabolism which, in turn, is influenced by ultrastructural effects. Thus Jones *et al.* (1977) discovered that the primary cause of decrease in dry matter yields and in regrowth potential in RMV-infected ryegrass swards was a corresponding decrease in the net rate of photosynthesis in the canopy of up to 50%. There was also an associated increase in dark respiration, also of the order of 50%. It is well known that virus infection often leads to modification and degeneration of organelles such as the nucleus, chloroplasts and mitochondria (Chamberlain & Catherall, 1976). Chloroplast degeneration includes loss of the thylakoid system, disruption of stroma lamellae and increase in starch and lipid material (Figure 16.3). Nuclear membranes may disintegrate and with severe disruption aggregates of virus particles may occur in the degenerate nucleus (Figure 16.4).

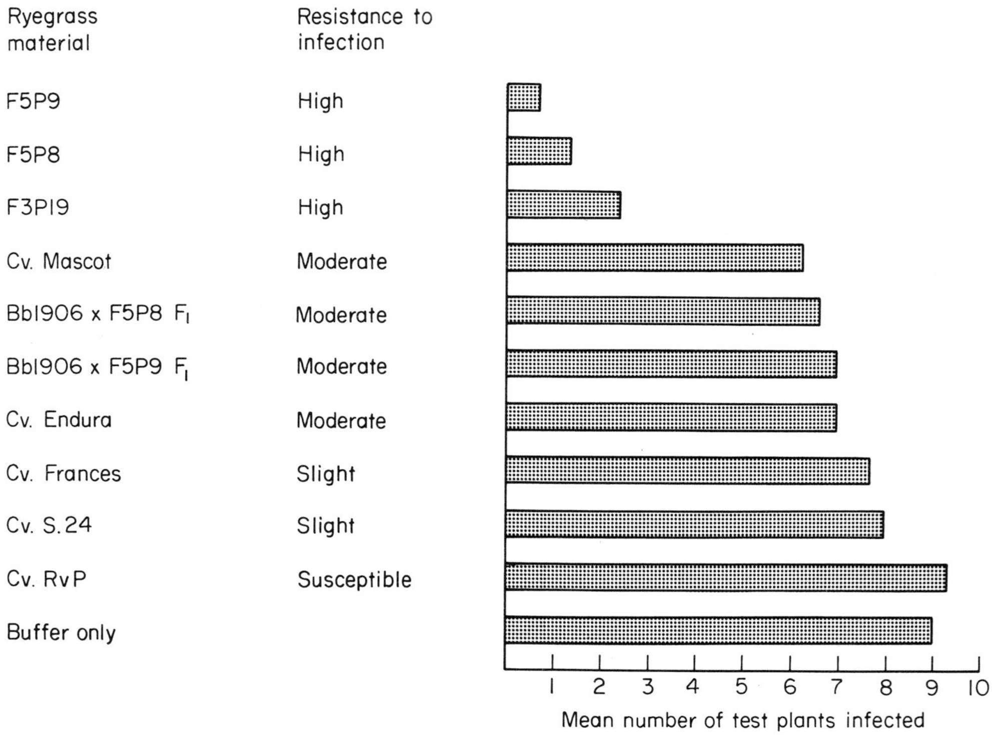

Figure 16.5. Capacity of sap from ryegrasses differing in resistance to infection to inactivate RMV. Data from Salehuzzaman & Wilkins, 1983.

The infection process and the role of resistance mechanisms

Viruses can only enter cells damaged by wounding, either by mechanical abrasion or by vector probing. Part of the process is mechanical, part biochemical. For example, Takebe & Otsuki (1969) have shown that negatively charged virus particles are aided in entry by positively charged molecules such as poly-L-ornithine, a technique now used in studying virus behaviour in protoplast systems. Uncoating of the virus occurs, followed by replication of both the nucleic acid core and the coat protein. Major adverse effects on the host occur during replication and it is here, and at assembly, that resistance mechanisms interfering with these processes may operate. Subsequently the virus may move from cell to cell via the plasmodesmata, and further resistance mechanisms limiting transport may come into effect. By studying these mechanisms we may come to understand the infection process better.

The role of inhibitors

Virus inactivators or inhibitors have often been implicated in mechanisms limiting infection. Thus Salehuzzaman & Wilkins (1983) found that ryegrass clones highly resistant to infection by RMV, when transmitted by the mite vector *Abacarus hystrix* and by manual inoculation, contained a constituent in healthy plant sap which, when that sap was mixed *in vitro* with sap from infected plants, reduced the infectivity of the inoculum when the mixture was tested on plants of a highly susceptible cultivar. This infectivity, as measured in terms of the number of test plants infected, was proportional to the innate resistance of the particular clone (Figure 16.5). Inhibition remained effective against a range of virus concentrations but was lost with a 16-fold dilution of sap from the most highly resistant clone. As inhibition increased the longer the virus was exposed *in vitro* to the healthy plant extract, it is logical to assume that the virus was inactivated rather than that resistance to it was induced in the test plants. Some specificity was indicated, because the extract was effective against another elongate particled virus, agropyron mosaic, but not against the isometric particled cynosurus mottle virus, in appropriate hosts.

This work is similar to that reported by Stirpe *et al.* (1981), who discovered that dianthins, glycoproteins produced by carnations, inhibited virus replication by binding to ribosomes. Binding occurred only with ribosomes from plants different from the glycoprotein source. While this might be the case in the work reported here, more recent evidence suggests it is not so.

The role of inclusion bodies

Inclusion bodies are a common feature of virus infection and take various forms. Structurally the most interesting are the so-called 'pinwheels' produced by a number of filamentous viruses including the potyvirus, RMV. Pinwheels have been shown by Edwardson (1974) to vary in precise morphology according to the inducing virus and by Shepard & Shalla (1969) to be composed of protein not serologically related to the inducing virus. They consist of a core from which radiates a number of curved plates and, sometimes, associated laminar aggregates and scrolls.

Chamberlain (1974, 1975, 1977) showed that the level of resistance in a number of ryegrass clones was correlated with the number of pinwheels occurring in a group within individual host cells, both mesophyll and phloem companion, and with their length and diameter. This was true whether resistance was assessed in terms of symptom expression or as a percentage of healthy plant yield (Figure 16.6). A similar situation existed with agropyron mosaic infection of wheat.

This association of resistance with pinwheel numbers leads to speculation that such inclusions may represent part of the host's defence mechanism. Possibly they absorb free viral components, or restrict virus movement. Pinwheels originate at the plasma membrane, close to plasmodesmata, and later move into the cytoplasm where they abut on the endoplasmic reticulum. One suggestion is that at the plasmodesmata pinwheels may interfere with intercellular metabolite transport and also delay systemic distribution of the virus particles, allowing time for other defence mechanisms to be activated. In resistant plants pinwheels were found at the plasmodesmata and in the cytoplasm, whereas in susceptible plants they were found

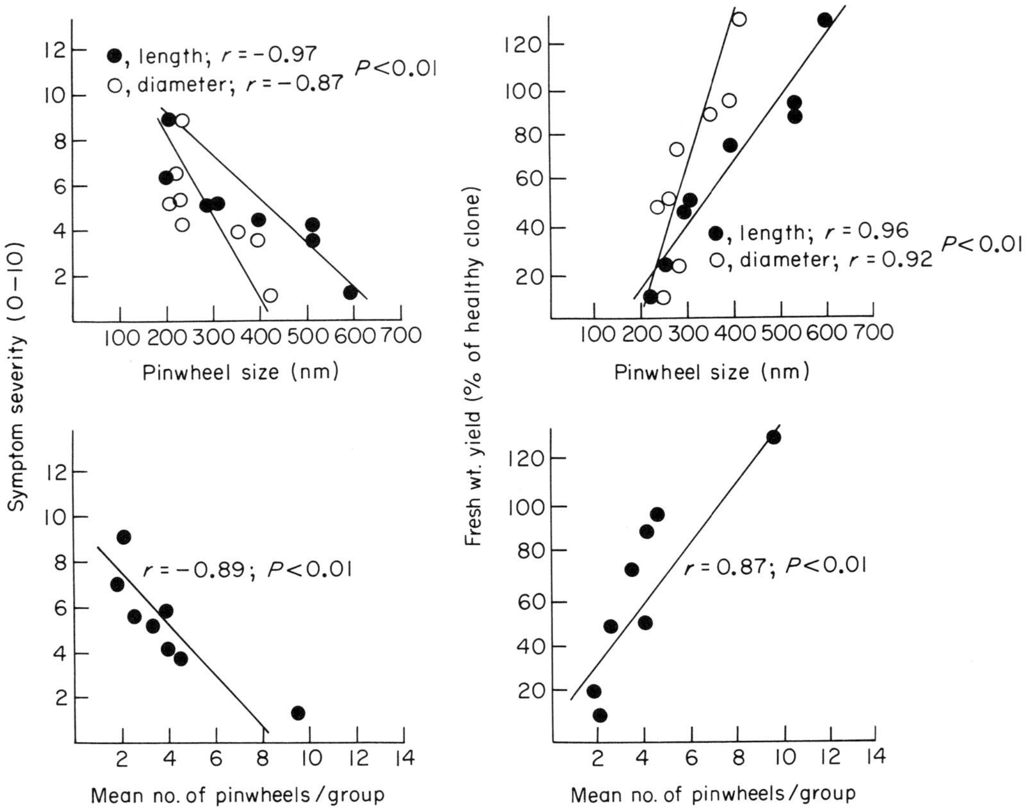

Figure 16.6. Relationship of yield and symptom severity with various characters of pinwheel cell inclusions in eight genotypes of RMV-infected Italian ryegrass. (From Chamberlain, 1975)

only in the cytoplasm in association with aggregates of virus particles and in cells well away from the lesion centres (Chamberlain *et al.*, 1977). This again suggests that pinwheels restrict virus movement. However, pinwheels could be detected in inoculated leaves only for the first 3 or 4 weeks after infection; thereafter they could be detected only in newly developing and presumably newly infected leaves. Possibly pinwheel production may cease when infection is under control, to be resumed if the virus becomes active again.

Limitation of virus concentration

Resistance may operate also by limiting virus concentration. Several allelic genes from wild Ethiopian types have been found to control resistance to BYDV in barley and these have proved useful in resistance breeding programmes. Jones & Catherall (1970) produced evidence to show that these genes operate by reducing virus multiplication. This reduction was particularly marked in resistant plants growing and maturing rapidly and less so in those which matured slowly either because of their genetic constitution or through manipulation of the environment. Fewer test plants acquired virus from aphids fed on resistant than on susceptible genotypes, again suggesting a lower virus concentration. More complete understanding of the mode of gene action here would clearly provide additional insight into the infection process.

Avoidance of loss

There are a number of ways in which losses caused by viruses may be avoided, or at least minimized.

Quarantine

While quarantine may be appropriate as a first line of defence, it is not always easy to define a non-indigenous virus, and new strains of an existing indigenous virus may unwittingly be introduced.

Good cultural practice

Sowing or planting clean stocks can do much to lessen the risk of spread of infection from within the crop. In this context, meristem tip culture following the work of Kassanis (1957) in freeing the potato cultivar King Edward from paracrinkle virus has proved very successful in a number of vegetatively reproduced crops and ornamentals.

Rogueing is commonly recommended to remove volunteers from previous crops harbouring viruses, although this has not always proved successful. Weed control can help to eliminate the sources of some viruses, e.g. removal of shepherd's purse (*Capsella bursa-pastoris*), chickweed (*Stellaria media*) and groundsel (*Senecio vulgaris*) can help in avoiding infection of sugar beet with virus yellows.

Cultivations can be planned to avoid peaks of vector infestation. For example, sowing short-term ryegrasses in autumn avoids the summer peaks of mites carrying RMV, late sowing of autumn cereals avoids the late season peaks of aphids carrying BYDV, and early sowing of sugar beet similarly avoids virus yellows.

Chemotherapy

Chemotherapy is probably currently the least successful method of control. Tetracyclines have been used to eradicate MLOs from plants. There are reports of Virazole (ribavirin), an antimetabolic nucleoside analogue, blocking virus replication and thus eradicating virus from plant tissue, e.g. potato virus X from cultured tobacco meristems (Shepard, 1977) and reducing alfalfa and and cucumber mosaic viruses in tissue cultures (Simpkins *et al.*, 1981). However, Virazole has not yet been shown to control viruses outside experimental systems.

An interesting example of a cosmetic effect of practical benefit derived from the use of chemicals is the amelioration of symptoms, though not the causal agent, of beet western yellows in lettuce by soil application of carbendazim (Tomlinson *et al.*, 1976). The action of this chemical was to delay senescence of the chloroplasts, leading to a healthy-looking and marketable product.

Control of vectors

Chemical control of vectors to avoid or limit introduction of virus and subsequent spread has been practised for some time and has proved successful in certain situations, notably in the control of persistent rather than stylet-borne viruses, and in the prevention of spread within rather than into a crop. The early warning system for BYDV control based on trapping information pioneered by ADAS is a good example of this.

Resistance breeding

Despite exceptions there are many examples of stable resistance to plant viruses achieved through plant breeding. Resistance to curly top of sugar beet has stood the test of time, and resistance in barley to BYDV introduced from Ethiopian sources has so far proved effective against all isolates and strains of the virus tested against it. Indeed, durable resistance seems more readily achievable to viruses than to many fungal pathogens. This is possibly because although viruses have considerable opportunities for mutation, the nature of the genome limits recombination and hence the ability to combine a resistance-breaking gene array with the overall fitness necessary for survival.

References

A'Brook J. (1975) Epidemiology and effects of grass viruses in relation to their insect vectors. *Annals of Applied Biology* **81**, 261–4.

A'Brook J. & Heard A.J. (1975) The effect of ryegrass mosaic virus on the yield of perennial ryegrass swards. *Annals of Applied Biology* **80**, 163–8.

Catherall P.L. (1966) The significance of virus diseases for the productivity of grassland. *Journal of the British Grassland Society* **21**, 116–22.

Catherall P.L. & Hayes J.D. (1970) Oat mosaic virus. *Plant Pathology* **19**, 78–81.

Chamberlain J.A. (1974) The relationship between tolerance and the production of pinwheel inclusions in plants infected with ryegrass mosaic virus. *Journal of General Virology* **23**, 201–4.

Chamberlain J.A. (1975) Electron microscopy of grass viruses in relation to possible resistance mechanisms. *Annals of Applied Biology* **81**, 264–6.

Chamberlain J.A. (1977) Pinwheel inclusions, their use in grass virus identification and possible role in host resistance. *Annales Phytopathologie* **9**, 249–53.

Chamberlain J.A. & Catherall P.L. (1976) Electron microscopy of some grasses and cereals infected with cocksfoot mottle, phleum mottle and cocksfoot mild mosaic viruses. *Journal of General Virology* **30**, 41–50.

Chamberlain J.A., Catherall P.L. & Jellings, A.J. (1977) Symptoms and electron microscopy of ryegrass mosaic virus in different grass species. *Journal of General Virology* **36**, 297–306.

Edwardson J.R. (1974) Some properties of the potato virus Y group. *Florida Agricultural Experiment Station Monograph* No. **4**, 398 pp.

Heathcote G.D. (1978) Effects of virus yellows on yield of some monogerm cultivars of sugar beet. *Annals of Applied Biology* **88**, 145–51.

Hull R. (1961) The health of the sugar beet crop in Great Britain. *Journal of the Royal Agricultural Society* **122**, 101.

Jones A.T. & Catherall P.L. (1970) The relationship between growth rate and the expression of tolerance to barley yellow dwarf virus in barley. *Annals of Applied Biology* **65**, 137–45.

Jones M.B., Heard A.J., Woledge J., Leafe E.L. & Plumb R.T. (1977) The effect of ryegrass mosaic virus on carbon assimilation and growth of ryegrass. *Annals of Applied Biology* **87**, 393–405.

Kassanis B (1957) The use of meristem tip culture to produce virus-free clones from infected potato varieties. *Annals of Applied Biology* **45**, 422.

Salehuzzaman M. & Wilkins P.W. (1983) Inhibitory activity in *Lolium perenne* associated with resistance to infection by ryegrass mosaic virus. *Physiological Plant Pathology* **22**, 199–207.

Shepard J.F. (1977) Regeneration of plants from protoplasts of PVX infected tobacco leaves. II. Influence of Viruzolen on the frequency of infection. *Virology* **78**, 261–6.

Shepard J.F. & Shalla T.A. (1969) Tobacco etch virus cylindrical inclusions: antigenically unrelated to the causal virus. *Virology* **38**, 185–8.

Simpkins I., Walkey D.G.A. & Neely N.A. (1981) Chemical suppression of virus in cultured plant tissue. *Annals of Applied Biology* **99**, 161–9.

Stirpe F., Williams D.G., Onyon L.J., Legg R.F. & Stevens W.A. (1981) Dianthins, ribosome damaging proteins with antiviral properties from *Dianthus caryophyllus* (carnations). *Biochemical Journal* **95**, 399–405.

Takebe I. & Otsuki Y. (1969) Infection of tobacco mesophyll protoplasts by toabacco mosaic virus. *Proceedings of the United States Academy of Sciences*, **64**, 843–8.

Tomlinson J.A., Faithfull F.M. & Ward C.M. (1976) Chemical suppression of the symptoms of two virus diseases. *Annals of Applied Biology* **84**, 31–41.

Uyemoto J.K., Claflin L.E., Wildon D.L. & Raney R.J. (1981) Maize chlorotic mottle and maize dwarf mosaic, effect of single and double inoculation on symptomatology and yield. *Plant Disease* **65**, 39–40.

Wilkins P.W. & Catherall P.L. (1977) Variation in reaction to barley yellow dwarf virus in ryegrass and its inheritance. *Annals of Applied Biology* **85**, 257–63.

Section 4
Damage and loss

17 World crop losses: an overview

I.M. SMITH
European and Mediterranean Plant Protection Organization,
1 rue Le Nôtre, 75016 Paris, France

L. CHIARAPPA and N.A. VAN DER GRAAFF
Plant Production and Protection Division, FAO,
Via delle Terme di Caracalla, Rome, Italy

Introduction

By the year 2000, if the world's population is to be fed as at present, world food production will probably need to increase by about 40%. Population projections, though perhaps unreliable in the longer term, can hardly fail to be realized in the short term unless some catastrophe intervenes. The present over-abundant younger generation will have their children before the age structure of the world's population can reach a more normal pattern, and all those children will have to be fed. The United Nations, through its Food and Agriculture Organization (FAO), sets as its targets a net improvement in nutrition, and a reduction in the inequalities between the different countries of the world. The task is formidable, but not impossible. Between 1950 and 1969, world agricultural production increased 1.7 times (in itself a considerable achievement), slightly more rapidly than world population (1.5 times). In the 1970s, the two are no more than keeping pace with each other.

An increase in agricultural production can, to a certain extent, be achieved by cultivating more land, especially in South America. However, resources of land and water are limited, and undoubtedly a large part of the increase must be made by progressive intensification of agriculture on the presently cultivated area. Yields can be increased directly by using better cultural techniques (e.g. mechanization), increasing inputs (especially fertilizers) and the breeding of higher yielding cultivars, well adapted to local conditions and responsive to fertilizers. These options are all demanding in terms of resources, energy and money, and success cannot be guaranteed. A cheaper alternative would be to prevent the high losses in production which occur at present by more efficient use of present techniques.

Analysis of crop losses

Major crop losses in many parts of the world frequently arise from causes which are practically unavoidable such as drought, flood, frost, hail. These lead to great

Wood R.K.S. & Jellis G.J. (1984) *Plant Diseases: infection, damage and loss.* Blackwell Scientific Publications, Oxford.

fluctuations in production from one year to the next. However, other causes of loss are to a great extent avoidable, at least ideally, and the losses due to pests, diseases and weeds clearly fall into this category.

The principles of loss assessment apply equally well to all biological agents causing crop losses, and it will be useful to consider the subject in the broadest way, and for convenience to use the word 'pest' in its wide sense, covering pests, diseases and weeds. We can consider following losses under three headings:

(a) actual losses
(b) potential losses
(c) the current cost of control measures.

These are illustrated in Table 17.1 which analyses losses in an individual crop (Zadoks, 1981a). It can be interpreted either in terms of all factors affecting yield, or in relation to pest losses only.

In relation to pest losses, the primitive yield is that which would be obtained without crop protection measures. The actual yield is that which is currently obtained by using certain crop protection measures. The attainable yield is that which experimentation shows could be achieved by full use of available crop protection methods under practical conditions. The economic yield, which gives the highest return on investment, is likely to be slightly less than the attainable yield, for each succeeding step in cutting losses is liable to cost more. The theoretical yield will be the yield obtainable in the absence of pests (not necessarily realizable under the practicable conditions of 'attainable yield').

In crop loss statistics, actual loss is the difference between actual and attainable yield; it is the figure usually quoted (FAO, 1967), and corresponds to the loss which could be made good by full use of existing knowledge. Potential loss is the difference between primitive and actual yield and represents the successes already achieved. The cost of crop protection measures means that the potential benefit must be less than the actual loss in financial terms. The belief that losses can be reduced is based on the assumption that the cost of achieving this objective is relatively low, but this has to be demonstrated for each particular case. New conditions (new techniques, new chemicals, higher market prices for a given com-

Table 17.1. Relation between actual and potential crop losses and yields.

Crop yield	Crop loss		Crop protection benefit
Theoretical		Unavoidable (Theoretical–Attainable)	
Attainable	Actual loss (FAO definition) (Attainable–Actual)	Avoidable (Attainable–Primitive)	
Economic			Potential benefit (Economic–Actual)
Actual	Potential loss (Actual–Primitive)		Actual benefit (Actual–Primitive)
Primitive			

From Zadoks (1981a)

modity) can affect the relationships shown in Table 17.1. In particular, if more economical and convenient control methods become available, the economic yield can more closely approach the attainable yield, and if more effective methods appear, the attainable yield can more closely approach the theoretical. If new methods are brought in where none previously existed, the actual yield will immediately exceed primitive yield.

In practical terms, units of yield and loss are usually obtained and expressed in tonnes per hectare, and losses are usually then expressed as a percentage of the attainable yield (rather than actual yield). This mode of expression is not necessarily the most suitable for a given purpose; from the standpoint of human nutrition, for example, it might be better to convert to kilojoules or to protein but, from the standpoint of disease management at the farm level, the loss must be expressed financially, although this can be difficult as the value of the crop varies with time, and in particular is affected by the amount produced. From the standpoint of agricultural statistics covering many crops, the losses may also be expressed as the additional land area which would need to be cultivated to compensate for them (although this may be very different, because the same absolute loss and also the land will not have the same value in different parts of the world).

Need for loss estimates

The scheme in Table 17.1 is a theoretical framework for the quantitative analysis of crop losses. In fact, such analyses have been done only to a limited extent, both at the level of the whole world and at that of a single farm. However, the need for such quantitative analyses has long been felt, and so many more or less accurate estimates have been made for different purposes. Governments need crop loss statistics to allocate material and research resources, and so do international agencies. Both need to be informed on the potential losses if pests spread or if new crops or cultural methods are introduced. Both need figures to inform the general public of the needs for crop protection measures. Farmers, and those who advise them, need specific figures to establish a basis for economic and effective control. Successful decision-making at all levels must depend on quantitative estimates of loss, and the level of accuracy of these estimates has to be tailored to the particular purposes.

Yield loss prediction in theory

How then are losses estimated? Firstly, the incidence or severity of damage caused by a pest can be determined directly. In an ideal loss-assessment scheme, research will establish a reliable system for estimating pest incidence or damage, which can be used at appropriate times through the growing season. The relationship between pest intensity and crop loss has to be established by detailed field experimentation. Yields are determined on standard plots from which different pest indices are

Table 17.2. Some examples of equations translating percentage disease to yield loss.

Disease	Equation	Reference
Single point		
Septoria tritici on wheat	$Y = -2.6943 + 0.6366 X$ X = % disease on foliage at G.S. 59	Eyal (1972)
Puccinia graminis on wheat	$Y = -25.53 + 27.17 \log X$ X = % disease at G.S. 75	Romig & Calpouzos (1970)
Piricularia oryzae on rice	$Y = 0.57 X$ X = % diseased plants 30 days after heading	Katsube & Koshimizu (1970)
Multiple point		
Rhynchosporium secalis on spring barley	$Y = (0.66 X_1 + 0.50 X_2) 12$ X_1 & X_2 = % disease on flag leaf & leaf 2 at G.S. 75	James *et al.* (1968)
Puccinia recondita on wheat	$Y = 5.3788 + 5.526 X_2 - 0.3308 X_5 + 0.5019 X_7$ X_2 = % disease on culms at G.S. 44 $X_5 + X_7$ = % disease on flag leaf – at G.S. 75 and 85	Burleigh *et al.* (1972)
Phytophthora infestans on potato (early infection)	$Y = 1.867 X_1 + 0.446 X_2 + 1.144 X_3 + 0.628 X_4 + 0.193 X_5 + 0.180 X_6 (+0.0 X_7) + 0.343 X_8 + 0.829 X_9$ where $X_1 \ldots X_9$ = % defoliation increments over 9 weeks	James *et al.* (1972)
Area under disease curve		
Cercospora spp. on cowpea	Area under curve of disease severity index (0–9) against time for at least 2 assessments per week from onset to harvest	
	% loss = 0.43 (Area) + 14.95	Schneider *et al.* (1976)

obtained at different times by suitable treatments (artificial inoculation, control measures, etc.). This experimentation gives a conversion function for predicting yield loss (with certain confidence intervals) from appropriate pest indices. The simplest such function is a linear regression equation based on the index at one point in time. However, the equation may relate to multiple points, or even to the area under a disease progress curve (Table 17.2). Computer simulation modelling of plant diseases will generally involve the incorporation of effects on yield.

Once such a conversion function has been obtained (or conversion functions, since one must ideally take account of varying agricultural and environmental conditions), it becomes possible to obtain loss estimates from the results of well-established survey schemes for pest incidence and current control measures. The percentage loss can be estimated for the whole area surveyed and, applied to the reference yield for the area (not always easy to estimate), can be converted to an absolute loss.

Yield loss prediction in practice

Good pest assessment schemes are widely available, although they could be better harmonized and more extensively tested for practicability and reliability under different conditions. Experimentation to establish conversion functions has been carried out by research establishments across the world, and compiled in the FAO Manual on the Evaluation and Prevention of Losses by Pests, Diseases and Weeds (Chiarappa, 1971). This manual consists of chapters on general principles (including statistical analysis), general methods, useful background information (e.g. crop growth-stage keys) and 136 individual assessment methods for particular pests, diseases and weeds. The manual is most complete and satisfactory for insect pests and weeds; there are fewer methods for diseases. The diseases which can be treated most easily are those where incidence virtually equals loss, i.e. an infected plant gives no yield (e.g. smuts, *Pyrenophora* spp. on cereals). Certain leaf diseases have also been covered, notably potato blight (*Phytophthora infestans*), several *Cercospora* leaf-spots, and Sigatoka disease (*Mycosphaerella musicola*) (see also Table 17.2). Developing intensity/loss relationships for fungi attacking the root system is much more problematical, although the manual covers one case (*Verticillium albo-atrum* on cotton).

In recent years, there has been a steady flow of research papers describing intensity/loss relationships for various diseases, e.g. *Pseudocercosporella herpotrichoides* (Clarkson, 1981), *Hymenula cerealis* (Bockus & Sim, 1982), *Puccinia coronata* (Singleton *et al.*, 1982), *Sphacelotheca reiliana* (Tuleen & Frederiksen, 1982), *Phytophthora cinnamomi* (Benson *et al.*, 1982), *Colletotrichum truncatum* on soybeans (Backman *et al.*, 1982). Some show that such a relationship may be nearly absent, e.g. in the case of dead-arm disease of grapevine (*Phomopsis viticola*) (Cucuzza & Sall, 1982).

Another possible source of data on intensity/loss relationships is experimentation carried out to assess the efficacy of pesticides, whether by industry or by registration authorities. This may incidentally provide data relating incidence to loss, but with the difficulties that trials are often carried out under conditions of deliberately and abnormally high disease levels, and that good yield data is not always obtained. The results of such experimentation are, unfortunately, often kept confidential.

The practical problems in providing ideal experimental conditions for determining intensity/loss relationships have been considered by Richardson (1981). The value of the results will depend on their statistical reliability and on their completeness. If the necessary different levels of attack at the right times cannot be obtained naturally, then it may be necessary to modify the epidemic by various devices not always achievable in practice, which in any case are liable to have undesirable interactions with other factors affecting yield. Richardson has examined the possibilities for alternative simpler experimental schemes, based, for example, on single plant comparisons, small area comparisons and half-field comparisons, and has shown where these may be valuable. In particular, the single-plant method makes it possible to obtain a wide range of disease levels.

Given a reliable intensity/loss relationship, the second element required in estimating actual losses is pest survey data. Surveys of pest incidence have been carried out very widely, but usually without the possibility of accurately estimating probable losses, except on the base of the surveyor's skill and experience. Experimentation on the determination of intensity/loss relationships has to lead to the definition of a sampling system for intensity which is simple and practicable for large-scale survey use. Although quite complex simulation models of pathosystems for forecasting purposes now exist, for example, in Europe (EPPO, 1979, 1980), a recent examination of current practice in Europe (EPPO, 1982) suggests that most practical forecasting by plant protection services is based on very simple empirical rules, requiring a minimum number of visits to a crop. Surveying in order to forecast losses may have to be linked with surveying for other purposes. Remote sensing (e.g. infra-red aerial photography) also has some promise as a survey method, especially if the conversion function is a simple one — e.g. detecting fields with very high loss.

In fact, it is only in a limited number of cases, e.g. the UK cereal disease survey (King, 1977a & 1977b; Cook & King, this volume) that extensive systematic surveys are combined with a tested intensity/loss relationship to give loss figures whose value can be critically assessed. In most cases, the results of a few trials on loss assessment may be combined with some simple estimates of disease incidence based on surveys, and may be scaled up to the region or country to provide what must only be approximate national loss figures. Van der Graaff (1981) considers in more detail the sources of information on which crop loss estimates could be based.

Crop loss assessment on a world scale

In 1967, FAO organized a Symposium on Crop Losses, at which the lack of hard data was emphasized, and the primary need for satisfactory methods was appreciated. An international network of correspondents was established, and the *Manual on the Evaluation and Prevention of Losses by Pests, Diseases and Weeds*, already described, was subsequently published (Chiarappa, 1971).

The FAO Symposium also stimulated the establishment of the FAO International Collaborative Programme on Crop Losses. This programme included the organization of crop loss workshops (held in Algeria, Italy, Nigeria, Brazil, Egypt and India), the establishment of *ad hoc* consultancies on crop loss appraisal and new developments in national and international programmes, including the establish-

Table 17.3. Pre-harvest losses of rice (in 1000 tonnes).

	Actual production	Potential production	Total loss (%)	% loss caused by insects	disease*	weeds
N. & C. America	4239	5369	20.5	3.5	6	11
S. America	7470	9396	20.5	3.5	6	11
Europe	1585	1865	15	2	3	10
Asia −China)	154 377	325 950	52	32	9	11
China	58 188	86 848	33	15	8	10
Africa	5610	8451	33	14	6	13
Oceania	35	40	13	2	3	8
	231 504	437 919	47.5	27.5	9	11

* Principally *Pyricularia oryzae*, *Cochliobolus miyabeanus* and root diseases
Data from Cramer (1967)

Table 17.4. Pre-harvest losses of citrus (in 1000 tonnes).

	Actual production	Potential production	Total loss (%)	% loss caused by insects	disease*	weeds
N. & C. America	8470	10 603	20	7	9	4
S. America	4210	5910	29	10	15	4
Europe	3996	4701	15	7	5	3
Africa	2273	2990	24	10	10	4
Asia	5166	6378	19	8	7	4
Oceania	280	355	21	8	9	4
	24 395	30 937	21	8	9	4

* Principally virus disease and *Phytophthora citrophthora*
Data from Cramer (1967)

ment of a Crop Loss Section in the Canada Department of Agriculture, projects on disease assessment for tropical crops in India, informal groups such as the Western Region Coordinating Committee 'Crop Loss Assessment' in the USA, and the International Committee on Crop Losses established by ISPP. In addition, crop loss assessment is now taught in plant protection courses in an increasing number of universities.

What has been the outcome? Cramer (1967) compiled crop loss estimates from available sources, good and poor, to cover world crops generally, worldwide and by continents. Tables 17.3 and 17.4 illustrate the kinds of data he collected. The figures are considered cautious and conservative, and are evidently of very variable quality, which makes it difficult to know how reliable the overall totals are. Nevertheless, no one has attempted to update this document as a whole (Walker, 1975; Cramer, 1975), and Cramer's work continues to be quoted as the best available source. FAO's activities have stimulated further research, and more reliable figures are progressively becoming available. However, their compilation is a major task which would need to be undertaken as a specifically planned project. It should be stressed that post-harvest food losses, in particular, have received increasing attention (National Academy of Sciences, 1978), as an area in which significant progress seems possible in reducing losses in developing countries. The United Nations General Assembly has set the target of a 50% reduction in post-harvest losses by 1985.

Consequences for the future

It may be questioned whether the refinement of world crop loss figures such as Cramer's will in itself bring any great benefits. It is possible that some of these estimates are grossly wrong, in which case better estimates are certainly needed. It is possible, in some areas, that new control methods have brought significant benefits; this requires confirmation. For example, in Western Europe, how much has treatment of cereals with new fungicides contributed in terms of increased yield, and is this all due to the control of known diseases? (EPPO, 1981). Or what have been the real consequences of the 'Green Revolution' in terms of increased potential or actual losses due to pests? The gap between actual and attainable yields may widen, while actual yields still increase. Full analysis of crop protection costs and residual yield losses on a country basis are also of great interest. Zadoks (1981a) has done this for the Netherlands, estimating that the total cost to the nation of plant pests can be broken down into 57% treatments costs, 11% collective costs and 32% residual losses. In general, large-scale crop loss figures serve as a general indication for policy-making, and it is doubtful whether policies would be greatly changed by their refinement.

There is scope for further work in the precise identification of situations where pests cause substantial losses which could be greatly reduced by control measures. There are extreme cases in which crops are hardly being cultivated because of a pest: for example, leaf blight (*Microcyclus ulei*) practically prevents economic rubber production in South America, and bacterial diseases practically prevent tomato production in certain seasons in the humid tropics. There may be no immediate practical solutions to these problems but they can at least be identified.

In other cases, actual yields being obtained are far short of the attainable yield, or else attainable yields are being obtained by using excessively costly control measures. In such cases, a pest management approach may ensure that available resources are used to best effect.

Effective pest management requires the establishment of a crop loss profile (Chiarappa, 1981). This implies an understanding of the damage/loss relationship, estimation of probable yield loss and damage based on estimates of pest incidence, and forecasts of how much of this loss can be avoided by appropriate control methods of known cost. This approach, preferably integrated to cover all pests of a crop, is widely preached but only practised in a few cases. It seems most promising for the control of animal pests and foliage diseases. Chiarappa (1974) has tried to assess for which of these such an approach could be expected to be successful or necessary. A number of conditions must be satisfied. There must be a significant economic saving in treating on the basis of an assessment, rather than treating routinely. Thus, if the costs of control are low (e.g. sulphur treatments against *Uncinula necator* on vine), a management scheme is probably not worth considering and preventive chemical control measures are more effectively used. Similarly, if the value of the crop per hectare is low (e.g. cereals), it may be difficult to find control measures which are sufficiently cheap. It must be possible to assess disease readily, which complicates the situation for many soil-borne pathogens. It must be possible to treat after symptom assessment and expect a yield benefit. There is no point in a 'supervised' control system if symptoms can only be assessed long after action should have been taken (e.g. *Taphrina deformans* on peach) or yield loss arises very soon after symptom appearance (e.g. *Sclerotinia laxa* causing blossom blight of apricot). Finally, the management scheme must be consistent with acceptable agricultural practice, both from the point of view of the farmer (can he apply control measures on demand?) and the consumer (who must be protected from residues).

In the intensive agriculture of many developed countries, this kind of evaluation has already been made for many diseases. Forecasting and warning systems exist for those which most benefit from a management approach (e.g. potato blight — *Phytophthora infestans*, fireblight — *Erwinia amylovora*, vine downy mildew — *Plasmopara viticola*, apple scab — *Venturia inaequalis*). As fungicides have become available to control cereal diseases economically, management systems have rapidly followed (such as EPIPRE; Zadoks, 1981b).

In developing countries, however, where FAO's attention is mostly concentrated, such evaluations may be lacking or be based on much less certain foundations. Integrated control is proposed as a lasting solution to the world's crop loss problems, but a careful appraisal of epidemiology, and in particular of intensity/loss relationships, in relation to available control measures should make it possible to single out the situations where different kinds of pest management approach are most suitable.

There are particular problems in establishing intensity/loss relationships for tropical plant pests. It may be difficult to obtain homogeneous areas for experimentation, and the single-plant approach proposed by Richardson (1981) as an alternative to working with populations in plots encounters difficulties if there is too much between-plant variation. Cassava mosaic is currently a very important disease attracting attention for the study of intensity/loss relationships (Ayanru & Sharma, 1982). Hau *et al.* (1980) used a single-plant type of assessment and could only find a correlation between yield loss and disease intensity (expressed as percentage loss of leaf area due to mosaic) if potential yield was calculated for each plant separately, owing to wide variation between plants. There is clearly a need for careful analysis of the experimental designs suitable for such work on tropical crops, and no certainty that models developed for temperate conditions can be transposed to the tropics.

Conclusions

The magnitude of world crop losses due to pests has now, to a certain extent, been analysed and understood on a world scale, and the international scientific community is well aware of the problem. However, it has become clear that to quantify losses does not itself provide solutions. Indeed, much of the work done on intensity/loss relationships has been done with pests already well controlled — the analysis follows rather than precedes the solution. We now face the more difficult task of deciding how losses can be reduced, especially in the Third World, and for which pests good understanding of the magnitude of losses can contribute to better management. The emphasis has shifted from collecting information to provide data on losses for countries or continents to the integration of intensity/loss relationships into management systems at a very practical farm level.

References

Ayanru D.K.G. & Sharma V.C. (1982) Effects of cassava mosaic disease on certain leaf parameters of field-grown cassava clones. *Phytopathology* **72**, 1057–9.

Backman P.A., Williams J.C. & Crawford M.A. (1982) Yield losses in soybeans from anthracnose caused by *Colletotrichum truncatum*. *Plant Disease* **66**, 1032–4.

Benson D.M., Jones R.K. & Barker K.R. (1982) Disease loss assessment for Azalea, Rhododendron, and Japanese holly in North Carolina nurseries. *Plant Disease* **66**, 125–8.

Bockus W.W. & Sim T.IV (1982) Quantifying Cephalosporium stripe disease severity on winter wheat. *Phytopathology* **72**, 493–5.

Burleigh J.R., Roelfs A.P. & Eversmeyer M.G. (1972) Estimating damage to wheat caused by *Puccinia recondita tritici*. *Phytopathology* **62**, 944–6.

Chiarappa L. (1971) *Crop loss assessment methods: FAO manual on the evaluation and prevention of losses by pests, diseases and weeds*, also *supplement 1* (1974) and *supplement 2* (1977). Commonwealth Agricultural Bureaux, Farnham Royal.

Chiarappa L. (1974) Possibility of supervised plant disease control in pest management systems. *FAO Plant Protection Bulletin* **22**, 65–8.

Chiarappa L. (1981) Establishing the crop loss profile. In: *Crop loss assessment methods: FAO manual on the evaluation and prevention of losses by pests, diseases and weeds, supplement 3* (Ed. by L. Chiarappa), pp. 21–4. Commonwealth Agricultural Bureaux, Farnham Royal.

Clarkson J.D.S. (1981) Relationship between eyespot severity and yield loss in winter wheat. *Plant Pathology* **30**, 125–31.

Cramer H.H. (1967) Plant protection and world crop production. *Pflanzenschutz-Nachrichten* 'Bayer'. Bayer, Leverkusen.

Cramer H.H. (1975) Economic aspects of plant protection. *Semaine d'étude, agriculture et hygiène des plantes, Gembloux, 8–12 September 1975.*

Cucuzza J.D. & Sall M.A. (1982) Phomopsis cane and leaf spot disease of grapevine: effects of chemical treatments on inoculum level, disease severity, and yield. *Plant Disease* **66**, 794–7.

EPPO (1979) Conference on systems modelling in modern crop protection. *EPPO Bulletin* **9**, 133–332.

EPPO (1980) Conference on forecasting in crop protection. *EPPO Bulletin* **10**, 3–275.

EPPO (1981) EPPO colloquium on the side-effects of fungicides in cereals. *EPPO Bulletin* **11**, 259–379.

EPPO (1982) Report of the first meeting of the Working Party on forecasting, Changins, 1982. *EPPO Publication Series C* No. **76**, pp. 1–29.

Eyal Z. (1972) Effect of Septoria leaf blotch on the yield of spring wheat in Israel. *Plant Disease Reporter* **56**, 983–6.

FAO (1967) *Report of the FAO Symposium on Crop Losses*. Food and Agricultural Organization, Rome.

Hau B., Kranz J. Dengel H.-J. & Hamelink J. (1980) On the development of loss assessment methods in the tropics. In: *Crop loss assessment: Proceedings of the E.C. Stakman Commemorative Symposium*. Miscellaneous Publication **7**, pp. 254–61, Agricultural Experiment Station, University of Minnesota.

James W.C., Jenkins J.E.E. & Jemmet J.L. (1968) The relationship between leaf blotch caused by *Rhynchosporium secalis* and losses in grain yield of spring barley. *Annals of Applied Biology* **62**, 273–88.

James W.C., Shih C.S., Hodgson W.A. & Callbeck L.C. (1972) The quantitative relationship between late blight of potato and loss in tuber yield. *Phytopathology* **62**, 92–6.

Katsube T. & Koshimizu Y. (1970) Influence of blast disease on harvests in rice plant. I. Effects of panicle infection on yield components and quality. *Bulletin of the Tohoku National Agricultural Experiment Station* **39**, 55–96.

King J.E. (1977a) Surveys of diseases of winter wheat in England and Wales, 1970–1975. *Plant Pathology* **26**, 8–20.

King J.E. (1977b) Surveys of foliar diseases of spring barley in England and Wales, 1972–1975. *Plant Pathology* **26**, 21–9.

National Academy of Sciences (1978) *Postharvest Food Losses in Developing Countries*. National Academy of Sciences, Washington.

Richardson M.J. (1981) The determination of yield-loss relationships by non-plot methods. In: *Crop loss assessment methods. FAO manual on the evaluation and prevention of losses by pests, diseases and weeds, supplement 3*. (Ed. by L. Chiarappa), pp. 59–63. Commonwealth Agricultural Bureaux, Farnham Royal.

Romig R.W. & Calpouzos L. (1970) The relationship between stem rust and loss in yield of spring wheat. *Phytopathology* **60**, 1801–5.

Schneider R.W., Williams R.J. & Sinclair J.B. (1976) Cercospora leafspot of cowpea: models for estimating yield loss. *Phytopathology* **66**, 384–8.

Singleton L.L., Moore M.B., Wilcoxson R.D. & Kernkamp M.F. (1982) Evaluation of oat crown rust disease parameters and yield in moderately resistant cultivars. *Phytopathology* **72**, 538–40.

Tuleen D.M. & Frederiksen R.A. (1982) Evaluating a crop loss model for head smut of sorghum. *Phytopathology* **72**, 1278–80.

Van der Graaff N.A. (1981) Increasing reliability of crop loss information: the use of 'indirect' data. In: *Crop loss assessment methods. FAO manual on the evaluation and prevention of losses by pests, diseases and weeds*. (Ed. by L. Chiarappa), pp. 65–9. Commonwealth Agricultural Bureaux, Farnham Royal.

Walker P.T. (1975) Pest control problems (pre-harvest) causing major losses in world food supplies. *FAO Plant Protection Bulletin* **23**, 70–77.

Zadoks J.C. (1981a) Crop loss today, profit tomorrow: an approach to quantifying production constraints and to measuring progress. In: *Crop loss assessment methods. FAO manual on the evaluation and prevention of losses by pests, diseases and weeds*. (Ed. by L. Chiarappa), pp. 5–11. Commonwealth Agricultural Bureaux, Farnham Royal.

Zadoks J.C. (1981b) EPIPRE: a disease and pest management system for winter wheat developed in the Netherlands. *EPPO Bulletin* **11**, 365–9.

(1983a). They are strongly affected by cultivar and exceptionally by cold winters.

Scab can cause premature defoliation but no studies have quantified the relationship between defoliation and disease. Scab can be particularly debilitating on shoots which can be killed directly, or indirectly by providing an infection site for cankers. No quantitative studies have been reported although a start has been made in estimating pustule numbers on shoots (Butt *et al.*, 1982). Wood scab is more important and possibly harder to control on pear than on apple.

Cankers obviously debilitate trees but there have been surprisingly few studies on the relationship between canker incidence or intensity and shoot or tree mortality apart from Wilson (1968) cited earlier.

Disease–quality relationships

The term quality is used in at least two senses in crop-loss studies: as a biochemical parameter such as protein or sugar content or as some aspect of appearance. The former is certainly more useful in that it can be operationally defined but the latter is in common parlance in horticulture and quality loss is used here to indicate some defect in shape, colour or skin finish. A major effect of mildew, discussed later, is to reduce mean fruit size and this, with some cultivars, leads to a downgraded crop. Mildew also affects the skin finish of both apples and pears through russeting, as do some of the fungicides used to control mildew, and this complicates the quantification of damage. Most studies have been largely anecdotal. In a 10-year study on the effects of scab and mildew incidence on fruit quality and yield, which will be referred to several times in this chapter, Kolbe (1981) estimates quality losses of 7–17% from mildew through russeting and smaller fruits.

Quality losses from scab are directly related to the incidence of scabbed fruits recorded in the grading process. Commercial grading systems, however, are not comparable to the disease assessment systems of research workers. Experience at East Malling (Butt *et al.*, 1981, 1982) indicated that small lesions on fruits, recorded in detailed assessments at harvest, were not noticed in conventional assessments, nor did they result in a downgraded crop when commercially graded. The relationship of fruit quality with the incidence of foliar scab has been noted by several workers: with scab on leaves at the beginning of the season (Shibkova, 1976); and with the proportion of shoots with scabbed leaves at harvest (Vanderscheer, 1980a). The intercepts of the equations reported by Jeger (1981) indicate that a small amount of foliar scab is present before fruit scab is observed. Kolbe (1981) reports quality losses of 35% due to scab.

Yield losses

Yield is the marketable portion of biomass production, the apple and pear fruits in this chapter. Losses in yield can be considered at the individual farm, or at the regional or national level. Estimates made for experimental farms are obviously of a different nature from those aggregated for a region, even when, as with fruit orchards in the UK (Butt & Jeger, 1982), the industry is reasonably well-defined geographically. This section is concerned with models of crop loss, components of yield, and losses in fruit orchards.

Models of yield loss

The need to consider the physiology of yield of diseased plants has increasingly been recognized (Teng & Gaunt, 1980). The diseases considered here certainly affect the physiology of orchard trees (Swinburne, 1975; Spotts & Ferree, 1979; Ellis *et al.*, 1981) but no attempts have been made to incorporate these effects into mechanistic yield loss models. Most models of yield loss have been descriptive equations of varying sophistication (James, 1974; Teng & Gaunt, 1980; Madden *et al.*, 1981) and predict the percentage yield loss from disease measured on a percentage scale, although the uses of these scales in yield-loss models have been criticized (MacKenzie & King, 1980; Lim & Gaunt, 1981). The most complete conceptual approach to yield loss–pest relationships has been given by Walker (1977) and could usefully be extended to diseases. No yield-loss models have been formulated specifically for perennial crops, where losses in any one year may depend on disease in previous years especially in those prior to cropping.

Components of yield

An important aspect of any yield-loss study is to determine the component of yield affected by disease, yet surprisingly few models of loss take this into account. Yield components often form a product; with orchard fruits the simplest is tree density x mean number of fruits/tree x mean fruit size, but this is a gross over-simplification. The main problems are that components are not independent. There are inter-temporal affects and complex relationships with leaf area that have been studied in detail by pomologists (Magness *et al.*, 1931; Jackson & Palmer, 1976; Hansen, 1977; Landsberg, 1979) and these make the formulation of a realistic yield-loss model for orchard fruits a difficult proposition.

Orchard losses

Although reports of total yield losses caused by mildew are common, e.g. 15–30% by Kolbe (1981), there are few quantitative studies of the effects on components of yield. Burth & Motte (1980) report a linear relationship between mean fruit weight and mildew incidence which suggests a 1g loss in weight for every percentage increment in mildew. In a comprehensive and continuing study at East Malling, Butt *et al.*, (1983) have quantified the dependence of total fruit weight/tree, fruit numbers/ tree and mean fruit weight of Cox's Orange Pippin, as well as the tree growth referred to earlier, on mildew incidence over an 8-year period using multiple regression analyses.

Reports of losses due to scab in the pre-war and immediate post-war years were not particularly useful because the amount of spray damage with the materials then available could be severe (McKay, 1938) although yield increases of up to 100% were claimed on occasion (Mills, 1947). More recently, Kolbe (1981) estimated yield losses of 30–50% but again this was not related to the levels of scab. Mendoza *et al.*, (1981) regressed the total weight of non-scabbed apples on the percentage of leaves scabbed, percentage of fruits scabbed and the date of assessment. All coefficients were negative in a multiple regression equation and, by comparing their magnitude, it appears that leaf scab was having a larger effect on yield than fruit scab (although details of the comparative ranges of disease were not given). The negative sign of the assessment date coefficient is counter-intuitive, (as early scab would generally lead to lower yields), and the equation is best considered as a convenient form of data summary rather than a realistic loss model.

Estimates of orchard losses caused by canker have never been accurately estimated (Swinburne, 1975), although Savidge (1946) considered that crop yield in a trial cider orchard over a 15-year period was linked more to freedom from canker than to any other orchard factor.

Economic losses

The economic losses suffered through disease are not simply the yield losses converted to monetary units. Loss implies a reference point and this may be quite different for the physical and economic cases. A rigorous economic analysis for pests was provided by Headley (Walker, 1977; Thompson & White, 1979) and can be adapted to diseases. At a minimum, the relationships between yield and disease and control measures and disease (both in monetary units) must be known. Plausible relationships for each are given in Figure 18.3: the effect of disease on yield is increasingly marked at high levels of disease, whereas it is increasingly costlier to reduce disease at low levels of disease. Where the two curves intersect, at either end of the scale, there is an absolute monetary loss. The maximum economic gain is where the difference between the two curves is greatest and this occurs at a level of disease which *does* cause a reduction in yield. Figure 18.3 is highly idealized but useful in that it provides the economic reference point and shows that excessive as well as inadequate control can lead to economic losses. The level of disease which provides the reference point has been termed an economic threshold which is unfortunate because the term has also been used in other senses and it is an optimum, not a threshold, that is implied. The only threshold in Figure 18.3 is where the yield revenue curve departs significantly from the horizontal.

This form of analysis has been attempted for fruit orchards and Thompson & White (1979) consider that more pragmatic action thresholds can be linked to the economic threshold. An attempt is being made at East Malling to characterize both the yield revenue and control cost functions (Butt & Jeger, 1982; Butt *et al.*, 1983). Results suggest that the economic threshold for apple powdery mildew on Cox's

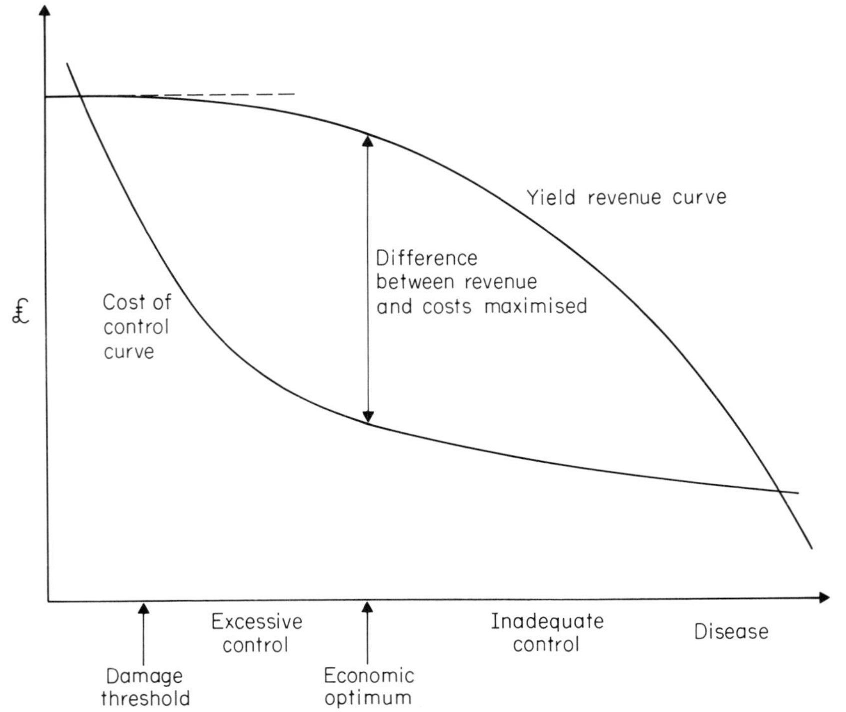

Figure 18.3. Relationships between yield revenue (£), the cost of disease control (£) and disease.

Orange Pippin may be very low indeed, largely because of the higher proportion of fruits in the premium size grade at low mildew levels.

Economic losses from apple scab have long been recognised, although rarely with the perception of Wallace (1913) who calculated monetary values for loss and net profits from spraying at the farm, regional and national levels. He estimated, conservatively, that losses of $47 million were suffered by growers in the USA in 1900. Interestingly, the few recent examples of regional losses have come from Kashmir (Malik *et al.*, 1976; Gupta, 1978) where a serious epidemic occurred, apparently for the first time, in 1973. A loss of Rs 54 lakhs (~ US$540 000) in a single season was estimated and has resulted in Government grants and loans of almost half that amount again as a 'calamity and relief' fund. Although it is rarely such a problem elsewhere, scab can account for half the *total* cost of pest management in the far north-east of the USA (Zuck & MacHardy, 1981) and the need for optimal control strategies is clear (MacHardy & Jeger, 1983). However, Kennel (1981) queries whether an optimum exists with scab and considers that the threshold is crossed with the slightest attack. This must depend, however, on the thoroughness or scale of the disease assessment as discussed earlier: the threshold may be very low indeed but it is better to recognize it as such than to maintain that scab is absent.

There have been virtually no data on orchard economic losses caused by canker.

Vanderscheer (1980b) estimated that the total annual cost of control measures and losses in the Netherlands came to upwards of 1 million guilders (~ US$350 000) for 1300 ha. Highly susceptible cultivars such as Spartan are being planted in the UK and, if chemical control becomes more common as a result, then a proper economic evaluation will be required given the high establishment costs and the delays before cropping (Butt & Jeger, 1982) that affect the profitability of new orchards.

A detailed procedure for establishing the regional costs of disease control for apple and pear has been given by Steiner in Aldwinckle (1978). A detailed model for evaluating economic losses in individual orchards, including discounting to allow for interest rate effects, is given by Heaton *et al.* (1981) who also demonstrated its use with losses due to Prunus ring spot virus. It also needs remembering that crop loss assessment itself costs money (Carlson, 1980) and should be included in the costings.

Implications for disease management

An aim of disease management is to optimize economic returns to the grower and this was stressed by James (1974) who argued that projected losses in yield, in monetary units rather than increases in disease, should be the basis of each and every control decision. The decision is helped, of course, if the crop price is known in advance. Similar comments have been made by MacKenzie & King (1980) and these have most relevance for epidemics proceeding logarithmically from low initial levels. However, some epidemics, and apple powdery mildew on vegetative shoots is a good example, do not proceed on this basis. Incidence is usually high initially because of the relationship between primary mildew and shoot phenology. The management objective is not to prevent incidence from reaching a threshold value but rather to maintain it within certain bounds or even to reduce it to a threshold value. These threshold values, linked to economic reference points, are being established at East Malling Research Station and can be incorporated into disease management schemes (Butt *et al.*, 1981, 1982) and with strategies such as integrating host partial resistance with use of fungicides (Jeger & Butt, 1983b). Most progress has been made with apple powdery mildew but the other diseases are also receiving attention.

Finally, it is arguable whether losses caused by individual disease or pests of orchard crops should be studied in isolation. Interactions between diseases, pests and control measures undoubtedly occur (Butt & Jeger, 1982) as in other crops but comprehensive appraisal of these remains a task for the future.

References

Aldwinckle H.S. Ed. (1978) Disease losses and cost of control — 1977. *Newsletter of the Apple and Pear Disease Workers* **1**, 296–313.

Burth V. & Motte G. (1980) Zum Aufreten und zur Bekämpfung des Apfelmehltaus in der industriernä-Bigen Apfelproduktion. *Nachrichtenblatt für den Pflanzenschutz in der DDR* **34**, 220–3.

Butt D.J. & Barlow G.P. (1979) The management of apple powdery mildew: a disease assessment method for growers. *Proceedings of the 1979 British Crop Protection Conference — Pests and Diseases*, pp. 77–86.

Butt D.J. & Jeger M.J. (1982) Decision-based management of orchard pathogens and pests in the United Kingdom. *Proceedings of the 1982 British Crop Protection Symposium — Decision Making in the Practice of Crop Protection* (Ed. by R.B. Austin), pp. 167–89.

Butt D.J., Souter R.D. & Swait A.A.J. (1983) Apple mildew crop loss study. *East Malling Research Station Report for 1982*, p. 81.

Butt D.J., Jeger M.J., Souter R.D. & Swait A.A.J. (1981) Management of apple orchard diseases. *East Malling Research Station Report for 1980*, p. 78.

Butt D.J., Jeger M.J., Souter R.D. & Swait A.A.J. (1982) Management of apple orchard diseases. *East Malling Research Station Report for 1981*, p. 79.

Byrde R.J.W., Clarke G.M. & Welford P.J. (1973) Negative binomial distribution of cankers on apple trees caused by *Nectria galligena*. *Transactions of the British Mycological Society* **60**, 580–3.

Carlson G.A. (1980) Economic aspects of crop loss information. In: *Crop Loss Assessment, Proceedings of E.C. Stakman Commemorative Symposium, Minneapolis, August 20–23*, pp. 208–17.

Crowdy S.H. (1952) Observations on apple canker IV. The infection of leaf scars. *Annals of Applied Biology* **39**, 569–80.

Ellis M.A., Ferree D.C. & Spring D.E. (1981) Photosynthesis, transpiration and carbohydrate content of apple leaves infected by *Podosphaera leucotricha*. *Phytopathology* **71**, 392–5.

Gupta G.K. (1978) Present status of apple scab (*Venturia inaequalis*) in Himachel Pradesh and strategy for its control. *Pesticides* **12**, 13–4.

Hansen P. (1977) Blad-frugt relationer og frugtud vikling hos aebletraeer. *Saertryk af Tidsskrift for Planteavl* **81**, 123–32.

Heaton C.R., Ogawa J.M. & Nyland G. (1981) Evaluating economic losses caused by pathogens of fruit and nut crops. *Plant Disease* **65**, 886–8.

Jackson J.E. & Palmer J.W. (1976) Effects of shade on the components of yield of apple. *Journal of the Science of Food and Agriculture* **27**, 395.

James W.C. (1974) Assessment of plant diseases and losses. *Annual Review of Phytopathology* **12**, 27–48.

Jeger M.J. (1981) Disease measurement in a study of apple scab epidemics. *Annals of Applied Biology* **99**, 43–51.

Jeger M.J. & Butt D.J. (1983a) The effects of weather during perennation on epidemics of apple mildew and scab. *EPPO Bulletin* **13**, 79–85.

Jeger M.J. & Butt D.J. (1983b) Using partial resistance in the integrated control of apple powdery mildew. *IOBC/WPRS Bulletin 1983/VI/4*, pp. 111–22.

Kennel W. (1981) Integrierte Bekämpfung des Apfelschorfes (*Venturia inaequalis*) *Erwerbsobstbau* **23**, 132–5.

Kolbe W. (1981) Überlegungen zur Schorf- und Mehltaubekämpfung im Obstbau. *Obstbau* **6**, 486–9.

Landsberg J.J. (1979) From bud to bursting blossom: weather and the apple crop. *Weather* **34**, 394–407.

Lim L.G. & Gaunt R.E. (1981) Leaf area as a factor in disease assessment. *Journal of Agricultural Science, Cambridge* **97**, 481–3.

MacHardy W.E. & Jeger M.J. (1983) Integrating control measures for the management of primary apple scab, *Venturia inaequalis* (Cke.) Wint. *Protection Ecology* **5**, 103–25.

McKay R. (1938) Spraying experiments for the control of apple scab in 1935 and 1936. *Journal of the Department of Agriculture (An Roinn Talmhaídheachta), Eire* **35**, 3–18.

MacKenzie D.R. & King E. (1980) Developing realistic crop loss models for plant diseases. In: *Crop Loss Assessment, Proceedings of E.C. Stakman Commemorative Symposium, Minneapolis, August 20–23*, pp. 85–9.

Madden L.V., Pennypacker S.P. & Kingsolver C.H. (1981) A comparison of crop loss models. *Phytopathologische Zeitschrift* **101**, 196–201.

Magness J.R., Overley F.L. & Luce W.A. (1931) Relation of foliage to fruit size and quality in apples and pears. *Washington State College of Agriculture Experiment Station, Bulletin* **249**, 26 pp.

Malik A.G., Sheikh Ghulam-ud-Din, Koul M.L. & Putoo B.L. (1976) Fighting apple scab epiphytotics in Kashmir. *Pesticides* **10**, 44–6.

Mendoza H.A., Bailey A.M. & Ibarra J.E. (1981) Evaluación de daños ocasionados por la 'roña del manzano' Venturia inaequalis (Cke.) Wint., en Santiago, N.L. *Boletiń de la Sociedad Mexicana de Micologiá* **15**, 225–52.

Mills W.D. (1947) Effects of sprays of lime sulfur and of elemental sulfur on apple in relation to yield. *Cornell University Agricultural Experiment Station, Memoir* **273**, 38 pp.

Savidge C. (1946) Cider apple production in Herefordshire: results from Burghill trial orchard. *Long Ashton Research Station Report for 1946*, pp.116–27.

Seem R.C. & Gilpatrick J.D. (1980) Incidence and severity relationships of secondary infections of powdery mildew on apple. *Phytopathology* **70**, 851–4.

Shibkova N.A. (1976) Vredonosnost' parshi yabloni v leningradshoí obl. *Mikologiya i Fitopatologiyi* **10**, 431–2.

Spotts R.A. & Ferree D.C. (1979) Photosynthesis, transpiration and water potential of apple leaves infected by *Venturia inaequalis. Phytopathology* **69**, 717–9.

Swinburne T.R. (1975) European canker of apple. (*Nectria galligena*). *Review of Plant Pathology* **54**, 787–99.

Teng P.S. & Gaunt R.E. (1980) Modelling systems of disease and yield loss in cereals. *Agricultural systems* **6**, 131–54.

Thompson P. & White G.B. (1979) Linking economic and action thresholds in pest management: implications for advisory services. *Acta Horticulturae* **97**, 97–104.

Vanderscheer H.A.Th. (1978) Overwintering of the pathogen and rating of 'secondary mildew' in relation to susceptibility of apple varieties to *Podosphaera leucotricha. Mededelingen Faculteit Landbouwwetenschappen Rijksuniversiteit, Gent* **43**, 901–7.

Vanderscheer H.A.Th. (1980a) Threshold of economic injury for apple powdery mildew and scab. In: *Integrated control of insect pests in the Netherlands* pp. 49–52. Pudoc, Wageningen.

Vanderscheer H.A.Th. (1980b) Kanker bij vruchtbomem. *Proefstation voor de Fruitteelt-Wilhelminadorp (Goes), Mededeling* No. **18**, 66 pp.

Walker P.T. (1977) Crop losses: some relation-ships between yield and infestation. *Mededelingen Faculteit Landbouwwetenschappen Rijksuniversiteit, Gent* **42**, 919–26.

Wallace E. (1913) Scab disease of apples. *Cornell University Agricultural Experiment Station, Bulletin* **335** pp. 542–624.

Wilson E.E. (1968) Control of European canker of apple by eradicative and protective fungicides. *Plant Disease Reporter* **52**, 227–31.

Zuck M.G. & MacHardy W.E. (1981) Recent experience in timing sprays for control of apple scab: equipment and test results. *Plant Disease* **65**, 995–8.

19 Loss caused by cereal diseases and the economics of fungicidal control

R.J. COOK[*]

Agricultural Development and Advisory Service, Welsh Office Agricultural Department, Ty Glas Road, Cardiff CF4 5ZB UK

J.E. KING

Agricultural Development and Advisory Service, Harpenden Laboratory, Hatching Green, Harpenden AL5 2BD UK

Introduction

Diseases of cereals are widespread in UK and each year some crops are severely damaged by one or more of the leaf, stem and root diseases. For this reason, fungicides have been developed and have become a standard part of cereal production, most crops receiving at least one spray treatment for the control of leaf and stem diseases. The Agricultural Development and Advisory Service (ADAS) of England and Wales have indicated that £60 million was spent on fungicides for cereals in 1982 and it is likely that this level of use will be maintained, at least for the foreseeable future. In view of the size of this market and the potential for misdirection of resources, it is essential that losses caused by diseases are periodically reviewed so that costs and benefits of fungicidal treatments can be assessed. This chapter reviews the extent of damage caused by diseases and the effects of fungicides on the financial and energy budgets of the cereal crop in England and Wales.

Disease survey results

The annual incidence and severity of diseases of cereals have been estimated by ADAS in surveys of randomly-selected commercial crops at the milky-ripe growth stage (GS), (GS 75, Zadoks *et al.*, 1974). Spring barley crops were surveyed during 1967-80, winter wheat during 1970–82 and winter barley in 1981 and 1982. A total of 3976 winter wheat crops, 3296 spring barley and 502 winter barley crops were examined. Of these, 1074 winter wheat, 797 spring barley and 372 winter barley crops had been treated with fungicidal sprays. Plant samples from each crop were assessed using standard keys (Anon., 1976). The severity of leaf diseases was recorded as the percentage area of the flag and second leaves with symptoms. Stem-base diseases were recorded as the percentage of tillers showing symptoms in each

[*] Present address: Agricultural Development and Advisory Service, Brooklands Avenue, Cambridge CB2 2DR

of three categories: slight, moderate and severe (Scott & Hollins, 1974). Precise details of the survey methods have been published elsewhere (King, 1980).

Disease severity and losses

In most seasons the most damaging leaf disease of both spring barley and winter wheat was mildew (*Erysiphe graminis*). Other fungal diseases, notably rhynchosporium (*Rhynchosporium secalis*) and brown rust (*Puccinia hordei*) on spring barley, net blotch (*Pyrenophora teres*) on winter barley and septoria (*Leptosphaeria nodorum* and *Mycosphaerella graminicola*), yellow rust (*Puccinia striiformis*) or brown rust (*Puccinia recondita*) on winter wheat, were common in most seasons but were generally less important than mildew. Diseases of the stem base, assessed on winter cereals since 1975, have been eyespot (*Pseudocercosporella herpotrichoides*), sharp eyespot (*Rhizoctonia cerealis*) which has become more common in recent years, and symptoms of infection by *Fusarium* spp.

Losses in grain yield caused by those diseases for which a relationship between severity and loss has been established (Table 19.1) have been calculated each year (King, 1977a, 1977b, 1977c). The average total loss caused by all diseases was estimated to be 10.3%/year for spring barley and 5.2%/year for winter wheat. Using 1982 grain values (£105/t for wheat, £100/t for barley) and production data from the Ministry of Agriculture, Fisheries and Food Statistics Division (which does not distinguish between winter and spring barley), these losses have been transformed to an estimate of financial losses averaging £150 million/year. Such an estimate

Table 19.1. Formulae used in yield loss estimates for ADAS cereal disease surveys.

Disease	Relationship	Reference
Spring barley		
mildew	$y = 2.5\sqrt{x_i}$	Large & Doling, 1962 King, 1973
brown rust	$y = 0.4x_{ii}$	King & Polley, 1976
rhynchosporium	$y = 0.5x_{ii}$	James *et al.*, 1968
Winter wheat		
mildew	$y = 2.0\sqrt{x_i}$	Large & Doling 1963 King, 1973
septoria	$y = x_{iii}$	King *et al.*, 1983
yellow rust	$y = 0.4x_{iii}$	King, 1976
eyespot	$y = 0.1x_m + 0.36x_s$	Clarkson, 1981
sharp eyespot	$y = 0.05x_m + 0.26x_s$	Clarkson & Cook, 1983

y – % loss in grain yield; x_i – % disease on leaf 3 at GS 58; x_{ii} – % disease on leaf 2 at GS 75; x_{iii} – % disease on flag leaf at GS 75; x_m – % tillers with moderate symptoms at GS 75; x_s – % tillers with severe symptoms at GS 75 takes account only of the direct effect of pathogens on crops and excludes the cost of any control measures which are taken in an attempt to reduce or avoid loss. The cost of fungicides and of their application and any damage done to crops by machinery during application may be legitimately regarded as an expense incurred as a result of the threat of disease. In the current context therefore, the £60 million spent on fungicides may be added to the estimated value of lost yield to give an estimated cost of diseases of about £210 million/year in England and Wales.

Effects of fungicides

The percentage of winter wheat crops sprayed each year has steadily increased from 1% in 1973 to about 80% in 1982, whereas in spring barley the percentage of sprayed crops has remained fairly stable at about 50%/year from 1974 until 1980. About 75% of winter barley crops surveyed in 1981 and 1982 were treated with foliar fungicides. Table 19.2 shows the increase in fungicide treatment of winter wheat and in particular the popularity of multiple spray programmes since 1979.

It is difficult to assess the effect of fungicide treatment on disease incidence and severity as recorded in the surveys. This is due, in part, to the tendency for fungicides to be applied to protect against subsequent losses of those crops which are most at risk from diseases. In addition, fungicide sprays serve to delay rather than prevent disease development as shown in experiments on fungicidal control of many diseases where the final severity of disease was often similar in treated and untreated crops. A yield benefit in such experiments results from retarded disease development during the critical shooting and heading phases of crop growth, even though diseases may progress rapidly towards the end of the season when the efficacy of fungicides is diminishing. It is therefore not possible to use the survey

Table 19.2. Annual percentage of winter wheat crops treated with fungicides 1973–82.

	1973	'74	'75	'76	'77	'78	'79	'80	'81	'82
Unsprayed	99	96	88	86	83	68	52	41	20	24
Single treatments										
MBC† early	0	0	3	10	6	14	17	18	12	12
late	0	0	2	0	1	3	7	3	4	4
Triazole early	0	0	0	0	1	2	2	5	5	2
late	0	0	0	0	1	5	13	9	8	6
Others	1	2	4	2	7	4	4	2	7	2
Programmes										
MBC/triazole	0	1	1	1	1	1	1	12	24	15
Others*	0	0	2	1	0	1	1	6	17	26

* includes programmes with MBC but without triazole
† includes all fungicides with an MBC component
Note No treated crops were sampled in 1970–72

data to examine the effects of fungicides on disease severity nationally at the later growth stages. Indeed, only rarely is the mean severity of disease on sprayed crops significantly less than that on unsprayed crops.

In the past, the frequencies of unsprayed crops with severities of diseases above particular thresholds, determined by the formulae given in Table 19.1, have been used to estimate the proportion of crops likely to yield an economic return on the cost of spray treatments (Cook *et al.*, 1981). However, it is now established that some of the fungicides commonly used on cereals, particularly carbendazim-generating (MBC) chemicals (Mukhopadhyay & Bandopadhyay, 1977; Tripathi *et al.*, 1982) and triazole compounds (Buchenauer & Grossmann, 1977), not only control disease but also reduce the rate of leaf senescence by virtue of growth regulatory activity. The control of low levels of unrecorded diseases and/or phylloplane microflora by fungicides also results in the delay of senescence (Dickinson, 1981). The occurrence of such effects in the field is evident in the results of variety trials where the use of standard fungicide programmes has increased green leaf area after anthesis to an extent which is greater than can be accounted for by disease control alone (Priestley, 1981). This effect can be detected in the survey data after removal of the annual variability in green leaf area by expressing the green leaf area for each crop as a percentage of the annual mean for all crops in the same year. After such transformation, the mean green leaf area for crops sprayed with triazole fungicides was greater than that for unsprayed crops in both winter cereals and in spring barley (Table 19.3). There was also an indication that green leaf area at GS 75 was

Table 19.3. Effect of fungicide spray treatment on relative green leaf area (%) (leaf 2 at GS 75).

	MBC* single	MBC† programme	Triazole≠ single	Other single	Unsprayed
Winter wheat					
Relative green leaf area (%)	99.9	101.2[b]	106.2[ab]	104.5	99.3[a]
Number of crops	531	109	177	226	2902
Spring barley					
Relative green leaf area	99.0	—	110.0[ab]	100.3[b]	99.6[a]
Number of crops	50	—	153	309	2784
Winter barley					
Relative green leaf area	98.8[b]	101.7	109.0[ab]	100.7	92.6[a]
Number of crops	107	51	108	37	132

* Any fungicide including a carbendazim generating material (MBC)
† Any programme in which at least one fungicide included an MBC
≠ Triazole fungicide alone or in mixture (except with MBC)
Figures, within each line, followed by the same suffix are significantly different ($P < 0.05$)

Table 19.4. Effect of fungicide spray date on relative green leaf area (%) (leaf 2 at GS 75). Number of crop.

	Date of Application					
	April or earlier	May 1–15	May 16–31	June 1–15	June 16–30	Untreated
Winter wheat	94.1^{ab}	101.3	103.8	102.5^{a}	108.7^{bc}	99.3^{c}
Number of crops	260	156	69	285	214	2902
Spring barley	99.9^{c}	96.5^{d}	103.3^{e}	107.9^{ad}	11.40^{bcde}	99.5^{ab}
Number of crops	311	120	144	167	55	2784
Winter barley	104.4	116.4	103.7	36.2	116.5	92.6
Number of crops	239	37	21	2	4	132

Figures, within each line, followed by the same suffix are significantly different ($P < 0.05$)

greatest for crops sprayed late, when any senescence-retardant effects of fungicides would have had least time to diminish before samples were taken (Table 19.4).

This effect of modern fungicides on the rate of senescence, together with the likelihood that the crops now left unsprayed are those least likely to give an economic return on spray costs, means that the use of disease severity thresholds could underestimate the proportion of surveyed crops likely to respond economically to fungicidal sprays. Such an underestimate is clear for winter wheat where the similarity between the estimates derived from the survey data and the frequency of economic responses recorded in fungicide experiments decreases with increasing costs and lateness of treatment, (Table 19.5). Such a discrepancy might be expected if the experiments were all in highly diseased crops where the effects of disease control on yield would be expected to be greater than the average. In practice, experiments were sited in commercial crops and disease levels recorded were within the range recorded on surveyed crops. Also, a range of cultivars was used in the experiments and the mean responses of individual cultivars to the full protective programmes were close to those recorded in experiments involving similar treatment conducted by the National Institute of Agricultural Botany in Cambridge during preparation of the list of recommended varieties (Anon., 1982).

These data support the contention that fungicides have some additional benefit over and above the control of obvious recordable diseases, presumably by reducing low levels of other diseases and also affecting green leaf area.

Fungicide costs and yield benefits

Fungicide experiments

During the past 5 years ADAS plant pathologists have co-operated in experiments

Table 19.5. ADAS fungicide programme experiments 1978–82.

Timing of treatment (GS)	No. of experiments	Mean yield (t/ha)		Percentage with profitable response*	
		Untreated	Response	Experiments	Survey†
Winter wheat	70	7.48	—	—	
31			0.28	70	72
39			0.47	65	48
31 + 39			0.63	82	27
31 + 39 + 59			0.89	63	13
Winter barley	48	5.53	—	—	—
Autumn			0.29	67	—
31			0.49	70	—
39			0.46	78	—
31 + 39			0.73	77	—
Spring barley	51	4.99	—	—	—
30 – 31			0.30	68	78

* Based on grain values of £105/t (wheat) and £100/t (barley). Fungicides costed at manufacturers' recommended prices with no addition for application costs. MBC/triazole (Spring barley) or MBC (Winter cereals) applied at GS 31. MBC/triazole applied at GS 39 and 59.

† From disease survey data — percentage of untreated crops with sufficient disease to repay treatment costs of £10/ha (GS 31), £20/ha (GS 39), £30/ha (GS 31 + 39) and £50/ha (GS 31 + 39 + 59) based on average yields of 5.84 t/ha for wheat and 4.93 t/ha for barley

using standard fungicide programmes designed to assess the frequency of economic yield response and the importance of treatment at different growth stages on wheat and barley. Table 19.5 shows the mean yield response for some of the fungicide programmes used in these experiments with the frequency of economic yield response for each crop. The increase in yield following application of fungicides ranged from 4% to 13% and, at the grain values quoted in Table 19.5 the greatest of these increases would support the use of the described two-spray or three-spray programmes on about 80% of winter cereal crops. On spring barley and on both winter cereals, a spray of an appropriate fungicide applied at the first-node growth stage (GS 31) would prove profitable on about 70% of crops.

Value of fungicidal sprays

The national value of fungicidal treatments may be estimated by multiplying the value per ha of the yield response following the application of fungicides in experiments (Table 19.5) by the national area of crops which are treated with similar fungicidal spray programmes. Such estimates indicate that the value of the increased yield resulting from the application of fungicidal sprays in 1982 was about £118 million to be set against fungicide costs of £60 million; a net benefit of £58 million. It is of interest to note that the percentage of surveyed crops sprayed with fungicides in 1982 was similar to the percentage of experiments in which profitable yield responses following fungicidal sprays were recorded (Table 19.5). This suggests that the area of commercial crops treated with fungicides is near to optimum and that any change in fungicide usage should not include an increase in the percentage of crops treated if the values of grain and chemicals are maintained at current levels.

Energy budget

The ratio of output to input in financial terms is about 2:1 for most spray treatments in the ADAS experiments and for the estimated values of the yield benefit and fungicide costs recorded in the surveys. These ratios underestimate the worth of fungicides expressed in energy terms.

Energy inputs for cereal production have been estimated by Leach (1975). These figures have been updated for the balance sheet shown in Table 19.6 using nitrogen inputs of 165 kg/ha for winter wheat and 145 kg/ha for winter barley and a mean energy requirement of 0.25 GJ/kg of active ingredient for fungicides (Lewis & Tatchell, 1979). The table shows that, in energy terms, the ratios of output to input relating to fungicides are 13:1 for wheat and 15:1 for barley; much more advantageous than the ratio of approximately 2:1 expressed in purely financial terms. The application of fungicides to cereals therefore is a productive use of natural as well as financial resources.

Discussion

This chapter indicates that the application of fungicides to cereals is associated with a high probability of a net financial benefit. Fungicide treatment is a cost effective method of increasing cereal yields in terms of grain and energy production. The estimates of losses caused by disease, derived from the ADAS disease surveys, suggest that diseases annually reduce production by £210 million (including £60 million for fungicide costs) at 1982 prices. By contrast, data from fungicide experiments suggest a substantially lower national benefit from fungicide treatment of only £118 million, even though the percentage of crops sprayed is similar to the percentage which the experiments indicate might repay the cost of spraying. The disparity between these figures suggests that although the yield response to sprays is cost-effective in a high percentage of crops, in many of these crops it can only be marginally so. There is thus considerable scope for identifying those crops which are most at risk from disease, and which therefore require most fungicide treatment. Work currently in hand in ADAS and commerce is aimed at improving the precision with which potentially responsive crops are identified.

Table 19.6. Estimated energy budget for winter cereals.

	Energy value (GJ/ha)		
	Wheat	Barley	Notes
TOTAL BUDGET			
Initial input (less sprays)	18.49	15.26	Leach, 1975
Additional nitrogen	0.92	1.30	0.026 GJ/kg of N (Leach, 1975)
Herbicides (2.5 kg ai)	0.63	0.63	0.25 GJ/kg ai (Lewis & Tatchell, 1979)
Fungicides:			
Wheat*	0.46	—	
Barley⁺	—	0.41	
Application costs (fuel, depreciation)	0.12	0.12	Leach, 1975
Total input	20.62	17.72	
Total Output	70.08	55.11	Metabolizable energy 11.18 GJ/t barley 12.00 GJ/t wheat
Output/Input ratio	3.40	3.11	
FUNGICIDE BUDGET			
Input	0.58	0.53	Fungicide and spray costs.
Output	7.56	8.16	Energy value of mean yield response in ADAS trials (0.63 t/ha wheat, 0.73 t/ha barley).
Output/Input ratio	13.03	15.40	

* MBC at GS 31 and triadimefon + captafol at GS 39
⁺ MBC + triadimefon at GS 31 and propiconazole at GS 39

References

Anon. (1976) *Manual of Plant Growth Stages and Disease Assessment Keys.* Ministry of Agriculture, Fisheries and Food (Publications), Alnwick, Northumberland.

Anon. (1982) Recommended varieties of cereals, 1982. Farmers Leaflet No. **8** National Institute of Agricultural Botany, Cambridge.

Buchenauer H. & Grossmann F. (1977) Triadimefon: mode of action in plants and fungi. In: *Internal Therapy of Plants* (Ed. by A. Fuchs), Netherlands Journal of Plant Pathology **83** supplement 1, 93–103.

Clarkson J.D.S. (1981) Relationship between eyespot severity and yield loss in winter wheat. *Plant Pathology* **30**, 125–131.

Clarkson J.D.S. & Cook R.J. (1983) Effect of sharp eyespot (*Rhizoctonia cerealis*) on yield loss in winter wheat and of some agronomic factors on disease incidence. *Plant Pathology* **32**, 421–8.

Cook R.J., Jenkins J.E.E. & King J.E. (1981) The deployment of fungicides in cereals. In: *Strategies for the control of cereal diseases* (Ed. by J.F. Jenkyn & R.T. Plumb), pp. 91–9. Blackwell Scientific Publications, Oxford.

Dickinson C.H. (1981) Leaf surface micro-organisms as pathogen antagonists and as minor pathogens. In: *Strategies for the control of cereal diseases* (Ed. by J.F. Jenkyn & R.T. Plumb), pp. 109–21. Blackwell Scientific Publications, Oxford.

James W.C., Jenkins J.E.E. & Jemmett J.L. (1968) The relationship between leaf blotch caused by *Rhynchosporium secalis* and losses in grain yield of spring barley. *Annals of Applied Biology* **62**, 273–88.

King J.E. (1973) Cereal foliar disease surveys. *Proceedings of the 7th British Insecticide and Fungicide Conference, Brighton 1973*, **3**, 771–80.

King J.E. (1976) Relationship between yield loss and severity of yellow rust recorded on a large number of single stems of winter wheat. *Plant Pathology* **25**, 172–77.

King J.E. (1977a) Surveys of foliar diseases of spring barley in England and Wales, 1972–75. *Plant Pathology* **26**, 21–9.

King J.E. (1977b) Surveys of diseases of winter wheat in England and Wales 1972–75. *Plant Pathology* **26**, 8–20.

King J.E. (1977c) The incidence and economic significance of diseases in cereals in England and Wales. *Proceedings of the 1977 British Crop Protection Conference — Pests and Diseases*, **3**, 677–87.

King J.E. (1980) Cereal survey methodology in England Wales. In: *Crop loss assessment. Miscellaneous publication* **7** pp. 124–33. University of Minnesota Agricultural Experimental Station.

King J.E. & Polley R.W. (1976) Observations on the epidemiology and effects on grain yield of brown rust on spring barley. *Plant Pathology* **25**, 63–73.

King J.E. Jenkins J.E.E. & Morgan W.A. (1983). The estimation of yield losses in wheat from severity of infection by *Septoria* species. *Plant Pathology* **32**, 239–49.

Large E.C. & Doling D.A. (1962) The measurement of cereal mildew and its effect on yield. *Plant Pathology* **11**, 47–57.

Large E.C. & Doling D.A. (1963) Effect of mildew on yield of winter wheat. *Plant Pathology* **12**, 128–30.

Leach G. (1975) *Energy and food production.* International Institute for Environment and Development. London and Washington.

Lewis D.A. & Tatchell K.A. (1979) Energy in UK Agriculture. *Journal of the Science of Food and Agriculture* **30**, 449–57.

Mukhopadhyay A.N. & Bandopadhyay R. (1977) Cytokinin-like activity of carbendazim *Pesticides* **11**, 24–5 & 8.

Priestley R.H. (1981) Fungicide treatment increases yield of cereal cultivars by reducing disease and delaying senescence. *EPPO Bulletin* **11**, 357–63.

Scott P.R. & Hollins T.W. (1974) Effects of eyespot on the yield of winter wheat. *Annals of applied Biology* **78**, 269–79.

Tripathi R.K., Tandon K., Schlösser E. & Hess W.M. (1982) Effect of fungicides on the physiology of plants. Part IV: Protection of cellular organelles of senescent wheat leaves by carbendazim. *Pesticides Science* **13**, 395–400.

Zadoks J.C., Chang T.T. & Konzak C.F. (1974) A decimal code for the growth stages of cereals. *Weed Research* **14**, 415–21.

20 Seedling diseases of cotton: damage, loss and control

J.G. WHITE
National Vegetable Research Station, Wellesbourne,
Warwick CV35 9EF UK

Introduction

Seedling diseases are of fundamental importance in crop production since they affect the initial establishment of crops. In a chapter on seedling diseases in his book, Wheeler (1969) noted that emergence may be poor, resulting in patchy seedling stands, and that emerged seedlings may subsequently die or fail to make satisfactory growth. Many fungi cause seedling diseases but they are distributed in relatively few genera e.g. *Pythium*, *Phytophthora*, *Rhizoctonia*, *Fusarium*, *Helminthosporium* and *Colletotrichum*. Additionally, there are individual species such as *Botrytis cinerea* which cause seedling diseases when environmental conditions are favourable for infection (Wheeler, 1969).

A number of non-pathological factors also influence seedling establishment (Hegarty, 1976) either indirectly by interaction with seedling diseases, or directly as in soil drying or capping during emergence. In attempting to establish satisfactory populations of crop plants, the farmer may deliberately sow too many seeds and then reduce the seedling stand to the required density or drill a number of seeds based on a 'field factor' calculation (Bleasdale, 1963) where he estimates the number of viable seeds expected to produce seedlings. However, seedling disease pathogens present on seed or in soil may introduce instability in that many seedlings fail to emerge. Soil sterilization will decrease the instability caused by soil-borne seedling diseases but for economic reasons it is rarely used commercially. A further approach to reducing such instability is the application of localized control measures such as seed or furrow applications of fungicides. Success in this approach relies upon the correct identification of the seedling diseases likely to be present and on the selection and application of the appropriate control measures.

By 1970 organomercury chemicals were used routinely in control of seedling diseases but because of environmental considerations they were replaced by non-mercurial protectant compounds such as thiram, quintozene and later by systemic fungicides such as benomyl, carboxin, chloroneb and thiabendazole. The possibility of replacing conventional fungicides by bio-control agents or novel chemicals was also under investigation. The present paper considers the progress made since 1970 in the control of seedling diseases with particular reference to the cotton crop in the

Wood R.K.S. & Jellis G.J. (1984) *Plant Diseases: infection, damage loss.* Blackwell Scientific Publications, Oxford.

USA. Most of the work relates to *Gossypium hirsutum* although the principles are relevant to the other commercially cultivated species of cotton, *G. arboreum*, *G. barbadense* and *G. herbaceum*. Because of the large number of organisms implicated in seedling diseases of cotton, reference to work on resistance breeding has been omitted.

Crop factors

Cotton is a row crop established entirely from dry seed. As with other row crops, the farmer in seeking optimum yields requires rows of plants to be correctly spaced for cultivation and harvesting operations with few long lengths of row devoid of plants. In traditional cropping systems this was achieved by sowing six or more seeds per station and then thinning by hand to the required plant population (Pothecary, 1970). This wasted seed and labour, and the disturbance caused by thinning could damage remaining plants. The act of deliberate oversowing could also aggravate the effects of the seedling diseases (Ebbels, 1980). In modern systems where the farmer must reduce to a minimum the number of operations per crop, the aim is to drill a number of seeds as close as possible to the number of plants finally required. In practice, American cotton farmers drill 24.5 kg seed/ha with the expectancy that 50% of seeds will produce seedlings capable of growing into mature plants (Ranney, 1971). Where the expected plant stand is not achieved, there is some insurance in that an inadequate stand of cotton may be offset by compensatory growth of the surviving plants. Thus seedling stands from 5.5/m row and 12.0/m row may give the same yield of lint (Minton *et al.*, 1982).

Seedling diseases of cotton

Seed may become contaminated with a range of fungi (Table 20.1), as well as with *Xanthomonas malvacearum* (bacterial blight), with the degree of contamination linked with the length of time the seed is exposed in the field (Roncadori *et al.*, 1972). Moist climates favour contamination of the maturing seed crop (Ebbels, 1980) while the effects on emergence of the seed produced vary with the degree of contamination and the environmental conditions at drilling. Roncadori *et al.* (1972) found that the temperatures most suitable for germination and emergence of cotton were also optimum for the growth of fungal contaminants on the seed. *X. malvacearum* which is ubiquitous and incompletely controlled by chemicals may, in wet conditions, rapidly increase and cause death of seedlings (Innes, 1982). The process of delinting seed with sulphuric acid destroys contaminants on the exterior of the seedcoat (Sinclair, 1965) but many pathogens may survive internally (Maier & Staffeldt, 1963).

Soil-borne fungi (Table 20.1) are potentially important in seedling diseases of cotton with the incidence and severity of infection determined by the species present, the location of the cotton growing area, previous cropping history and the

Table 20.1. Seed- and soil-borne fungi which may act in seedling diseases of cotton.

Fungi isolated from cotton seed	Soil-borne fungi
Alternaria alternata	*Fusarium* spp.
A. macrospora	*Phytophthora* spp.
A. tenuissima	*Pythium* spp.
Aspergillus niger	*Sclerotium rolfsii*
A. flavus	*Rhizoctonia solani*
Diplodia gossypina	*Trichocladium* (= *Thielaviopsis*) *basicola*
Fusarium oxysporum	
F. roseum	
Glomerella gossypii	
Nigrospora sphaerica	
Rhizopus arrhizus	

sanitation measures adopted. In contrast to many temperate crops, seedling diseases of cotton are more severe at relatively low (*c.* 19°C) than at high (*c.* 27°C) temperatures (Ebbels, 1980) although they are favoured by abundant soil moisture (Sinclair, 1965). The species most commonly associated with seedling disease problems is *Rhizoctonia solani*, the causal agent of 'soreshin'. The literature contains numerous papers such as those of Paulus *et al.* (1970, 1973, 1976, 1977) describing the benefits to be derived from using fungicides that control this pathogen. These results must be interpreted with caution however, since the presence of artificial inoculum of *R. solani* may override the effects of other fungi such as *Pythium ultimum*, *P. aphanidermatum* and *Trichocladium* (= *Thielaviopsis*) *basicola* which may be locally significant and may cause seedling disease when air and soil temperatures are low (DeVay *et al.*, 1982).

In the USA, yield losses of cotton due to diseases have been monitored annually. Of an average 15.0% loss caused by all diseases, *X. malvacearum* accounted for 1.0% and other seedling diseases for an average 3.6% in the years 1965–71 (Ranney, 1973). Reporting of these surveys continued until 1975 and in this later period the scale of losses remained essentially the same. This suggests that over the years substantial losses of the cotton crop were avoided, possibly by the adoption of effective control measures. However, in spite of oversowing, thousands of hectares of cotton in which target seedling stands are not achieved are replanted each year (Pinckard & Melville, 1975). Thus the full significance of seedling diseases in cotton cropping is rather more than yield loss statistics would suggest (Table 20.2).

Chemical control of seedling diseases of cotton

Chemicals may be applied to seed for the control of seed and soil-borne pathogens.

Table 20.2. The choice the farmer must make when cotton seedling diseases are a problem.

Poor stand due to seedling diseases: Replanting	Poor stand due to seedling diseases: Not replanting
Additional seed	Plants not uniform
Additional wages	Skippy stand
Additional use of equipment	Loss of fertilizer
Loss of weed control chemical	Additional costs of weed control
Loss of soil moisture	Weak plants more susceptible to other diseases and insects
Late planting	Oil and flaming more difficult
Late harvest	Effect on mechanical harvesting
Additional insect control	
Increased boll rots	

From Sinclair (1965)

Where soil-borne fungi are expected to cause poor seedling stands, incorporation of fungicides in the furrow is recommended. However, the area of cotton planted in which only the seed is treated far exceeds that in which soil treatment is used (Sinclair, 1965).

Chemicals for the control of *X. malvacearum* are few compared with those for the control of fungal pathogens. Organomercury compounds such as methylmercury dicyandiamide and 2-methoxy-ethylmercury chloride have been replaced by non-mercurial chemicals such as bronopol, Cufram Z (mixed metal ethylene bisdithiocarbamate) and Busan II (60% 2-(thiocyanomethylthio) benzothiazole). None of these chemicals completely controls *X. malvacearum* but they are considered ecologically more acceptable than organomercury compounds. Chemical control of *X. malvacearum* has recently been reviewed by Innes (1982).

For many years, the standard fungicide for incorporation into the furrow was quintozene. Although a number of techniques were used, the basic aim was to place the fungicide in the soil close to the emerging hypocotyl, the site of greatest activity of *R. solani* (Sinclair, 1965). While quintozene gave good control of *R. solani* its presence enhanced infection by *Pythium* spp. (Gibson *et al.*, 1961) necessitating the treatment of seed with fungicides effective against these fungi.

Organomercury compounds were introduced as seed treatments in the 1950s and for a time nearly 90% of seed sown in the USA was treated with this group of chemicals (Ranney, 1971). Although non-mercurial compounds such as thiram and captan had been available for many years, and were known to equal organomercury compounds in their effectiveness, their use remained insignificant until the latter were withdrawn in 1971. This was partly because of the ease of application of organomercury compounds in liquid sprays, as compared with the more difficult slurry applications of the non-mercurials (Ranney, 1971).

The introduction of systemic fungicides widened the possibilities for control of seedling diseases. When used alone, these fungicides offered no real improvement but in combination with protectant fungicides they markedly improved seedling stands. Thus, excellent seedling stands could be obtained in the presence of *R. solani* and *Pythium* spp. with combined seed treatments of thiram + chloroneb, fenaminosulf + chloroneb, or captan + chloroneb, although chloroneb alone was ineffective (Paulus *et al.*, 1973). Paulus *et al.* (1976) also demonstrated that acceptable seedling stands (>50%) could be obtained with seven combination treatments with different fungicides.

As fungicides with activity against specific fungi became available, the scope for combination seed treatments was broadened. Thus, captan + carboxin gave disease control in soil contaminated with *Fusarium* spp. (Minton & Green, 1980) while carboxin + benomyl, or carboxin + carbendazim was effective against *T. basicola* (Papavizas & Lewis, 1977). The logical extension of this approach was to load seed with as many as four fungicides active against the range of seedling diseases likely to be encountered (Garber & Carter, 1979).

Loading more than one fungicide on to cotton seed required a knowledge of the compatibility of the compounds selected and of the need to apply the chemicals in the correct sequence as Ranney & Heartley (1972) found that some sequences of application resulted in phytotoxicity.

With the advent of pneumatic seed drills, it became unacceptable to have large amounts of loose fungicide in seedlots, and this led to the widespread use of the organic solvent infusion technique (OSIT) whereby fungicides were applied to seed in solvents such as acetone (Papavizas & Lewis, 1977) or methylene chloride (Halloin *et al.*, 1978). The latter authors however, found that the OSIT did not improve the performance of fungicides compared with other methods of application, and also could reduce seedling stands as the solvents killed embryos of seeds damaged during processing.

In recent years, highly specific compounds suitable for seed and furrow treatment have been introduced. BAS 389 (*N*-cyclohexyl-*N*-methoxy-2,5-dimethyl-3-furanocarboxamide) applied alone was effective against *R. solani* (Papavizas *et al.*, 1979) and when used in combination with the Phycomycete specific fungicides propamocarb and metalaxyl gave high quality seedlings in glasshouse and field tests (Papavizas *et al.*, 1981; Paulus *et al.*, 1977).

In the literature quoted so far the authors have in general investigated the effects of fungicides on seedling stand in isolation from other chemical treatments. However, pesticides and herbicides also are used extensively in the cultivation of cotton and they can modify the effects of seedling diseases. Nematicides (Kappelmann, 1977, 1980) and insecticides (Pinckard & Melville, 1975) may alleviate diseases, while herbicides such as trifluralin may aggravate them (Neubauer & Avizohar-Hershenson, 1973; Grinstein *et al.*, 1976; Miller *et al.*, 1979).

Biological control of seedling diseases of cotton

In spite of much effort in the USA to develop biological control agents, none have been successfully applied to cotton drilled in the field. In the glasshouse, suppression of *R. solani* by mycophagous collembolans (Curl, 1979) and of *P. ultimum* by *Pseudomonas fluorescens* (Howell & Stipanovic, 1980) has been achieved. In the latter case control was mediated by the chemical pyoluteorin which was ineffective against *R. solani* and was inactivated when added direct to soil. More promising was the production of UV-induced biotypes of *Trichoderma harzianum* active against *R. solani* on cotton and *P. ultimum* on peas (Papavizas *et al.*, 1982). The biotypes, which could survive in soil, also had tolerance to benomyl, and although work was at an early stage, the possibilities for integrated disease control were apparent.

Novel chemicals for the control of seedling diseases of cotton

In recent years, there has been a move towards the introduction of agents which are not manufactured specifically for disease control. Industrial mycelial residues (Miller *et al.*, 1973), fatty acids (Hilderbrand & Weinhold, 1976), L-sorbose (Howell, 1978) and natural products from cotton seedlings (Hunter, 1978) have been screened for activity against *R. solani* but so far these have been less effective than conventional fungicides.

Conclusions

The model of the USA cotton crop demonstrates the inherent problems associated with the provision of effective control measures for seedling diseases in a crop which can only be established from dry seed. The involvement of both seed-borne and soil-borne pathogens is a common problem and emphasises the importance of carefully selecting control measures. Given that seedling diseases are not a problem in every field at every sowing, forecasting of the presence of the pathogens is important, hence the recommendation for expensive furrow treatments in fields with a history of seedling loss. Because the grower cannot forecast accurately for every drilling, insurance in the form of effective seed treatments is required.

Loss of the organomercury fungicides was made good by substitution with equally effective fungicides whose performance was enhanced by their use in combination with systemic fungicides active against *R. solani* but largely ineffective against Phycomycetes. These combinations were superseded by the introduction of BAS 389 (*N*-cyclohexyl-*N*-methoxy-2,5-dimethyl-3-furanocarboxamide), which was highly active against *R. solani* and the Phycomyete specific fungicides propamocarb and metalaxyl. Used singly, these chemicals gave good seedling stands; in combination, they produced seedlings with less disease than had hitherto been possible.

Attempts to control seedling disease complexes by biological agents or novel chemicals have not yet been successful; agents active against individual diseases in the glasshouse must retain this activity in the field in the presence of a range of micro-organisms and so far this has has not been achieved.

A large research input has provided cotton growers with a wealth of information on effective control measures for seedling diseases. Such information may well prove of value in the control of similar problems on other crops.

References

Bleasdale J.K.A. (1963) The bed-system of carrot growing. *Short Term Leaflet No.* **27**, *Ministry of Agriculture Fisheries & Food.*

Curl E.A. (1979) Suppression of *Rhizoctonia solani* and damping-off in cotton by mycophagous insects of the order Collembola. *Phytopathology* **69**, 526.

DeVay J.E., Garber R.H. & Matheron D. (1982) Role of *Pythium* species in the seedling disease complex of cotton in California. *Plant Disease* **66**, 151–4.

Ebbels D.L. (1980) Cotton diseases. *Outlook on Agriculture* **10**, 176–83.

Garber R.H. & Carter L.M. (1979) Influence of cultural practices on cotton seedling emergence rate and disease. *Phytopathology* **69**, 915.

Gibson I.A.S., Ledger M. & Boehm E. (1961) An anomalous effect of pentachloronitrobenzene on the incidence of damping-off caused by a *Pythium* sp. *Phytopathology* **51**, 531–3.

Grinstein A., Katan J. & Eshel Y. (1976) Effect of dinitroaniline herbicides on plant resistance to soilborne pathogens. *Phytopathology* **66**, 517–22.

Halloin J.M., Minton E.B. & Petersen H.D. (1978) Fungicide application to cottonseed using methylene chloride carrier. *Crop Science* **18**, 909–10.

Hegarty T.W. (1976) Field establishment of some vegetable crops: response to a range of soil conditions. *Journal of Horticultural Science* **51**, 133–46.

Hilderbrand D.C. & Weinhold A.R. (1976) Effect of nutrients upon the sensitivity of *Rhizoctonia solani* to fatty acids and use of undecylenic acid to control *Rhizoctonia* damping-off. *Plant Disease Reporter* **60**, 7–11.

Howell C.R. (1978) Seed treatment with L-sorbose to control damping-off of cotton seedlings by *Rhizoctonia solani*. *Phytopathology* **68**, 1096–8.

Howell C.R. & Stipanovic R.D. (1980) Suppression of *Pythium ultimum* — induced damping-off of cotton seedlings by *Pseudomonas fluorescens* and its antibiotic, pyoluteorin. *Phytopathology* **70**, 712–5.

Hunter R.E. (1978) Effects of catechin in culture and in cotton seedlings on the growth and polygalacturonase activity of *Rhizoctonia solani*. *Phytopathology* **68**, 1032–6.

Innes N.L. (1982) Bacterial blight of cotton. *Biological Reviews*, (In press).

Kappelman A.J. Jnr. (1977) Effect of fungicides and insecticides applied at planting on cotton emergence, seedling survival and vigour. *Plant Disease Reporter* **61**, 703–6.

Kappelman A.J. Jnr. (1980) Effect of fungicides and insecticides, and their combinations on stand establishment and yield of cotton. *Plant Disease* **64**, 1076–8.

Maier C.R. & Staffeldt E.E. (1963) Cotton seedling disease complex in New Mexico. *New Mexico Agricultural Experiment Station Bulletin*. No. **474**, 41 pp.

Miller P.M., Sands D.C. & Rich S. (1973) Effect of industrial mycelial residues, woodfibre wastes and chitin on plant-parasitic nematodes and some soil-borne diseases. *Plant Disease Reporter* **57**, 438–42.

Miller J.H., Carter C.H., Garber R.H. & Devay J.E. (1979) Weed and disease responses to herbicides in single and double row cotton (*Gossypium hirsutum*). *Weed Science* **27**, 444–9.

Minton E.B. & Green J.A. (1980) Germination and stand with cottonseed treatment fungicides: formulations and rates. *Crop Science* **20**, 5–7.

Minton E.B., Papavizas G.C. & Lewis J.A. (1982) Effect of fungicide seed treatments and seed quality on seedling diseases and yield of cotton. *Plant Disease* **66**, 832–5.

Neubauer R. & Avizohar-Hershenson Z. (1973) Effect of the herbicide trifluralin on *Rhizoctonia* disease of cotton. *Phytopathology* **63**, 651–2.

Papavizas G.C. & Lewis J.A. (1977) Effect of cottonseed treatment with systemic fungicides on seedling disease. *Plant Disease Reporter* **61**, 538–42.

Papavizas G.C., Lewis J.A. & O'Neill N.R. (1979) BAS 389, A new fungicide for control of *Rhizoctonia solani* in cotton. *Plant Disease Reporter* **63**, 569–73.

Papavizas G.C., Lewis J.A., Minton E.B. & O'Neill N.R. (1981) New systemic fungicides for the control of cotton seedling disease. *Phytopathology* **70**, 113–8.

Papavizas G.C., Lewis J.A. & Abd-El Moity T.H. (1982) Evaluation of new biotypes of *Trichoderma harzianum* for tolerance to benomyl and enhanced biocontrol capabilities. *Phytopathology* **72**, 126–32.

Paulus A.O., Shibuya F., Osgood J., Dewolfe T., Cudney D. & House J. (1970) Controlling *Rhizoctonia* seedling disease of cotton in Southern California. *California Agriculture* **24**, 12–4.

Paulus A.O., Nelson J., Dewolfe T., House J. & Shibuya F. (1973) Nonmercury fungicides for control of seedling disease of cotton. *California Agriculture* **27**, 9–10.

Paulus A.O., Nelson J.A., Shibuya F. & Garber R.H. (1976) Treating cottonseed cuts losses to damping-off. *California Agriculture* **30**, 15.

Paulus A.O., Nelson J., Harvey O. & Shibuya F. (1977) Damping-off in cotton controlled with combination seed treatment fungicides. *California Agriculture* **31**, 7–8.

Pinckard J.A. & Melville D.R. (1975) Some new developments in cottonseed dressings. *Plant Disease Reporter* **59**, 262–6.

Pothecary B.P. (1970) The development of mechanisation for cotton growing in the Sudan Gezira. In: *Cotton Growth in the Gezira Environment* (Ed. by M.A. Siddig and L.C. Hughes), pp. 67–77. Sudan Agricultural Research Corporation.

Ranney C.D. (1971) Effective substitutes for alkyl mercury seed treatments for cotton seed. *Plant Disease Reporter* **55**, 285–8.

Ranney C.D. (1973) Dynamics of cotton disease losses. *Plant Disease Reporter* **57**, 325–8.

Ranney C.D. & Heartley W.H. (1972) Multiple cottonseed treatments: effects of sequence of application of pesticides on germination, seedling growth and survival. *Crop Science* **12**, 847–50.

Roncadori R.W., Brooks O.L. & Perry C. (1972) Effect of field exposure on fungal invasion and deterioration of cotton seed. *Phytopathology* **62**, 1137–9.

Sinclair J.B. (1965) Cotton seedling diseases and their control. *Louisiana Agricultural Experiment Station Bulletin* No. **590**, 35 pp.

Wheeler B.E.J. (1969) *An Introduction to Plant Diseases*. Wiley, London.

21 Damage and loss caused by potato diseases

G.J. JELLIS AND R.E. BOULTON
Plant Breeding Institute, Maris Lane, Trumpington,
Cambridge CB2 2LQ, UK

Introduction

Potatoes are prone to attack by diseases caused by fungi, viruses and viroids, bacteria and mycoplasmas. These pathogens may infect the foliage, roots or tubers causing non-emergence, weakened plants, premature death and rotten or poor quality tubers. Because potatoes are vegetatively propagated, many diseases are carried over from one crop to the next and it is necessary to ensure that crops are initially healthy by replacing seed stocks frequently. This significantly increases the cost of growing the crop. Seed potato growers have to ensure the health of their stocks by careful management; a small amount of disease can lead to considerable losses in revenue as a result of downgrading.

Losses caused by disease are not therefore simply a matter of a decrease in total yield of tubers at harvest. Pathogens can affect the growing plant so that tubers are oversized or undersized, or are unsuitable for processing because of their dry-matter content. They can also cause rots several months after harvest or disfigure the tubers making them unsuitable for quality markets.

Annual losses in production caused by disease, pests and weeds were calculated by Cramer (1967). At that time it was estimated that 32.3% of the crop grown worldwide was lost by disease, 9.7% by pests and 6% by weeds. Comparative figures for the UK were in turn 15%, 3% and 4%.

In this chapter the significance of diseases in crops destined for the seed, ware and processing markets is considered, with particular reference to the UK.

Diseases causing yield reduction at harvest

A number of pathogens affect yield to some extent although often these effects are small under normal growing conditions in the UK. Much research has been done to relate damage caused by blight and by various virus diseases to loss of yield and this section will be restricted to these diseases.

Blight

Blight (*Phytophthora infestans*) is probably the single most important disease of

Wood R.K.S. & Jellis G.J. (1984) *Plant Diseases: infection, damage and loss.* Blackwell Scientific Publications, Oxford.

potatoes worldwide and losses occur nearly everywhere the crop is grown. The disease spreads rapidly in suitable weather. In Britain, complete destruction of the foliage rarely takes less than a month from the time infection is first observed; in some years the disease is arrested by the onset of a dry period but spreads to the rest of the foliage if suitable conditions recur. Unlike diseases affecting plant establishment, blight directly affects tuber yield, as an infected plant is not usually surrounded by healthy neighbours which can compensate for it. Methods for calculating losses in crop yield from disease incidence have been published by a number of workers including Large (1952), Olofsson (1971) and James *et al.* (1972).

Large, working in Britain, developed a technique based on reading blight progress curves in conjunction with bulking curves for the tubers. The bulking curve was based on the results of 60 crops of the cultivar Majestic at intervals of 2 weeks over six growing seasons. Increase in tuber yield was taken to cease when lesions affected 75% of the haulm. By recording the date when this stage is reached it is possible to calculate the potential loss in yield. During 1944–57 a number of fungicide trials were carried out to check the accuracy of this technique (Cox & Large, 1960). It was concluded that at the time when the maincrop area was dominated by the cultivars Majestic and King Edward, the method could be used to give a reasonable estimate of average blight losses in a particular region. It was not claimed to be accurate for assessing losses in a particular crop because rate and timing of tuber production may vary considerably depending on a number of factors.

Cox & Large (1960) calculated that in England and Wales during 1950–60 losses due to premature defoliation by blight accounted for about 8% of the potential production from an unsprayed crop. Since that time new cultivars have been introduced and are now widely grown but most are no more resistant to foliage blight than Majestic. The major factor which has influenced national yield loss resulting from blight is fungicidal control. In 1960, only about 30% of crops were sprayed (Cox & Large, 1960) and it was estimated that less than half of the yield loss could be prevented by spraying with the chemicals commonly in use (copper oxychloride and cuprous oxide preparations). By 1977, 87% of the crop area in Great Britain was sprayed against blight, on average three times, using dithiocarbamates (Taylor & Rowe, 1979). More recently, these protectant fungicides have been partially replaced by systemic fungicides belonging to the acylalamine group, marketed as acylalamine/dithiocarbamate mixtures. At present good control of blight is being achieved using these mixtures, although strains of the pathogen insensitive to the acylalanine component have been reported (Dowley & O'Sullivan, 1981).

In Sweden, Olofsson (1971) related yield loss caused by blight to the number of days between planting and first observed infection. James *et al.* (1972) reported that neither the Large nor the Olofsson methods were applicable in eastern Canada. They used regression analysis techniques to develop a multiple point method in which weekly changes in the progress of epidemics were used to determine yield loss (see Table 17.2, this volume). This method gave accurate estimates for the cultivars

Green Mountain and Katahdin but the particular equation used may not be suitable for some other cultivars with different rates of bulking, or in other countries where the growing season is longer or shorter than in Canada. A development of the James equation was the production of crop loss prediction tables for given apparent infection rates (Latin *et al.*, 1978).

Recently Rotem *et al.* (1983), working in Israel, have demonstrated that the influence of seasonal and cultural factors on the growth of foliage and the yield of tubers can alter the relationship between infection and crop loss. Whereas previous work related yields to defoliation of the host plant, they used healthy haulm area, because different amounts of healthy foliage may persist with the same percentage defoliation. Furthermore they considered that expression of yield losses as a percentage derived by comparison with the healthiest plot can be misleading as this does not give an indication of the actual weight loss; in their experiments the same percentage loss of yield corresponded to actual yields which differed by up to a factor of two when the same cultivar was grown in different environments.

None of the methods described for estimating crop losses takes account of the incidence of tuber blight. Assessment of tuber blight is not possible in the growing crop because susceptibility to foliage and tuber blight are not related and environmental factors such as soil type and amount of soil covering the tubers affect the incidence of tuber blight. Cox & Large (1960) considered that losses caused by tuber blight, which they considered to be 3% per annum in England and Wales between 1950 and 1960, probably affected the national economy more than losses in yield resulting from foliage blight. This is because in the main ware-growing areas blight is usually only serious in wet years, when yields are high, and the effect nationally is to reduce a surplus production.

Viruses

Serious losses caused by viruses and viroids can sometimes occur in the year of primary infection, especially if infection occurs early in the growing season. The concentration of potato virus Y (PVY) shortly after primary infection can be as high as that following secondary infection (Gugerli, 1978), and the Y^o strain in particular may cause severe systemic necrosis, leaf drop and death of the plant. The concentration of the phloem-confined potato leaf-roll virus (PLRV) is often relatively low in the foliage of plants with primary infection and the virus may not become fully systemic (Tamada & Harrison, 1980). However early infection can reduce the efficiency of the plant. Potato spindle tuber viroid (PSTV) can also have serious effects on yield and quality in the year in which primary infection occurs (Pfannenstiel & Slack, 1980). Most potato viruses are of major economic importance, however, because of losses caused by secondary infection which occurs when infected seed tubers are planted.

Potato viruses X, A, S and M are generally termed 'mild mosaic' inducing viruses, as visual symptoms are often mild or absent in infected plants. Some cultivar/virus strain combinations, however, result in severe mosaic and dwarfing, and some induce hypersensitivity which can be lethal. Bald (1943) reported yield losses of 11–45%, and Bonde & Merriam (1951) losses of up to 18%, in field-grown potatoes with secondary PVX infection. The cultivar King Edward showed a marketable yield increase of 10% when freed from PVM (paracrinkle) infection (Bawden & Kassanis, 1965). The mild mosaic viruses often show interactive effects when they occur in combination. For example, field plots of the cultivar Netted Gem with dual PVS and mild PVX infection showed a decrease in marketable yield of 25%, but no significant decreases with single infection by each virus (Wright, 1977). Dual infection with PVA and PVX causes a severe rugose mosaic, which was common on the cultivar Désirée in the UK in 1977. The incidence of mild mosaics is relatively low in the UK, largely because of virus testing, seed crop inspection, and resistant cultivars. They remain potentially damaging however, especially in seed crops and breeders' multiplication stocks.

PVY and PLRV are the most damaging potato viruses in the UK. Their incidence tends to fluctuate over a period of years (Figure 21.1) reflecting a number of interacting factors (Hill, 1978), the most significant being the incidence of viruliferous aphid vectors early in the growing season. Depending on cultivar and growing conditions, secondary infection with strains of PVY can produce a range of symptoms including mild or severe mosaic, stipple streak, rugosity, mild or severe necrosis and dwarfing; individual plants may incur total loss of yield. Dual infection with PVY and PVX can cause a severe rugose mosaic. PLRV causes leaf rolling, yellowing, necrosis and dwarfing, with a reduction in photosynthetic efficiency and in the translocation of assimilates. Number of stems per plant, number of tubers and mean tuber weight are also reduced (Killick, 1979). Dual infection with PVY and PLRV can be lethal.

Harper *et al.* (1975) reported yield losses of up to 60% in field plots planted with a high proportion of PLRV infected tubers. Kolbe (1981) reported that, after successively replanting the progeny of plots initially grown from virus-free seed, the mean incidence of PVY and PLRV infected plants was 39, 72, 88 and 88% in the first, second, third and fourth years respectively. The mean yield decreased by 29, 62, 76 and 83% relative to virus-free seed in each year and there was a progressive increase in the proportion of small tubers.

Many of the viruses and virus-like organisms reported from potato growing areas in other parts of the world, for example Latin America (Fribourg, 1980), are not indigenous to western Europe. Imports of breeding material must be constantly monitored for these pathogens (Howell, 1981) since many are considered potentially damaging. Pfannenstiel & Slack (1980) reported that, over 3 years of replanting, many cultivars primarily inoculated with a severe strain of PSTV showed a progressive decrease in tuber numbers and an increase in growth cracking; by the third year many showed a decrease in tuber weight per plant of over 80%.

Virus infection can also influence host plant interactions with other pathogens. For example, Jones & Mullen (1974) found that PVX-infected tubers were less

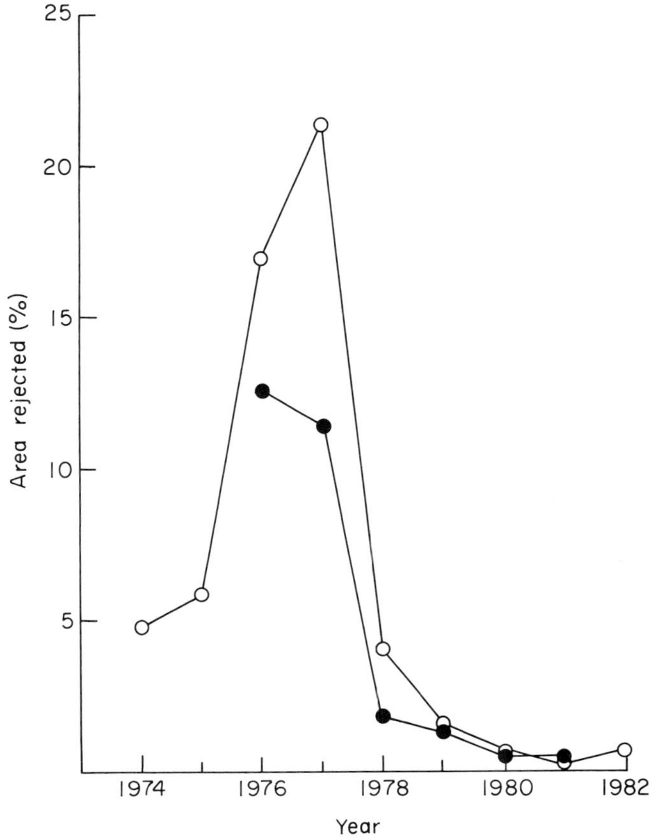

Figure 21.1. Percentage (all entered grades) of total area which was rejected due to virus diseases in seed potato inspection schemes in England and Wales (1974–82) and Scotland (1976–81). o, England and Wales; •, Scotland.
Data from the Ministry of Agriculture, Fisheries and Food, London and the Department of Agriculture and Fisheries for Scotland, Edinburgh.

susceptible to *Fusareum roseum*; by contrast PVX infection can significantly increase colonization of plants by *Verticillium dahliae* particularly in fields continuously cropped with potatoes (Goodell *et al.*, 1982). Mycelium of *Phytophthora infestans* spread more quickly in PVS-infected than in virus-free tubers (Schöber & Weidemann, 1982). Richardson & Doling (1957) found that the incidence of blight in tubers from plants infected with PLRV was twice that in tubers from virus-free plants and suggested that rolled leaves favour blight by retaining moisture. By contrast Pietkiewicz (1971) reported that infection with PVX or PVY enhanced the field resistance of potato leaves to blight.

Crop losses as a result of diseases caused by non-viral pathogens are often directly related to disease severity (expressed as a percentage of plant tissue affected), disease incidence and the growth stage of the crop. However, viruses often infect individual plants systemically and different viruses or strains can have widely different physiological effects. Accordingly, Bos (1981) suggested the use of the term 'disease intensity' rather than 'severity' to describe the effects of virus disease on individual plants. Yield losses attributed to virus diseases are often reported in terms of loss in individual plants or in uniformly infected plots. Crop loss assessment is usually more complex, however, since the incidence of infected plants can be very variable and for many crops, including potatoes, compensatory growth of non-infected plants must be taken into account.

Assessment of loss caused by virus diseases in the potato crop has been extensively studied and critically reviewed by Reestman (1970), and more recent work includes that of van der Zaag (1977) and Killick (1979). Reestman (1970) defined conditions allowing optimal compensation for poor growth as high fertility and sufficient water supply, and either normal planting density with good sprout development or high planting densities. Plant spacing both within and between rows is important. Sensitive cultivars with secondary infection by PLRV or severe strains of PVY have poor growth, often from emergence, and the foliage of healthy neighbouring plants will grow to occupy the space available. Often no yield loss will be incurred with virus incidence of less than about 10%; yields may actually be increased in some cases. However, further aphid-borne primary virus infection, other pests and diseases, poor growing conditions or low planting densities allow less compensation and thus increase loss.

The mild mosaic viruses, which can reduce yield but do not reduce the foliar growth of infected plants to any great extent, allow very little compensation by neighbouring healthy plants. Loss will increase in a linear fashion with the incidence of infection; for example, infection with PVX was shown to reduce crop yield by 0.3% for each 1% incidence of diseased plants (Bonde, 1953).

Unclassified, home-saved seed accounts for about one third of the seed planted annually for the UK ware crop (van Loon & Crosnier, 1982) and provided that satisfactory yields can be attained this offers considerable cost benefit to the grower. A higher incidence of virus infection, potentially capable of causing high losses in the subsequent ware crop, is much more likely in home-saved seed since it is usually grown in areas of high aphid incidence. Growers in England and Wales can reduce the risk of loss by taking advantage of the tuber testing scheme offered by the Agricultural Development and Advisory Service.

The British classification scheme, as specified in the Seed Potato Regulations 1978, Statutory Instruments 1978, No. 215, specifies low tolerances for the incidence of PVY and PLRV at inspection. A higher incidence of virus may be found in ware crops grown from classified seed, however, due to undetected primary infection in the seed crop, particularly within susceptible cultivars (Hill, 1978).

The costs of administering classification schemes for large-scale seed potato production must be regarded in part as a loss due to virus diseases. The costs are outweighed by the advantages of healthy seed for ware production, and other factors are also considered besides viruses, e.g. blackleg (*Erwinia carotovora* subsp. *atroseptica*), trueness to type and isolation of the seed crop. Even so, financial loss to seed growers occurs when crops are downgraded or rejected for exceeding the strict virus tolerance levels specified. For example, in Scotland in 1976 and 1977, an average of 3514 ha of basic seed potatoes was downgraded from the 'FS' to the 'AAI' category because of excessive virus infection, mainly PLRV. This represented 19% of the mean total area inspected in those years. Losses to seed growers in Great Britain in years of high virus incidence are illustrated in Fig. 21.1. It can be seen that virus incidence has declined overall since the peak of 1976–77. This is because of years less favourable to vector activity and strict implementation of the classification scheme. However, given the required combination of factors, incidence could again rapidly increase.

Diseases causing post-harvest losses

In this section, diseases which disfigure tubers or cause rotting during storage are considered. The effects of some diseases may be apparent at harvest, e.g. common scab (*Streptomyces scabies*), powdery scab (*Spongospora subterranea*), black scurf (*Rhizoctonia solani*) and spraing (tobacco rattle virus and potato mop top virus). Other diseases only show their main effects after a period of storage, e.g. gangrene (*Phoma exigua*), dry rot (*Fusarium* spp.), soft rot (*Erwinia carotovora*), skin spot (*Polyscytalum pustulans*) and silver scurf (*Helminthosporium solani*). The storage rots cause direct losses due to tuber decay; other diseases cause blemishes which can result in either a failure of the seed or ware crop to reach the prescribed grading standards or a reduction in the selling price.

The Seed Potato Regulations 1978 make tuber standards an important part of the classification scheme in Great Britain. Tolerances for specified diseases and pests, damage and defects are laid down in Schedule 5 of the regulations. Prior to this, seed potato inspections were inaugurated in England and Wales as a training exercise during 1974–77 and the incidence of tuber diseases has been reported by Ebbels (1983). In three of the four years, gangrene was the disease most frequently responsible for crops failing to reach the required standards. Stubbs (1981) reported the results of inspections of seed tubers carried out in England and Wales in 1978 and 1979. The failure rate was similar in both years, 7.6% in 1978 and 7.4% in 1979, with tuber rots (Group II of Schedule 5) accounting for 42% and 54% of the rejections in these two years.

Latent diseases in particular are of great concern to the seed producer and merchant as grading standards based on visual symptoms can give only partial control. The fungi causing gangrene and dry rot are wound pathogens, capable of remaining dormant until tubers are wounded at lifting, grading or during transit. Thus although contamination of the tubers may occur at the site where the seed was grown, conditions of handling and storage can be critical in determining whether or not substantial losses will occur. The problems associated with dissatisfaction of the quality of Scottish seed in the 1960s were discussed by Hay (1969). He reported that in surveys carried out during the seasons 1966/67 and 1967/68, over half of the reasons given for complaints to seed merchants concerned damage caused by latent diseases. More recently, fungicides effective in controlling a number of these latent diseases have been introduced. In Scotland, fumigation with 2-aminobutane is favoured. If applied within 14 days of harvest this gives good control of gangrene and skin spot and some control of silver scurf but is not effective against dry rot (Graham *et al.*, 1973). In Northern Ireland thiabendazole applied as an ultra low volume spray at harvest is widely used (Logan *et al.*, 1975). This controls dry rot and silver scurf as well as gangrene and skin spot.

Some tuber diseases (e.g. common scab) appear to have little effect on yield (Adams & Hide, 1981). Others, particularly gangrene, dry rot, black scruf, skin spot and blackleg may lead to seed decay, sprout or shoot death and consequently patchy crops. However the potato plant has a considerable ability to compensate (Hirst *et al.*, 1973, and section on virus diseases) so unless seed losses are great or the incidence of disease is high, total yield may not suffer greatly. What frequently appears to be more important is the effect of the disease on the proportion of marketable tubers. In California, Weinhold *et al.* (1982) found that stem canker (*Rhizoctonia solani*) had little effect on total yield but a considerable effect on the proportion of tubers which was misshapen or which exhibited growth cracks. In England stem canker, gangrene and skin spot all have an effect on tuber size; stem canker and gangrene tend to increase, and skin spot to decrease the proportion of small tubers (Hide *et al.*, 1973a, 1973b; Griffith *et al.*, 1974).

Grading standards are also enforced for potatoes sold for human consumption by registered producers and licensed ware merchants (The Grading Standards Prescription, 1980 and The New Potatoes Prescription, 1980). These specify that tubers must be free from:

(a) disease or rot affecting the flesh of the potato,

(b) disease, other than common scab, affecting the skin of the potato which renders the potato unsuitable for human consumption and

(c) common scab affecting more than 25% of the surface area of the potato.

Tubers which are shrivelled or wizened, damaged to a depth greater than 3 mm, greened, bruised, frosted, misshapen or affected by growth cracks, second growth or hollow heart, or tainted or affected by growth shoots must also be excluded, and there are specifications for tuber size. In practice there is a tolerance of faults of 5% tubers by weight. A summary of the main faults found by inspections carried out at wholesalers' premises in 1980 and 1981 is shown in Table 21.1. The major fault in both years was excessive mechanical damage and this is reflected in the incidence of disease.

Severity and type of damage are important factors in determining the incidence of a number of storage diseases. Gangrene, for example, occurs most frequently after crush wounding (Griffith, 1970), particularly if the wounds are deep and penetrate the vascular ring (Jellis, 1975). In a survey of six cultivars grown at four sites and harvested commercially, Jellis & Haslam (1980) found an average of 26 crush wounds on each 100 tubers sampled and 18% of the wounds penetrated the vascular ring.

Table 21.1. Faults reported by Potato Marketing Board inspectors following inspections of wholesalers' premises in 1980/81 and 1981/82.

Inspections are categorized according to the *main* fault found.

a) Overall summary of faults.

Year	No. of inspections	Total substandard	Fault (% total substandard)				
			Common scab	Other* disease	Mechanical damage	Pest	Other
1980	41 814	1629 (3.9%)	0.9	17.2	43.2	7.0	19.9
1981	39 017	2070 (5.3%)	4.3	13.4	36.0	4.1	17.6

* see Table 1(b)

b) The incidence of diseases other than common scab (% total disease).

Year	Blight	Gangrene	Dry rot	Bacterial rots	Spraing	Not stated
1980	23.2	13.6	11.8	48.2	1.4	1.8
1981	20.9	23.0	9.0	43.5	3.2	0.4

Data from the Potato Marketing Board, London.

Some blemishing diseases, such as silver scurf and black dot (*Colletotrichum coccodes*), do not render the potato unsuitable for human consumption but detract from the appearance of washed potatoes. Severe infection can completely obscure the pigmentation of coloured cultivars such as King Edward, reducing their market value (Jellis & Taylor, 1977). The diseases are also important because they can be responsible for excessive fresh weight losses during storage (Mooi, 1968; Hunger and McIntyre, 1979).

Conclusions

With improved management of the major diseases in the UK, by breeding and classification (certification) schemes to control wart disease (*Synchytrium endobioticum*), classification schemes and insecticides to control virus diseases, and fungicides to control blight, losses caused by the lesser known tuber-borne fungal and bacterial diseases have become relatively more important. In the last two decades considerable effort has gone into the study of the prevalence, epidemiology, effects and control of these diseases.

Hide (1981) examined seed potato tubers for evidence of fungal diseases over a period of 14 years. He found that although few tubers appeared unfit to plant because of decay or failure to sprout, most tubers planted were affected by one or more diseases. Pérombelon (1972) reported that most Scottish seed tubers were contaminated with *Erwinia carotovora*. In an attempt to reduce this high incidence of tuber-borne pathogens a number of strategies have been developed. These include the virus-tested stem cutting (VTSC) scheme, the introduction of compulsory grading standards, the use of fungicides (no effective bactericide is yet available), and the development of screening tests for use by plant breeders in the production of resistant cultivars. There have been recent developments in methods for detecting viruses and also producing seed which should lead to improved seed health. Enzyme-linked immunosorbent assay (ELISA) was first introduced for plant viruses in the mid-1970s (Clark & Adams, 1977) and its potential in potato seed testing was quickly appreciated. In particular, unlike other serological techniques, it can be used for detecting PLRV (Casper, 1977). More recently, nucleic acid hybridization techniques have been used to detect PSTV (Owens & Diener, 1981), which lacks the antigenic protein coat characteristic of viruses, as well as to detect a number of conventional potato viruses (Jellis, 1983).

Although the VTSC scheme produces foundation seed free from tuber-borne diseases, subsequent seed crops can become rapidly reinfected (Hide, 1978). It would therefore be desirable to be able to propagate material more rapidly to reduce the risk of reinfection of seed before release to the ware grower. Tissue culture methods provide this opportunity. They have the potential of producing many thousands of plants within a single year, eliminating the need for at least the first three years of vegetative propagation (Hussey & Stacey, 1981). Another new approach to seed production is the use of true seed. This is of particular interest in developing countries where the high cost of seed tubers is a major limiting factor in potato production. It has been demonstrated in China and more recently at the International Potato Center, Peru, that potato production from true seed is feasible and this has led to a considerable amount of research throughout the world (International Potato Center, 1979).

In conclusion, it is difficult to quantify total losses in the UK caused by potato diseases because their effects are so diverse. Frequently they influence quality rather than yield so that loss has to be related to current market prices and premiums paid for high quality. More information on losses will become available with the introduction of the grading scheme for seed tubers and this will enable the effects of new control measures to be monitored.

References

Adams M.J. & Hide G.A. (1981) Effects of common scab (*Streptomyces scabies*) on potatoes. *Annals of Applied Biology* **98**, 211–6.

Bald J.G. (1943) Potato virus X: mixtures of strains and the leaf area and yield of infected potatoes. *Australian Council for Scientific and Industrial Research, Bulletin* **165**, 32 pp.

Bawden F.C. & Kassanis B. (1965) The potato variety King Edward VII and paracrinkle virus. *Report of Rothamsted Experimental Station for 1964*, 282–90.

Bonde R. (1953) Potato X-virus causes large losses; better seed is the answer. *Maine Farm Research* **2**, 10–2.

Bonde R. & Merriam D. (1951) Control of latent mosaic. *67th Annual Report. Maine Agriculture Experiment Station Bulletin* **491**, 47–8.

Bos L. (1981) Assessment of crop losses caused by viruses. In: *Crop loss assessment methods: FAO manual on the evaluation and prevention of losses by pests, diseases and weeds, Supplement 3* (Ed. by L. Chiarappa), pp. 79–84. Commonwealth Agricultural Bureaux, Farnham Royal.

Casper R. (1977) Detection of potato leafroll virus in potato and in *Physalis floridana* by enzyme-linked immunosorbent assay (ELISA). *Phytopathologische Zeitschrift* **90**, 364–8.

Clark M.F. & Adams A.N. (1977) Characteristics of the method of enzyme-linked immunosorbent assay for the detection of plant viruses. *Journal of General Virology* **34**, 732–42.

Cox A.E. & Large E.C. (1960) Potato blight epidemics throughout the world. *United States Department of Agriculture, Handbook* **174**, 230 pp.

Cramer H.H. (1967) *Plant protection and world crop production. Pflanzenschutz — Nachrichten* 'Bayer'. Bayer, Leverkusen.

Dowley L.J. & O'Sullivan E. (1981) Metalaxyl-resistant strains of *Phytophthora infestans* (Mont.) de Bary in Ireland. *Potato Research* **24**, 417–21.

Ebbels D.L. (1983) Incidence of tuber diseases in classified seed potatoes harvested in England and Wales, 1974–77. *Plant Pathology* **32**, 145–50.

Fribourg C.E. (1980) Historia y distribución de los virus de papa en América Latina. *Fitopatologia* **15**, 13–24.

Goodell J.J., Powelson M.L. & Allen T.C. (1982) Interrelations between potato virus X, *Verticillium dahliae*, and *Colletotrichum atramentarium* in potato. *Phytopathology* **72**, 631–4.

Graham D.C., Hamilton G.A., Nash M.J. & Lennard J.H. (1973) Fumigation of bulk-stored potatoes with 2-aminobutane for control of gangrene, skin spot and silver scurf diseases. *Potato Research* **16**, 234–43.

Griffith R.L. (1970) Pathogen, wound type, temperature and gangrene infection. *Report of Rothamsted Experimental Station for 1969*, Part 1, pp. 165–6.

Griffith R.L., Hide G.A., Hirst J.M. & Stedman O.J. (1974) Effects of gangrene (*Phoma exigua*) on potatoes. *Annals of Applied Biology* **77**, 237–50.

Gugerli P. (1978) The detection of two potato viruses by enzyme-linked immunosorbent assay (ELISA). *Phytopathologische Zeitschrift* **92**, 51–6.

Harper F.R., Nelson G.A. & Pittman U.J. (1975) Relationship between leaf roll symptoms and yield in Netted Gem potato. *Phytopathology* **65**, 1242–4.

Hay F.C. (1969) *Report on the marketing of Scotch seed potatoes.* The National Farmers' Union of Scotland, 139 pp.

Hide G.A. (1978) Incidence of pathogenic fungi on Scottish potato seed stocks derived from stem cuttings. *Potato Research* **21**, 277–89.

Hide G.A. (1981) Fungus diseases on potato seed tubers planted in England and Wales, 1963–76. *Annals of Applied Biology* **98**, 377–93.

Hide G.A., Hirst J.M. & Stedman O.J. (1973a) Effects of black scurf (*Rhizoctonia solani*) on potatoes. *Annals of Applied Biology* **74**, 139–48.

Hide G.A., Hirst J.M. & Stedman O.J. (1973b) Effects of skin spot (*Oospora pustulans*) on potatoes. *Annals of Applied Biology* **73**, 143–50.

Hill S.A. (1978) Current virus infection of seed potatoes in Britain. In: *Plant Disease Epidemiology* (Ed. by P.R. Scott & A. Bainbridge), pp. 229–34. Blackwell Scientific Publications, Oxford.

Hirst J.M., Hide G.A., Stedman O.J. & Griffith, R.L. (1973) Yield compensation in gappy potato crops and methods to measure effects of fungi pathogenic on seed potatoes. *Annals of Applied Biology* **73**, 143–50.

Howell P.J. (1981) The risks from exotic potato viruses. *EPPO Bulletin* **11**, 243–9.

Hunger R.M. & McIntyre G.A. (1979) Occurrence, development, and losses associated with silver scurf and black dot on Colorado potatoes. *American Potato Journal* **56**, 289–306.

Hussey G. & Stacey N.J. (1981) *In vitro* propagation of potato (*Solanum tuberosum* L.). *Annals of Botany* **48**, 787–96.

International Potato Centre (1979) Production of potatoes from true seed. *Report of a Planning Conference, Manila, Philippines, September 13–15 1979*, 172 pp.

James W.C., Shih C.S., Hodgson W.A. & Callbeck L.C. (1972) The quantitative relationship between late blight of potato and loss in tuber yield. *Phytopathology* **62**, 92–6.

Jellis G.J. (1975) The susceptibility of potato tuber tissues to infection by *Phoma exigua* var. *foveata*. *Potato Research* **18**, 116–9.

Jellis G.J. (1983) Some current research topics at the Plant Breeding Institute, Cambridge. *The Seed Potato. Journal of the National Association of Seed Potato Merchants* **23**, 16–9.

Jellis G.J. & Haslam M. (1980) The incidence and importance of wounds which penetrate the vascular ring. *Annals of Applied Biology* **96**, 371–4.

Jellis G.J. & Taylor G.S. (1977) The development of silver scurf (*Helminthosporium solani*) disease of potato. *Annals of Applied Biology* **86**, 19–28.

Jones E.D. & Mullen J.M. (1974) The effect of potato virus X on susceptibility of potato tubers to *Fusareum roseum* 'Avenaceum'. *American Potato Journal* **51**, 209–15.

Killick R.J. (1979) The effect of infection with potato leaf roll virus (PLRV) on yield and some of its components in a variety of potato (*Solanum tuberosum*). *Annals of Applied Biology* **91**, 67–74.

Kolbe W. (1981) Ergebnisse eines Kartoffel — Nachbauversuches. *Der Kartoffelbau* **32**, 352–3.

Large E.C. (1952) The interpretation of progress curves for potato blight and other plant diseases. *Plant Pathology* **1**, 109–17.

Latin R.X., MacKenzie D.R. & Cole H. Jr (1978) Crop loss tables for potato blight epidemics. *American Potato Journal* **55**, 541–8.

Logan C., Copeland R.B. & Little G. (1975) Potato gangrene control by ultra low volume sprays of thiabendazole. *Annals of Applied Biology* **80**, 199–204.

Mooi J.C. (1968) De aantasting van de aardappel door zilverschurft (*Helminthosporium solani*). *Mededeling van het Instituut voor Plantenziektenkundig Onderzoek* **482**, 62 pp.

Olofsson B. (1971) In: *Crop loss assessment methods: FAO manual on the evaluation and prevention of losses by pests, diseases and weeds* (Ed. by L. Chiarappa). Commonwealth Agricultural Bureaux, Farnham Royal.

Owens R.A. & Diener T.O. (1981) Sensitive and rapid diagnosis of potato spindle tuber viroid disease by nucleic acid hybridization. *Science* **213**, 670–2.

Pérombelon M.C. (1972) The extent and survival of contamination of potato stocks in Scotland by *Erwinia carotovora* var. *carotovora* and *E. carotovora* var. *atroseptica*. *Annals of Applied Biology* **71**, 111–7.

Pfannenstiel M.A. & Slack S.A. (1980) Response of potato cultivars to infection by the potato spindle tuber viroid. *Phytopathology* **70**, 922–6.

Pietkiewicz J. (1971) Influence of viruses X and Y on the reaction of potatoes to *Phytophthora infestans* de By. *Ziemniak* 1971, 99–108.

Reestman A.J. (1970) Importance of the degree of virus infection for the production of ware potatoes. *Potato Research* **13**, 248–68.

Richardson D.E. & Doling D.A. (1957) Potato blight and leaf-roll virus. *Nature* (London) **180**, 866–7.

Rotem J., Bashi E. & Kranz J. (1983) Studies of crop loss in potato blight caused by *Phytophthora infestans*. *Plant Pathology* **32**, 117–22.

Schöber B. & Weidemann H-L. (1982) Der Einfluss von Infektionen mit Kartoffelvirus S (PVS) auf die Anfalligkeit der Kartoffelknollen gegenüber *Phytophthora infestans* (Mont.) de Bary. *Nachrichtenblatt des Deutschen Pflanzenschutzdienstes* **34**, 8–10.

Stubbs J. (1981) Seed tuber dressing standards. *The Seed Potato. Journal of the National Association of Seed Potato Merchants* **21**, 17–8.

Tamada T. & Harrison B.D. (1980) Factors affecting the detection of potato leafroll virus in potato foliage by enzyme-linked immunosorbent assay. *Annals of Applied Biology* **95**, 209–19.

Taylor J.A.H. & Rowe D.G. (1979) *Maincrop potato production techniques 1977–8*. Potato Marketing Board, London.

van der Zaag D.E. (1977) Betekenis van infectie van poters met virus op de opbrengst bij aardappelen. *Landbouwkundig Tijdschrift* **89**, 154–7.

van Loon C.D. & Crosnier J.C. (1982) 1955–1980, 25 ans de culture de la pomme de terre en Europe. In: *EAPR. 25 Jahre Europäische Gesellschaft für Kartoffel-forschung 1957–1982* (Ed. by B. Schöber, A. Specht, N.E. Foldo & J.H.W. Holden), pp. 35–53. European Association for Potato Research, Wageningen.

Weinhold A.R., Bowman T. & Hall D.H. (1982) Rhizoctonia disease of potato: effect on yield and control by seed tuber treatment. *Plant Disease* **66**, 815–8.

Wright N.C. (1977) The effect of separate infections by potato viruses X and S on Netted Gem potato. *American Potato Journal* **54**, 147–9.

22 Damage and loss caused by disease in hardy ornamentals

E.J. EVANS
Agricultural Development and Advisory Service, Government Offices, Coley Park, Reading RG1 6DT, UK

S.C. MELVILLE
Agricultural Development and Advisory Service, Staplake Mount, Starcross, Exeter EX6 8PE, UK

Introduction

Hardy ornamentals in England and Wales have a farm-gate value of approximately £110 million (Anon., 1982) of which bulb crops account for £31 million. The major crops in order of importance are narcissi (valued at £19 million), tulips, anemones, gladioli and irises. Narcissi and tulips are grown for both bulbs and flowers; the remainder are grown almost exclusively for their blooms. In recent years containerized plants have proved to be the most profitable sector of the nursery-stock industry. Production is valued at £30 million, considerably in excess of trees (£19 million), shrubs (£14 million), roses (£8 million) and herbaceous plants (£2 million).

The term hardy ornamentals encompasses a diversity of plant species, many of which are susceptible to a wide range of diseases. Some of these diseases can cause severe damage and plant loss whereas others gradually affect the growth and appearance of their hosts, often over several seasons. Modern techniques such as containerization further complicate assessment of loss by introducing a degree of mobility into the production cycle. Dwarf conifers, for example, may be raised by a specialist propagator, grown-on by a nurseryman, held and sold by a retail outlet and finally planted for amenity purposes. The value of the plant can be influenced by disease at any stage of its journey from propagator to customer. Assessment of crop loss attributable to a specific disease is especially difficult under these circumstances. Nevertheless, in this chapter an attempt has been made to estimate damage and loss caused by selected diseases of bulb crops and hardy-ornamental nursery stock in the UK.

Bulb crops

The UK is the world's largest producer of narcissi. Some 60% of the total narcissus bulb crop is forced by growers in the UK for flowers, about 25% sold as bulbs for the

Wood R.K.S. & Jellis G.J. (1984) *Plant Diseases: infection, damage and loss.* Blackwell Scientific Publications, Oxford.

home market and 15% exported as dry bulbs and flowers. The establishment of a thriving export trade in recent years has highlighted the importance of maintaining and improving the health of stocks. Stringent phytosanitary requirements imposed by importing countries are a major constraint to further expansion of exports.

Basal rot (*Fusarium oxysporum* f.sp. *narcissi*) is an important problem of narcissus in the UK, particularly in relation to bulb exports, since the main importing countries require that consignments should be almost or entirely free of the disease. It occurs typically as a soft brown rot of the bulb extending upwards from the basal plate. Affected bulbs may fail to produce a plant or give rise to poorly developed shoots, so that losses occur both in the growing crop and in bulbs after harvest. *F. oxysporum* is a wound parasite and penetrates the bulb through dying roots towards the end of the growing season. Infection can also occur through wounds made by emerging roots shortly after planting or by way of tissues damaged at lifting. Losses due to basal rot vary greatly from field to field and from season to season. Bulb infection and the subsequent development of the disease are very dependent on temperature, the fungus being most active at relatively high temperatures (optimum 25°C). At sites and in seasons favourable for basal rot up to 40% of bulbs of a susceptible cultivar such as Golden Harvest may be affected and unmarketable. Severe attacks during the last 12 years have probably been associated with high summer temperatures. Analysis of temperature records at Rosewarne Experimental Horticulture Station (EHS), Camborne, Cornwall, UK, supports this view. Work at the EHS indicates that any factor leading to higher soil temperatures tends to increase basal rot. Thus, more disease occurs on south-facing than on north-facing slopes, more in ridges than in beds and more in bare than in weed-covered ground. The disease is also encouraged by mechanical damage and high storage temperatures (Tompsett, 1979). Control by cultural measures and the use of a fungicide in the hot water treatment can appreciably reduce commercial losses, often to below 5% bulbs affected. The export standard however, requires less than 2% basal rot and current practices can only just achieve this under experimental farm conditions. Data from pre-export inspections show that basal rot is a major cause of rejection. Failure to meet the export standard is penalized by loss of an attractive premium as high-quality export samples can command up to twice the home market price for bulbs.

Other bulb rots such as neck rot and smoulder (*Botrytis narcissicola*) can also reduce home ware sales and limit exports. Neck rot occurs at the base of the old flower stalk and surrounding scales and may extend to the base of the bulb. The cause of this disorder has not been established satisfactorily. Several fungi have been isolated from affected bulbs, including *F. oxysporum*, *B. narcissicola* and *Penicillium* spp. and although the basal rot fungus is often implicated, isolation and inoculation studies have failed to show that it is the sole cause of the problem.

Most commercial bulb stocks are infected to a greater or lesser degree by viruses. In recent years production of virus-free stocks by meristem tip culture (Stone *et al.*, 1976) has provided an opportunity to assess losses due to these diseases Multiplication and evaluation of virus-free bulbs are now in progress; initial results suggest they could substantially improve bulb and flower yields. Commercial stocks of *Narcissus tazetta* cv. Grand Soleil d'Or on the Isles of Scilly are extensively infected with two or more viruses (Brunt, 1971), mainly narcissus degeneration virus and arabis mosaic virus. Results from small-scale comparative trials (Anon., 1982) showed that virus-free stock from the Glasshouse Crops Research Institute, Littlehampton, Sussex, UK, was more productive than a virus-infected commercial stock (937 flowers compared with 646, produced in each case from 300 bulbs). Virus-tested bulbs of many of the major narcissus cultivars have now been obtained and it is anticipated that these will gradually replace existing commercial stocks during the next decade. The introduction of virus-free stocks is likely to improve the competitive position of UK bulb exports.

Prospects for control of most leaf diseases have been greatly improved by the use of modern fungicides, either as bulb dips or as foliar sprays. Severe losses are now relatively infrequent on commercial holdings although diseases such as tulip fire (*Botrytis tulipae*) and narcissus leaf scorch (*Stagonospora curtisii*) can be locally damaging especially in amenity plantings. Tulips are often planted annually in borders in formal bedding schemes and sclerotia of *B. tulipae* from the previous season's planting can cause infection which spoils the aesthetic effects. Narcissus leaf scorch and white mould (*Ramularia vallisumbrosae*) can both cause premature leaf death and reduced bulb weight with a consequent decrease in flower production the following season. These diseases are potentially very damaging, especially in the wetter south-western areas of England. Following trials by Jenkins & Hawken (1969) on white mould control, large scale commercial losses have been avoided by the use of two or three (mainly post-flowering) sprays of a dithiocarbamate fungicide or, more recently, a mixture of a dithiocarbamate and a benzimidazole fungicide. Similarly, the debris-borne leaf spot pathogen (*Mycosphaerella macrospora*) and bulb or debris-borne ink disease pathogen (*Drechslera iridis*) can be major constraints to profitable long-term production of bulbous iris. These diseases tend to increase in severity with the age of the plantation caused by accumulation of infected debris. Whereas crops can often be grown on annually without fungicide sprays it is virtually impossible to produce an economic long-term crop without an efficient spray programme. Results of a recent trial at Rosewarne EHS (Table 22.1) illustrate loss from ink disease through spotting of flowering stems and its prevention by a programme of seven, three-weekly (November-April) fungicide sprays (Tompsett, 1980). Long-term (up to 6 years) iris production is cheaper to operate than annual cropping, demanding less soil preparation and less labour for cleaning, planting and lifting bulbs. The economic benefits from applying a fungicide programme and burning debris annually (the latter also stimulates earlier and improved flowering) are presented in Table 22.2.

Anemone leaf curl (*Colletotrichum* spp.) is one disease which is not as yet amenable to chemical control. Leaf curl was first recorded in the UK in 1968 on crops grown from Dutch corms. The symptoms are a down-curling of the leaves

Table 22.1. The effect of a fungicide spray programme on ink disease (*Drechslera iridis*) and flower quality of long-term bulbous iris (cv. Wedgwood), 1980.

	Ink disease (0–4 scale)	Class I flower stems (%)	Class III unmarketable flower stems (%)
Untreated	3.13	4	62
Mancozeb	0.50***	35***	6***
Chlorothalonil	0.25***	93***	0***

*** Significantly different from the untreated control at P = 0.001
Data from Tompsett (1980)

Table 22.2. Relative crop value of long-term bulbous iris (cv. Wedgwood) in relation to control measures for ink disease (*Drechslera iridis*).

Treatment	Relative crop value (marketable flowering stems)
Annual burning of debris + fungicide spray programme	100
Spray programme only	63
Annual burning only	41
Untreated	30

Data from Tompsett (1980)

followed by twisting and coiling of petioles and flower stalks and, sometimes eventual death of affected plants. A postal survey conducted by the Agricultural Development and Advisory Service (ADAS) in southwest England in 1978 indicated that approximately 60% of corms were affected with average plant losses of 10–15% (G.W. Griffin, personal communication). Similar losses have been recorded in successive years with the worst affected crops proving entirely unmarketable. To date the cultivar St. Piran has remained virtually free from severe infection and the continued success of the anemone industry depends on maintaining its health.

Hardy nursery stock

Despite recent improvements in chemical and cultural control, die-back and wilt diseases caused by soil-borne *Phytopthora* spp. are still the main disease problems. *Phytophthora* spp. affect a wide range of hardy ornamentals both in field soil and in containers. *P. cryptogea*, *P. citricola* and especially *P. cinnamomi* are the species most frequently responsible for damage. In the UK *P. cinnamomi* causes major loss of *Chamaecyparis lawsoniana* and species of *Cupressus*, *Rhododendron*, *Erica* and

Calluna. Infection can lead to a root and foot rot, wilting, leaf discoloration and death of affected plants. Foliar symptoms seldom appear until the plant is substantially affected. Plants only slightly infected show no foliar symptoms and the disease can remain undetected and is therefore maintained on the nursery or spread if the plant is sold. Advisory experience in south-east England suggests plant losses of up to 80% in susceptible species when these are grown on infested land. Losses have declined in recent years through improvement of field drainage, use of fungicides and containerization. Growing plants in containers enables them to be isolated from contaminated field soil. This technique has reduced the risk of infection on nurseries where standards of hygiene and of management are high. Where standards are lower, introduction of the pathogen through contaminated compost or irrigation water has led to widespread damage.

Other soil-borne pathogens are less important but nevertheless can be locally damaging where constraints on land lead to intensive cropping of susceptible species. Downy mildew (*Peronospora sparsa*) can reduce the stand of rose seedlings on long-established rose land. Wilt (*Verticillium dahliae*) is often a problem in *Acer platanoides* grown in fields recently cropped with strawberries or potatoes. Similar problems can occur with herbaceous plants. Bacterial wilt (*Xanthomonas campestris*) is normally a disease of minor economic importance but in areas where wallflowers are grown intensively such as in the seedlings' beds of parks' departments, accumulation of infected trash has led to loss of up to 90% of transplants (Griffin & Baker, 1976).

Fungal pathogens can cause major problems during propagation, especially with plants that take a long time to establish. A wide range of fungi can induce rotting and failure to root under these circumstances. Symptoms may be caused by a complex of fungi, including *Botrytis cinerea*, *Cylindrocarpon destructans*, *Glomerella cingulata*, *Monochaetia karstenii*, *Pestalotiopsis* spp., *Pythium* spp. and *Rhizoctonia solani*. The extent of loss varies greatly from nursery to nursery and is determined by many factors including host susceptibility, initial health of cuttings, rooting medium, propagation regime (temperature and humidity) and fungicide usage. Mist propagation is the most commonly used method of rooting softwood cuttings but the use of a polythene tent with bottom heat has increased in popularity, especially for autumn and winter work. Improved rooting of some of the more difficult subjects has been achieved using the latter technique although constant high humidities tend to encourage disease. It is impossible to estimate total loss during propagation because of the multiplicity of plant species and diseases. Nevertheless losses recorded as caused by one or more diseases of specific hosts can be substantial (Humphreys-Jones, 1977, 1980). Recent fungicide trials in south-east England have demonstrated that a complex of root pathogens (mainly *Cylindrocarpon destructans* and *Pestalotiopsis sydowiana*) can reduce the numbers of hybrid rhododendrons rooted under polythene by as much as 40% (Smith, 1977) and root rot (*Rhizoctonia solani*) causes similar losses in *Calluna*. Damage of this order of magnitude is not uncommon on commercial holdings where it has a marked influence on the profitability of the enterprise, reducing income from sales and increasing labour and equipment costs.

Container production increased rapidly during the late 1970s in response to an increase in the popularity of garden centres and their customers' demands for high-quality stock. Leaf and stem diseases usually disfigure plants and decrease their retail value. Many foliar diseases are now amenable to chemical control so that major loss of income due, for example, to powdery mildew (*Sphaerotheca pannosa*) and black spot (*Diplocarpon rosae*) of roses or scab (*Venturia inaequalis*) of *Malus* spp. is relatively rare. Other pathogens are not so easily controlled. Leaf spot and stem blight *Pestalotiopsis* spp. and *Monochaetia karstenii* have increased in importance so that on some nurseries they regularly reduce quality of stock. These pathogens are well adapted to modern growing regimes, are favoured by high temperatures and humidity and spread rapidly in tightly-packed container beds especially where overhead watering is used. Species of coniferous trees (e.g. *Juniperus*) and *Erica*, *Camellia* and *Rhododendron* spp. are most commonly affected.

Diseases can reduce the quantity and quality of ornamentals. The qualitative aspect is becoming increasingly important as the industry attempts to meet the demands of a more discriminating public. Looking to the future, there are prospects of decreasing damage and loss by introducing improved propagating material, free of major pathogens. Such material is gradually being released for multiplication either by conventional or by micro-propagation techniques. Its success relies on subsequent protection through rational use of cultural and chemical disease-control measures.

References

Anon. (1982) *Commission review, hardy nursery stock and bulbs 1982.* Ministry of Agriculture, Fisheries and Food, London.

Brunt A.A. (1971) Virus diseases of narcissus. In: *The Daffodil and Tulip Year Book 1971*, pp. 18–37. The Royal Horticultural Society, London.

Griffin M.J. & Baker L.A.E. (1976) Bacterial wilt of wallflowers caused by *Xanthomonas campestris. Plant Pathology* **25**, 108–14.

Humphreys-Jones D.R. (1977) Leaf and shoot death (*Coniothyrium fuckelli*) of *Juniperus communis* L. var. *compressa* Carr. *Plant Pathology* **26**, 47–8.

Humphreys-Jones D.R. (1980) *Myrothecium roridum* leaf blotch on *Kilkwitzia amabalis. Plant Pathology* **29**, 55.

Jenkins J.E.E. & Hawken R.H. (1969) The control of narcissus white mould. *Plant Pathology* **18**, 122–9.

Smith P.M. (1977) Rotting during rooting of *Rhododendron* cuttings. *Annual Report of the Glasshouse Crops Research Institute for 1976*, pp. 121–2.

Stone O.M., Hollings M. & Pawley R.R. (1976) Production of virus free plants: *Narcissus. Annual Report of the Glasshouse Crops Research Institute for 1975*, p. 127.

Tompsett A.A. (1979) The control of narcissus basal rot (*Fusarium oxysporum* f.sp. *narcissi*). *Annual Review of the Rosewarne and Isles of Scilly Experimental Horticulture Stations for 1978*, pp. 13–23.

Tompsett A.A. (1980) Recent experimental work on Dutch iris. *Annual Review of the Rosewarne and Isles of Scilly Experimental Horticulture Stations for 1979*, pp. 17–23.

23 Damage and loss caused by *Phytophthora* species in forest crops

GRETNA WESTE

Botany School, University of Melbourne, Parkville, Victoria 3052, Australia.

Introduction

Fungi of the genus *Phytophthora* have caused disease and death of forest trees in such diverse parts of the world as America, Asia, Australia and Europe. The most destructive species, *Phytophthora cinnamomi*, was originally described from *Cinnamomum burmanni* in a tropical forest on the west coast highlands of Sumatra (Rands, 1922). The disease may be tropical in origin, gradually extending under the influence of a milder climate or as a result of increased travel and immigration before the introduction of plant quarantine.

This chapter reviews damage and loss due to *Phytophthora* spp. in world forest crops. The detailed analysis of host damage is based on research into *Phytophthora cinnamomi* in Australian eucalypt forests (Weste, 1983).

Damage and loss in softwood forests

Most of the destruction recorded from America and Europe has been due to *Phytophthora* infections of softwood crops. For example, *P. cinnamomi* was responsible for little-leaf disease of short leaf pine (*Pinus echinata*), and to a lesser extent of loblolly pine (*P. taeda*), which affected forests on 6 million ha of south and south-eastern USA (Campbell & Copeland, 1954). The same pathogen caused root-rot and death of Douglas fir (*Pseudotsuga menziesii*) in the forest of the Pacific north-west. The pathogen was introduced on ornamental plantings and spread by water along valleys and elsewhere by forestry. In Douglas fir the symptoms were short internodes, short needles and eventual death. Other timber species of the same forest were also susceptible, including firs, *Abies* spp. and western hemlock, *Tsuga heterophylla* (Roth & Kuhlamn, 1966). *Phytophthora lateralis* caused epidemic disease in Port Orford cedars (*Chamaecyparis lawsoniana*) in commercial forests in Oregon and Portland, resulting in extensive disease and death. In addition to root rot, the pathogen caused aerial infections and stem cankers which destroyed the crown of the tree, and which arose from sporangia carried by driving rain and wet winds. All infections apparently originated from infested soil.

Damage to softwood crops in Australia and New Zealand has been confined to

Wood R.K.S. & Jellis G.J. (1984) *Plants Diseases: infection, damage and loss.* Blackwell Scientific Publications, Oxford.

seedling crops, usually in forest nurseries and crops in areas liable to water-saturated soils. *P. cinnamomi*, together with species of *Pythium* and sometimes other fungi, cause 'damping off' in forest nurseries where hygiene is lax, plants are too crowded or are maintained in too humid an atmosphere. In Australian nurseries *P. cinnamomi* has become a major problem. *Pinus radiata* was susceptible when grown on well-drained, fertile soil for the first 5 years after planting out but then became tolerant. During the early years mortalities occurred in 11% of the pine seedlings planted on a fertilized and well-drained soil. No deaths occurred later although the pathogen continued to be isolated from samples of soil and fine roots (Weste, 1983).

In wetter areas the situation is different and *P. cinnamomi* has caused severe damage and deaths in *P. radiata*. In northern New Zealand, Newhook (1959) recorded that many thousands of mature trees died in shelterbelts of *P. radiata* in 1953 and 1956. During these epidemics abnormally high summer and autumn rainfall was associated with massive infection and death of feeder rootlets. The cold wet winters which followed retarded rootlet regeneration and shelterbelt trees with large crowns became defoliated and died whereas adjacent plantation pines with sparse crowns survived. Adjacent trial plantations that had been treated with superphosphate also survived. Since all soils had similar pathogen populations Newhook concluded that the difference depended on the relationship between transpiration demand and number of healthy absorbing roots. Whereas isolated shelterbelt trees became waterlogged in wet periods but had relatively large transpiration demands during dry periods, plantation trees removed excess water from wet soils and their sparse crowns made less transpiration demand.

Damage and loss in hardwood forests

Damage from *Phytophthora* species has also affected hardwood forests of the world, for example beech and chestnut in America and Britain (Brasier & Strouts, 1976). In the 19th century the American chestnut (*Castanea dentata*) was virtually eliminated from the southern portion of its range (Zentmyer, 1980). *P. cinnamomi* was also largely responsible for the decline in forests of Ohia (*Metrosideros collina* spp. *polymorpha*) on Hawaii (Kliejunas & Ko, 1974). Ohia is a pioneer species on new lava flows and provides valuable watershed protection, so that the decline and death of large areas of forest is a serious problem.

Dead stands of *Nothofagus* spp. in the New Guinea highlands at 1000–2000 m were also associated with *P. cinnamomi* (Shaw *et al.*, 1972). The rain forests of central Queensland have areas of dieback and death where muddy puddles made by feral pigs provide ideal conditions for zoospore production (Brown, 1976). Parts of Australian coastal forests as far apart as Tasmania and Queensland, Western Australia and Victoria, are devastated by the same pathogen.

Probably the most severe disease ever observed in any mixed species forest is that caused by *P. cinnamomi* in the native forests of south-western and to a lesser extent south-eastern Australia. In West Australia 220 000 ha of jarrah (*Eucalyptus marginata*) forest have become diseased, resulting in the destruction of the prime timber species, the endangering of the Perth water supply and the annihilation of the beautiful wildflowers for which Western Australia has become well known.

Australian forests are dominated by the genus *Eucalyptus* of which there are about 500 species adapted to various environmental conditions found from near the equator to about 43°S. Associated with the eucalypt forests are lower strata of *Acacia* spp. and mixed species of woody shrubs. The eucalypts and acacias grow rapidly and are drought tolerant. Much of this flora is susceptible to *P. cinnamomi*, including the principal hardwood forest trees belonging to the *Eucalyptus* subgenus *Monocalyptus*, such as the West Australian jarrah and the Victorian silvertop (*E. sieberi*). The first sign of *P. cinnamomi* in the forest is often a massive death of the understorey to be followed 1–3 years later by dieback and death of the susceptible trees (Figure 23.1, 23.2).

The symptoms of dieback disease in Western Australia were noticed in small areas as far back as about 1920 but widespread disease was not apparent until after World War II when greatly increased construction and improvement of roads provided more opportunities for infection of forests. Combined with ignorance of the cause of the disease, this led to an ideal situation for new infections over a large

Figure 23.1. Disease caused by *Phytophthora cinnamomi* in a dry sclerophyll forest, Brisbane Ranges, Victoria, Australia. Note the diseased shrubs and the sharp boundary between diseased and unaffected vegetation.

Figure 23.2. Tree death caused by *Phytophthora cinnamomi* in a dry sclerophyll forest, Brisbane Ranges, Victoria, Australia. The trees are stringy bark eucalypts, *Eucalyptus macrorhyncha*, *E. obliqua* and *E. baxteri*. Trees may die 1–3 years after death of shrubs.

area of State forest and the disease spread very rapidly from about 1950 to 1975, particularly during the wet years of 1964 and 1973.

In south-eastern Australia, Victorian coastal forests dominated by *E. sieberi* became infected with *P. cinnamomi* after 1950 and 12 000 ha of poor sites are now severely diseased with large areas of dead trees (Weste & Marks, 1974).

The dry sclerophyll shrubby forest of the Brisbane Ranges in south-western Victoria was a source of useful timber and the impact of *P. cinnamomi* on these poor quality stringy bark forests is horrific. In northern Australia *P. cinnamomi* has killed the dominant species, *E. tetradonta*, in monsoonal native forests adjacent to mining settlements and this is a potential disaster because Northern Territory forests border vast deserts. Eucalypt forests of New South Wales and Queensland are less severely affected, possibly because the summer rainfall alleviates host water stress in coastal forests of northern but not southern Australia.

In parts of coastal north Queensland large rivers are bordered with a rich and varied flora consisting of about 50 species of mangrove. Recent investigations have shown mangrove deaths to be associated with *Phytophthora* spp. Mangrove forests near Gladstone were damaged by *Phytophthora operculata* in association with industrial pollution and soil reclamation (Pegg *et al.*, 1980). The mangroves on Cape York, far north Queensland, were killed by *P. nicotianae* in a region free from habitation, industry or pollution (Weste *et al.*, 1982). *P. nicotianae* has been consistently associated with plant disease and death in the Northern Territory (Blowes & Pitkethly, 1981).

Site factors affecting damage by *Phytophthora*

Site interaction with host and pathogen is of overriding importance in determining the degree of forest destruction due to *Phytophthora* spp. Sites which are a major disease hazard are those with soil temperatures greater than 10°C and a soil matric water potential higher than -9 bar. High rainfall during spring and early summer promotes production and dispersal of zoospores and subsequent infection. Disease risks are increased when dry periods during summer or autumn lead to water stress in forest trees which are unable to replace water lost by transpiration when the root system is infected.

The disease has made tremendous impact on dry sclerophyll shrubby forests which grow in regions with a dry summer and autumn climate, often in shallow soils. During winter and spring such soils are periodically saturated, a condition to which the roots of stringy bark eucalypts are particularly sensitive. During wet periods in spring, production, release and dispersal of zoospores are maximal. Zoospores are attracted toward host roots on which they encyst prior to penetration.

While the pathogen induces early symptoms of dieback by causing decay of the fine roots, it normally kills by spreading into the main roots and sometimes into the collar, thus girdling the plant. In very wet areas zoospores may penetrate the collar through wounds or grafts, for example in spray-irrigated orchard trees. In native forests understorey species die within a few months but death of trees may take 1–3 years and may occur suddenly or by gradual dieback, depending on environmental stress. Widespread death of trees along gullies or in shallow soil on ridges is illustrated in Figure 23.2.

Invasion by the pathogen does not invariably result in destruction of susceptible forest species. The degree of destruction depends on concomitant environmental factors. Forest-invading *Phytophthora* spp. have made little impact where trees are grown on deep, fertile, well-drained soils with reliable rainfall, and when understorey species are field resistant. The worst affected sites are those with shallow, poorly drained soils which become waterlogged in wet seasons and droughted during dry seasons.

In south-eastern Australia major timber production comes from wet sclerophyll forests grown on deep friable soils with high organic content, a large microbial population, a reliable rainfall and a field-resistant understorey. Consequently *P. cinnamomi* has caused little disease in these forests. However the dominant tree and major timber species, *Eucalyptus regnans*, is susceptible when grown in less favourable sites or in compacted soil such as on the margin of forest parks.

The inoculum potential of *Phytophthora* depends to a great extent on soil temperature and soil-water potential. When soil temperatures fall below 10°C

Phytophthora has proved difficult to isolate, although this temperature does not kill the pathogen. In spring, pathogen populations increase rapidly with increasing soil temperatures and disease potential becomes high. The pathogen overwinters in host roots or in large clumps of organic matter. During wet periods in spring and summer when soil temperatures reach 15°C or higher a population explosion may occur with the formation of sporangia and zoospores resulting in the infection of new roots. During periods of water stress, when the matric soil-water potential is less than -9 bar, the pathogen is again difficult to isolate but during the same dry period symptoms appear and a zone of dead vegetation may mark the path travelled by the spring zoospore population. Where the pathogen is spread by water run-off, a clearly defined boundary or disease front may separate infected from healthy vegetation (Figure 23.1). Where there is no definite water flow, dying vegetation may form a mosaic scattered amongst healthy forest (Weste, 1983).

Extension of disease due to *P. cinnamomi* has been measured in Victorian eucalypt forests and in three epidemic years was recorded as 171 m per annum along the contour. Rapid disease extension by zoospore dispersal on wet downhill slopes has been measured in several forests as 400 m per annum (Weste, 1983). There are few records of aerial spread of those *Phytophthora* spp. which infect trees but wind-driven rain and splash droplets may carry the sporangia of some species such as *P. cactorum* and *P. lateralis* and cause collar rot and crown dieback in peach and apple orchards.

In Australia the impact of *P. cinnamomi* on the forest is clearly visible where understorey species are susceptible because these become chlorotic and die rapidly *en masse*. The grass tree, *Xanthorrhoea australis* (Liliales, Victoria) and *Banksia grandis* (Proteaceae, Western Australia) are particularly susceptible and being large make useful disease indicators in the south-east and south-west respectively. These two species acquire massive root infections that are largely responsible for the very high levels of inoculum of the pathogen in disease forests. In West Australia it is proposed to reduce the inoculum level by using fire to change the understorey from *B. grandis* to more resistant legumes (Shea *et al.*, 1978).

As a result of infection with *P. cinnamomi*, parts of dry sclerophyll shrubby eucalypt forest in Victoria were changed to a grassy forest or to a sedge woodland. When the susceptible shrubs died out, bare ground increased from 10% to 90% in severely diseased forests such as those of the Brisbane Ranges. This bare ground was later colonized by resistant sedges or grasses. After an epidemic the population density of the pathogen declined gradually (Weste, 1983) but regeneration of seedlings of susceptible shrubs has rarely been observed on any diseased site. Alterations to the whole ecosystem following invasion by *P. cinnamomi* are being monitored biennially for each type of forest. Changes in species diversity, population density, ground coverage and the whole ecosystem are being measured and recorded. There is no record of major ecological changes in response to a pathogen apart from those being studied in Australian eucalypt-dominated forests.

Disease potential in infected eucalypt forests depends on the following factors:
(a) susceptibility of the dominant tree and understorey species

(b) a soil temperature range between 15 and 28°C with a soil water potential greater than -9 bar
(c) periodic free water within this temperature range for zoospore production and dispersal
(d) high density of host roots
(e) a low soil microbial population density.

Disease potential is increased in forests with marginal conditions for host survival, such as shallow soils which may become waterlogged in winter or spring and may dry out during summer and autumn to water potentials less than -9 bar (actually -60 to -80 bar in the Brisbane Ranges). Disease potential is inversely related to the numbers of soil micro-organisms present. Many microbes are either competitive, antagonistic or predatory to fungi. Suppressive soils are those with high populations of bacteria and actinomycetes which lyse the hyphae and suppress formation of sporangia (Broadbent & Baker, 1974). In Australian forests, soils with an extremely low population of actinomycetes and aerobic bacteria are prone to severe disease, whereas in suppressive soils the micro-organisms keep the pathogen in check and symptoms do not appear. Soils remain suppressive only while their matric water potential is -1 bar or more.

What characteristics make *Phytophthora* species so destructive?

The genus includes about 50 species which range from host-specific biotrophs such as *P. infestans*, *P. erythroseptica* and *P. megasperma* to necrotrophs with an extraordinarily wide host range such as *P. cinnamomi* and *P. cactorum*. There are also aquatic saprophytic species which may not cause disease. There are probably less than a dozen species which cause severe disease in forests. All are disseminated
(a) in flowing water
(b) on infected nursery stock
(c) by movement of infected soil and gravel *per se* and
(d) on off-road vehicles and equipment.

The general life cycle of those *Phytophthora* species which infect forest crops is particularly suited to warm temperate forests in many parts of the world. During spring and summer when soils are warm and wet, maximum numbers of free swimming zoospores are released; in autumn when soils dry out and become colder chlamydospores are formed (Figure 23.3). The former spores are motile and the latter are resistant to drying and to bacterial lysis and both act as infective agents.

Chlamydospores of *P. cinnamomi* have been observed to increase in number in non-sterile, host-free soils producing mycelium and more chlamydospores or sporangia and zoospores depending on water or nutrient levels (Weste, 1983). This demonstrates a competitive saprophytic ability. Oospores are extremely resistant and serve to maintain the population when growth of the pathogen has almost ceased due to adverse environmental factors. They appear to be rare, however, in heterothallic species.

The short life cycle, rapid production of zoospores, efficient dispersal in soil

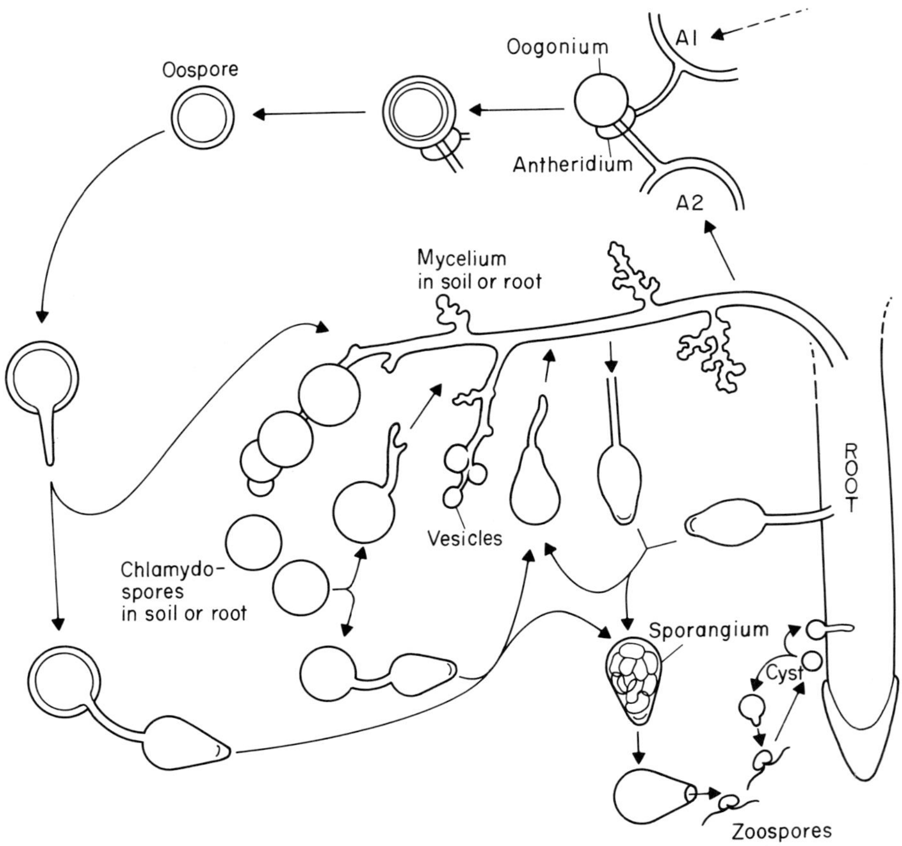

Figure 23.3. The life cycle of *Phytophthora cinnamomi*, illustrating alternative stages in the life cycle by which the fungus adapts to environmental changes. Drawing by H.J. Swart.

water, attraction to plant roots, encystment and germ tube orientation, are all factors favouring rapid and efficient infection and therefore epidemics. Survival and reproduction of *Phytophthora* spp. are enhanced by a number of variants in the life cycle which are utilized as adaptations to changes in environmental factors. Most propagules germinate to produce either a germ tube or a sporangium which in turn may germinate directly or produce zoospores (Figure 23.3); behaviour depends on prevailing nutrient or moisture conditions. Non-specific, root-infecting *Phytophthora* species are therefore extraordinarily well adapted for survival and infection of forest trees. Their greatest hazards in soils are low water content and high populations of antagonistic or competitive soil microorganisms.

How do *Phytophthora* species damage their hosts?

P. cinnamomi penetrates unsuberized root tissue, usually the absorbing and elongating region of young roots. It infects and penetrates many, probably all, plant species (Weste, 1983). Once inside the root cortex the extent of further penetration and damage varies with host susceptibility. Evidence therefore suggests that resistance is expressed after penetration. The whole root system may be invaded in species which are fully susceptible such as *Banksia grandis*, *Isopogon ceratophyllus* (both Proteaceae) and *Xanthorrhoea australis* (Liliales). In such hosts, the fungus destroys the main roots and may cause a collar rot so that death occurs within a few weeks. With less susceptible but sensitive eucalypt hosts such as *E. marginata* (jarrah) and *E. sieberi* (silvertop), a site conducive to disease development is necessary before mature forest trees are killed. In such host species on high risk sites, the same total invasion of the vertical root system occurs as in susceptible species, and leads to mass death of trees.

In all field-resistant species examined, *P. cinnamomi* infected and penetrated the roots, stopped root growth, grew for *c.* 96 h and produced a small necrotic lesion in which the pathogen remained viable and confined, and which therefore provided a source of infection for roots of other species. Field-resistant species develop no secondary symptoms such as dieback or wilt, except under very wet conditions. Field-resistant species include some eucalypts such as *E. calophylla* (marri) and *E. maculata* (spotted gum), the grasses and sedges which invade diseased forests, and crop plants such as wheat and maize.

When wheat seed is germinated on infected soil *P. cinnamomi* can be isolated from surface-sterilized roots when the grain is harvested. Plants therefore range from complete susceptibility to field resistance but no complete resistance has been recorded.

Recent research, testing a range of susceptible and field-resistant species, has shown that susceptible species wilt and die when infected by *P. cinnamomi* because of failure in water transport through the roots. When 8–15% of the root system is infected, root conductance may be reduced to 4% of uninfected controls. Root conductance is unaffected when field-resistant species are infected (Dawson & Weste, unpublished research).

Management of a diseased forest

The management of forests containing dieback disease with the aim of ultimately allowing continuation of normal forest activities such as forest improvement and timber harvesting, is difficult. Activities must be conducted within constraints designed to prevent infection in new areas and minimize disease extension and severity within those forests already infected. The West Australian Forests Department has adopted procedures commencing with a quarantine period, followed by aerial detection of disease, mapping, laboratory testing to confirm infection, and then culminating in strict site management.

(a) Quarantine: One of the main obstacles to forest hygiene has been the inability to detect all infection loci because of the time lapse between infection and the appearance of disease. The West Australian Forests Department imposed a three year quarantine period during which activities, including logging, in the forest were severely restricted, in order to allow disease to become manifest. In areas not quarantined, stringent hygiene procedures were imposed to prevent or minimize dispersal of the pathogen. These included: confinement of forest activities to summer, disease-safe periods (when soils are too dry for pathogen activity) and washing-down vehicles before leaving an infested forest (McKinnel, 1981).

(b) Detection and location of the pathogen: In large forests such as occur in Victoria and Western Australia small foci of disease may only be detected from the air. Dead trees can be detected readily but trees may not die until several years after pathogen invasion so that aerial photography is required to detect symptoms in extremely susceptible indicator species of the understorey. A system of colour photography at a scale of 1:4000 has been developed which enables a single sick Banksia or grass tree (Bradshaw, 1974) to be identified and located on a grid. If tests for *P. cinnamomi* are positive, the plant can be isolated so that disease is not spread. Sophisticated techniques of navigation and calibration allow precise location of diseased shrubs in aerial photographs in relation to the ground locations.

(c) Laboratory tests: In order to establish the cause of the symptoms observed in aerial photographs, the pathogen must be isolated and identified in the laboratory because trees also die from other causes such as fire and drought.

(d) Mapping: Tests are also required to determine the extent of soil infestation before a diseased area can be accurately mapped. Maps are printed in colour code to indicate the locations of infection and disease status of the forest.

(e) Management plans: The management of a forest ecosystem containing infested soils aims to create conditions less favourable for the pathogen and more favourable for the trees.

Advantage is taken of the overriding importance of site in contributing to disease severity. Moist lowland sites contribute to high mortality whereas forests on well-drained upland sites may remain relatively healthy. Management therefore plans to improve drainage and to use controlled burning to replace the highly susceptible species such as *Banksia grandis* with legumes which improve the nitrogen status of forest soils and reduce the inoculum potential because of their lower susceptibility. *Acacia pulchella* is favoured in Western Australia because the roots produce chemicals that inhibit mycelial growth and zoospore production (Whitfield *et al.*, 1981). Other plans include improvement of forest quality, avoidance of soil puddling which encourages zoospore production, the use of pathogen-free water supplies and the insistence that all nursery stock is pathogen-free (McKinnel, 1981). Dieback disease has infested West Australian jarrah forests for more than 60 years (Podger *et al.*, 1965) and it is considered unlikely that many sites of maximum disease potential remain. With careful forest management the dramatic destruction of the entire forest ecosystem on a grand scale will not recur.

The use of fungicides is hardly practicable for large areas of uncultivated native forest but is certainly feasible in forest nurseries. Acylalanine soil drenches produce successful control of *P. cinnamomi*.

There are two other methods of control which may prove useful. The first is to plant resistant species or resistant strains of the desired forest species. For example, occasional trees of *E. marginata* show resistance and clones from such trees may well provide a forest of resistant trees in the future. Such clones have already been established (Grant, unpublished research). The second is based on the observation that soils containing a high microbial population density normally suppress *P. cinnamomi*. Australian avocado groves have been maintained on moist soils kept suppressive by the addition of fowl manure and organic matter (Broadbent & Baker, 1974). Such additions to uncultivated forest soils are not yet practicable but offer scope for possible future control.

One complication to such biological control is the requirement of suppressive organisms for a relatively high soil-water potential. Another is that there may be seasonal differences in maximum population densities. For example, in dry sclerophyll forests in Victoria soil microorganisms attain greatest numbers in late autumn, whereas the pathogen population reaches its maximum in late spring or early summer, thereby avoiding biological control.

References

Blowes W.M. & Pitkethly R.N. (1981) The occurrence of *Phytophthora* spp. in the Northern Territory, *Australasian Plant Pathology* **10**, 8–10.

Bradshaw F.J. (1974) Large scale 70 mm colour photography as an aid to detecting jarrah dieback. In: *International Society for Photogrammetry, Commission VII Proceedings. Symposium on remote sensing and photointerpretation, Banff Alberta, Canada*, pp. 439–48.

Brasier C.M. & Strouts R.G. (1976) New records of *Phytophthora* on trees in Britain. *European Journal of Forest Pathology*, **6**, 129–36.

Broadbent P. & Baker K.F. (1974) Behaviour of *Phytophthora cinnamomi* in soils suppressive and conducive to root rot. *Australian Journal of Agricultural Research* **25**, 121–35.

Brown B.N. (1976) *Phytophthora cinnamomi* associated with patch death in tropical rain forests in Queensland, *Australasian Plant Pathology* **5**, 1–4.

Campbell W.A. & Copeland Jr. O.L. (1954) Little leaf disease of short leaf and loblolly pines. *Circular US Department of Agriculture* No. **940**, 41 pp.

Kliejunas J.T. & Ko W.A. (1974) Deficiency of inorganic nutrients as a contributing factor to Ohia decline. *Phytopathology* **64**, 891–6.

McKinnel F.H. (1981) Review of the dieback disease situation. *Technical Papers, Forests Department of Western Australia*, 8 pp.

Newhook F.J. (1959) The association of *Phytophthora* spp. with mortalities of *Pinus radiata* and other conifers. 1. Symptoms and epidemiology in shelterbelts. *New Zealand Journal of Agricultural Research* **2**, 808–43.

Pegg K.G., Gillespie N.C. & Forsberg L.I. (1980) *Phytophthora* spp. associated with mangrove death in central Queensland. *Australasian Plant Pathology* **9**, 6–7.

Podger F.D., Doepel R.F. & Zentmyer G.A. (1965) Association of *Phytophthora cinnamomi* with a disease of *Eucalyptus marginata* forest in Western Australia. *Plant Disease Reporter* **49**, 943–7.

Roth L.F. & Kuhlman E.G. (1966) *Phytophthora cinnamomi*, an unlikely threat to Douglas fir forestry. *Forest Science* **12**, 147–59.

Rands R.D. (1922) Streepkanker van Kaneel, veroorzaaktdoor *Phytophthora cinnamomi* n.sp. *Mededelingen van het instituut voor plantenziekten*. Batavia. Indonesia.

Shaw D.E., Cartledge E.G. & Stamps D.J. (1972) First records of *Phytophthora cinnamomi* in Papua New Guinea. *Papua New Guinea Agricultural Journal* **23**, 46–8.

Shea S.R., Gillen K.S. & Kill R.J. (1978) Variation in sporangial production of *Phytophthora cinnamomi* Rands on jarrah (*E. marginata* Sm.) forest sites with different understorey compositions. *Australian Forest Research* **8**, 219–26.

Weste G. (1983) Population dynamics and survival of *Phytophthora*. In: *Phytophthora, Its Biology, Taxonomy, Ecology and Pathology* (Ed by D.C. Erwin, S. Bartnicki-Garcia & P.H. Tsao) pp. 237–59. American Phytopathology Society, St. Paul, Minnesota.

Weste G. & Marks G.C. (1974) The distribution of *Phytophthora cinnamomi* in Victoria, *Transactions of the British Mycological Society* **63**, 559–74.

Weste G., Cahill, D. & Stamps D.J. (1982) Mangrove dieback in north Queensland, Australia. *Transactions of the British Mycological Society* **79**, 165–7.

Whitfield F.B., Shea S.R., Gillen K.J. & Shaw K.J. (1981) Volatile components from the roots of *Acacia pulchella* R.Br. and their effect on *Phytophthora cinnamomi* Rands. *Australian Journal of Botany* **29**, 195–208.

Zentmyer G.A. (1980) *Phytophthora cinnamomi* and the disease it causes. *American Phytopathology Society Monograph* No. **10**, 96 pp.

24 Damage and loss caused by dwarf mistletoes in coniferous forests of western North America

FRANK G. HAWKSWORTH

USDA Forest Service
Rocky Mountain Forest and Range Experiment Station
Fort Collins, Colorado 80526, USA

CHARLES G. SHAW, III[1]

USDA Forest Service
Pacific Northwest Forest and Range Experiment Station
Juneau, Alaska 99802, USA

Introduction

Dwarf mistletoes (genus *Arceuthobium*, family Viscaceae) are a unique group of forest tree pathogens that attack conifers in many parts of the northern hemisphere. In much of western North America, they are considered to be the single most damaging disease agent for many commercially important conifers. In North America, all six genera of the Pinaceae are parasitized: *Pinus*, *Picea*, *Abies*, *Tsuga*, *Larix*, and *Pseudotsuga*. Oddly enough, *Juniperus*, which is the main host genus of *Arceuthobium* in the Old World, Europe, Asia and Africa, is not attacked in North America. However, trees in this genus are hosts for several species of the related mistletoe genus *Phoradendron*.

In southern Asia (India, Pakistan, China) *Juniperus* and other conifers (*Pinus*, *Abies*, *Picea* and *Keeteleria*) are damaged by *Arceuthobium*. Although the dwarf mistletoes have a wide geographic distribution, they are considered serious forest parasites primarily in North America (western Canada, western USA, and Mexico).

The genus *Arceuthobium*

The genus *Arceuthobium* is unique among the mistletoes as it is the only genus found in both the Old and New Worlds. The genus contains approximately 40 taxa, of which thirty-two have been discussed previously (Hawksworth & Wiens, 1972). Eight more are now known: one in the Azores, one in East Africa (Hawksworth & Wiens, 1976), four in Mexico and Central America (Hawksworth & Wiens, 1977,

[1] This manuscript was written while Dr. Charles G. Shaw was on study leave with the British Forestry Commission on a trip partially funded by the American Philosophical Society.

Wood R.K.S. & Jellis G.J. (1984) *Plant Diseases: infection, damage and loss.* Blackwell Scientific Publications, Oxford.

1980), and two in south-western China (Kiu & Ren, 1982). Of the 40 known taxa, thirty-two occur in the New World and only eight in the Old World.

Dwarf mistletoe shoots are typically small (3–10 cm tall) and leafless (Figure 24.1). A unique feature of the genus is an explosive mechanism within the fruit which enables them to eject their seeds for distances of up to 30 m (average 5–15 m) (Figure 24.2). The dwarf mistletoe root system extends through the host phloem and xylem,

Figure 24.1. Aerial shoots of dwarf mistletoe (*Arceuthobium vaginatum*) on ponderosa pine.

and provides a means for transferring water and nutrients from the tree.

The dwarf mistletoes have a relatively long life cycle compared to most tree pathogens. It typically takes 5 years or more from the time of infection to the production of mature fruits. Several years after initial infection, infected branches in most host trees become distorted and develop into witches' brooms (Tinnin *et al.*, 1982) (Figure 24.3). The witches' brooms appropriate water and nutrients from the uninfected parts of the tree; the tree declines in vigour and dies prematurely.

The dwarf mistletoes generally have narrow host ranges, usually attacking only a single host species or a group of closely related hosts. The same hosts are usually attacked throughout the geographical range of the parasite, but there are notable exceptions. For example, *Arceuthobium tsugense* severely attacks *Tsuga mertensiana* in California but is rare on this tree in Alaska (Shaw, 1982a).

The dwarf mistletoes are one of the few forest pathogens that can be controlled by silvicultural methods, as summarized by Scharpf & Parmeter (1978). Of the sixteen dwarf mistletoes found in western North America, the seven listed in Table 24.1 are the most serious because they are widespread and parasitize commercially important tree species.

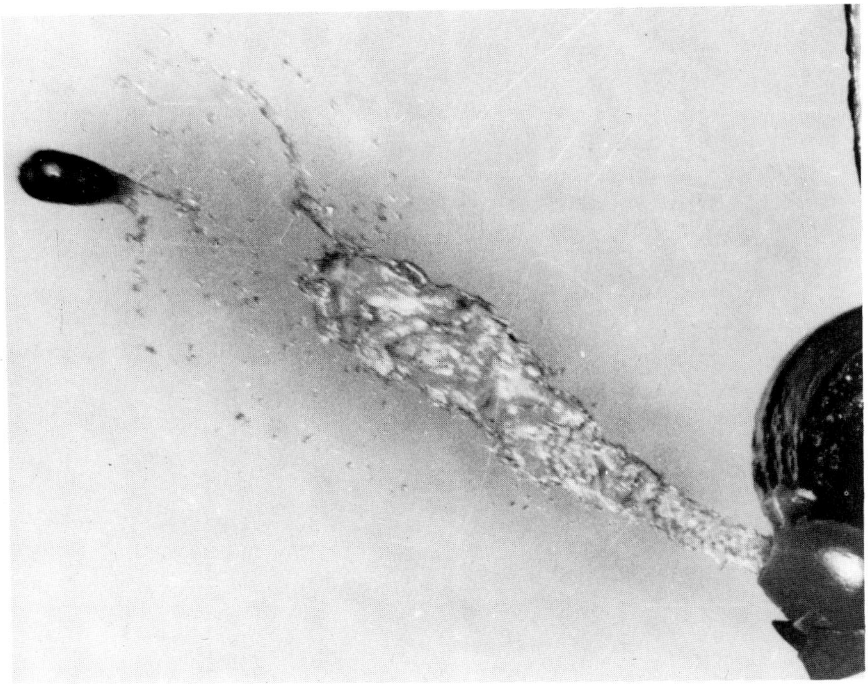

Figure 24.2. A dwarf mistletoe seed just after being forcibly ejected from the fruit. Photograph taken at 5 millionths of a second to capture image of seed flying at 30 m/s (100 km/h). Seeds can be expelled to distances of up to 30 m.

Figure 24.3. Abnormal branches (witches' brooms) in ponderosa pine induced by the dwarf mistletoe, *Arceuthobium vaginatum*. The tree, and the one in the left background, have been killed by the parasite.

Types of damage caused by dwarf mistletoes

Dwarf mistletoes affect forests in many ways (Hawksworth, 1978a), including
(a) reduction of tree height and diameter growth rates,
(b) increased mortality,
(c) decreased seed production,
(d) reduced wood quality,
(e) predisposition of affected trees to attack by insects and fungi causing decay, and
(f) ecological effects.

Growth loss

The most common result of dwarf mistletoe infection is a reduction in the growth rate of affected trees. The amount of reduction depends on the intensity of attack. It

Table 24.1. Dwarf mistletoes which cause most damage to commercial free species.

Dwarf mistletoe species	Principal host species	Distribution
1 *Arceuthobium abietinum*	*Abies concolor*, *A. magnifica*	California, Oregon
2 *A. americanum*	*Pinus contorta*	Western USA British Columbia Alta., Canada
3 *A. campylopodum*	*P. ponderosa*, *P. jeffreyi*	North-western USA California
4. *A. douglasii*	*Pseudotsuga menziesii*	Western USA British Columbia Canada*
5 *A. laricis*	*Larix occidentalis*	North-eastern USA British Columbia Canada
6 *A. tsugense*	*Tsuga heterophylla*	South-eastern Alaska to Northern California
7 *A. vaginatum*	*P. ponderosa*	South-western USA

* But not in the main commercial Douglas-fir forests west of the Cascade Mountains in Oregon, Washington, or British Columbia

Table 24.2. The effect of dwarf mistletoe (*Arceuthobium vaginatum*) on the radial growth rate of *Pinus ponderosa*.

Dwarf mistletoe[†] rating	No. of trees	Mean radial growth for last 5 years (cm)	Growth rate as percentage of that for non-infected and lightly infected trees (classes 0–3)
0	86	0.53	100
1	52	0.53	100
2	44	0.53	100
3	39	0.53	100
4	44	0.48	91
5	63	0.41	77
6	108	0.25	47

Data from Hawksworth (1961).

[†] Dwarf mistletoe intensity is usually assessed by using a 6-class rating system (Hawksworth, 1977). In this system, the live crown is visually divided into thirds and each third is then rated as: 0 (no infection); 1 (light infection—less than half the branches infected); or 2 (heavy infection—more than half of the branches are infected). The ratings for each third are then added to obtain a tree rating, which ranges from 0 (no infection) to 6 (heavily infected in each third). Ratings for all live trees can be averaged to obtain a stand rating.

ranges from essentially no measurable effect in lightly infected trees to severe retardation in heavily infected trees. In general, height growth rates are more markedly reduced than diameter growth rates. For example, in *Pinus hartwegii* infected by two species of *Arceuthobium* in Mexico, height growth was reduced by

Table 24.3. The effect of dwarf mistletoe on tree mortality in New Mexico.

Dwarf mistletoe rating	Percentage of trees killed in previous 10 years	
	Arceuthobium vaginatum on *Pinus ponderosa*	*A. douglasii* on *Pseudotsuga menziesii*
0	1	1
1	1	1
2	2	4
3	4	8
4	9	13
5	12	24
6	38	45

Data from Hawksworth & Lusher (1956).

47%, more than twice the 22% reduction in diameter growth (Andrade & Tovar, 1981). The effects of varying degrees of dwarf mistletoe infection on recent radial growth in 140-year-old dominant trees of *Pinus ponderosa* are shown in Table 24.2. Growth was significantly reduced only in trees in infection classes 4, 5, and 6. As a general rule, when the upper portions of a tree's crown become infected, tree growth rates are significantly reduced.

Mortality

Mortality rates differ for various host-parasite combinations but, as with growth reduction, are generally related to the intensity of infection (Table 24.3). Note that the dwarf mistletoe on *Pseudotsuga menziesii* is more lethal than the one on *Pinus ponderosa*.

Decreased seed production

Although few quantitative data are available, dwarf mistletoes adversely affect cone and seed production of infected host trees. In heavily infected trees of many species, seed production may be negligible. Munns (1919) found that *Pinus jeffreyi* affected by reduced by 20%. Korstian Long (1922) studied *Pinus ponderosa* infected by dwarf mistletoe in Arizona and found that the 'reproductive value' (yield of tree seed per pound x number of seeds per pound x germination percent) of heavily infected trees was 75% less than that for comparable uninfected trees. Recently, Schaffer *et al.* (1983) reported that lodgepole pines heavily infected by dwarf mistletoe in Colorado had significantly smaller cones and seeds. As a result of these effects, Myers (1974) suggested that trees with dwarf mistletoe ratings over class 3 should not be left as seed sources for natural regeneration of forest stands.

Reduced wood quality

The anatomy of host wood within dwarf mistletoe stem infections is markedly altered. Affected wood has shorter and distorted tracheids and a greatly increased proportion of ray tissue, which results in reduced strength and poor pulping characteristics (Piirto *et al.*, 1974; Hunt, 1971). The practical significance of the strength reduction has not been determined but it is presumably slight (Dobie & Britneff, 1975; Wilcox *et al.*, 1973). Most mistletoe-affected wood is near the outside of the stem and tends to be removed when the log is squared. Another noticeable, but again not quantified, effect of dwarf mistletoe is the increased size of knots in wood and the frequent swellings and distortions of the stem.

Predisposition to attack by insects and fungi causing decay

The concept that a single pathogen or insect alone seldom kills trees is gradually becoming accepted. Rather, tree killing typically involves a complex of pathological, entomological, and environmental factors.

In California, a Pest Damage Inventory system has been developed to examine the reasons for tree mortality including pathogens, insects and abiotic factors (Byler, 1978). The type of analysis resulting from this system is given in the following example for *Pinus jeffreyi* and *P. coulteri* in Southern California: root pathogens were involved in 52% of the tree mortalities and dwarf mistletoe in 60%; root pathogens *and* dwarf mistletoes were jointly involved in 28% of the tree mortalities (Fig. 24.4).

The relationship between dwarf mistletoes and bark beetles (*Dendroctonus* spp.) as causes of tree mortality has long been debated. Until recently, there had been few quantitative data on the subject. It now appears that, for some complexes (e.g., *Arceuthobium vaginatum*/*Dendroctonus ponderosae* in ponderosa pine), susceptibility of mistletoe-infected trees to attack by the beetle is increased (McCambridge *et al.*, 1982). For other complexes (e.g., *Arceuthobium americanum*/*Dendroctonus ponderosa* in lodgepole pine), mistletoe-infected trees are no more, and perhaps even less, susceptible to beetle attack than are non-infected trees (McGregor, 1978).

Dwarf mistletoe infections on the main stems of true firs and hemlock, particularly those with exposed wounds, frequently provide entry courts for decay fungi (Aho, 1982; Etheridge, 1973). However, this is not a problem in pines, spruce, or Douglas fir, presumably because of the higher resin content in these species.

Ecological effects

The dwarf mistletoes have very many ecological effects in infected forests. Frequently, forest pests are regarded as foreign to, and not an integral part of, the total forest ecosystem. It should be remembered, however, that ecosystems have evolved over the millenia, with forest pests having an active role in their development. For example, dwarf mistletoes are actively involved in forest succession. In general, seral species, such as *Pinus contorta*, are severely parasitized whereas the climax species, such as various species of *Picea* and *Abies*, are generally immune, or nearly so. The overall result of dwarf mistletoe parasitism,

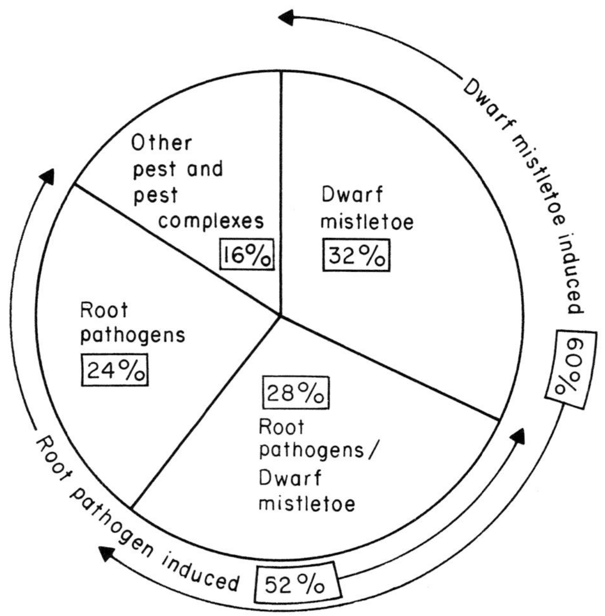

Figure 24.4. Mortality factors in *Pinus jeffrey* and *P. coulteri* in the Laguna Mountain Area, Cleveland National Forest, California (1977). Mortality factors are dwarf mistletoe (*Arceuthobium campylopodum*), root pathogens (mainly *Fomes annosus*), and insects (mainly *Melanophila* and *Ips*). Data from Byler (1978).

then, is a tendency to hasten succession toward the less-damaged climax species (Hawksworth, 1975).

The fire history of forests is an exceedingly important factor governing the distribution of dwarf mistletoes (Alexander & Hawksworth, 1975). In natural stands, fire has been the primary control agent. Fires severe enough to kill large areas of infected trees essentially eliminate the parasite from the stand except where an occasional infected tree may survive. Trees typically re-establish into burned-over areas much faster than the mistletoe. With a general increase in fire protection over the last half century or so, there has been a gradual increase in areas infested by mistletoe. In a broader, long-term sense, fires also favour mistletoe in that, when less-susceptible climax forests are burned, they are often replaced by more susceptible seral species.

The effects of fire on the ecology of some dwarf mistletoes have been investigated, particularly in the northern Rocky Mountains (Wicker & Leaphart, 1976). However, details of interactions between fires, hosts, and dwarf mistletoes are poorly known, particularly the possibility of using controlled burns as a management technique to reduce infestations.

Estimates of damage caused by dwarf mistletoes

Dwarf mistletoes are the most serious disease agents in many western coniferous forests. In many areas they cause greater losses than forest fires, insects, and other

Table 24.4. Estimates of commercial forest types infected by dwarf mistletoes in the USA.

Host and *Arceuthobium* sp.	Percentage of area affected	Location	Reference
Lodgepole pine			
(*A. americanum*)	50	Colorado, Wyoming	Johnson *et al.* (1981)
(*A. americanum*)	42	California, Oregon, Washington	Bolsinger (1978)
(*A. americanum*)	35	Montana	Dooling & Eder (1981)
(*A. americanum*)	60	Southern Idaho, Utah	Hoffman & Hobbs (1979)
Western larch			
(*A. laricis*)	47	Oregon, Washington	Bolsinger (1978)
Ponderosa and Jeffry pine			
(*A. campylopodum*)	26	California, Oregon, Washington	Bolsinger (1978)
Ponderosa pine			
(*A. vaginatum*)	36	Arizona, New Mexico	Andrews & Daniels (1960)
(*A. vaginatum*)	21	Southern Utah	Hoffman (personal communication)
Douglas fir			
(*A. douglasii*)	47	Arizona, New Mexico	Andrews & Daniels (1960)
(*A. douglasii*)	58	Southern Idaho, Utah	Hoffman (personal communication)

pathogens combined. Table 24.4 gives examples of the amount of dwarf mistletoe in some important commercial tree species. In some National Forests, as much as 80% of a forest type may be affected. Dwarf mistletoes are estimated to affect at least 8.5 million ha of commercial forests in the USA (Drummond, 1982).

Rather conservative estimates of direct losses (growth loss and mortality) caused by dwarf mistletoes are 12 million m^3 of wood per year in the United States (Drummond, 1982) and 4 million m^3 of wood per year in western Canada (Baranyay & Smith, 1972). This amount of wood lost is equivalent to that needed to build nearly 300 000 average-sized homes, and represents a loss of more than $200 million annually.

Silvicultural control of dwarf mistletoe

The basic principles for effective silvicultural control of dwarf mistletoe have long been known (Weir, 1916); however, only in the last decade have these principles been generally applied as forest management practices in western North America. Several features make the dwarf mistletoe amenable to silvicultural control (Hawksworth, 1978a; Knutson & Tinnin, 1980).

(a) They are obligate parasites; that is, they need a living host to survive. Hence, once an infected tree or branch is cut, the dwarf mistletoe is no longer a threat.

(b) They generally have narrow host ranges; that is, they are usually confined to a single host or a group of closely related species. It is frequently possible to favour an immune or lightly-infected tree species to minimize damage from dwarf mistletoe.

(c) Dwarf mistletoes have long life cycles. The time from infection to seed production is typically 5–6 years and sometimes even longer. This long life cycle means that build-up of populations is relatively slow.

(d) They have a slow rate of spread. Due to the long life cycle and limited seed dispersal, spread of dwarf mistletoes is relatively slow. On the average, dwarf mistletoes progress through stands at about 0.3–0.6 m/year.

The type of silvicultural control recommended depends on the growth characteristics and ecology of the host tree. In most situations, particularly for pines, mistletoes spread and intensify rapidly in young stands (Hawksworth & Graham, 1963). However, in other instances, e.g., *Abies* and *Tsuga*, dwarf mistletoe intensification in young stands is relatively slow and the problem is less severe than in old-growth stands (Scharpf & Parmeter, 1982; Shaw, 1982b).

For some forest types, clear-felling is a satisfactory method of control. This procedure is particularly suited to forests of *Pinus contorta*, because they are adapted to reproduction in open areas (frequently from seed retained in cones in the branches left after logging); other tree species cannot be treated by clear-felling, because they do not reproduce well under these conditions. For these species it is often possible to use shelterwood or seed tree harvests in diseased stands. For these methods, some of the original trees are retained to provide seed to regenerate this area. However, use of these methods in mistletoe-infested stands can be successful only if great care is taken to remove diseased residual trees before the new young stand becomes infected. In most species, diseased residuals should be removed before the young stand is about 1 m high (usually 5–10 years old depending on species and site factors) as such small trees usually have little infection because of their small 'target area' for interception of mistletoe seeds (Wicker & Shaw, 1967). If a stand is already generally infested, the silvicultural options are quite limited, and one is usually faced with the alternatives, harvest the entire stand and begin anew, or do nothing.

In diseased stands that are not yet of harvest age, thinning is possible, if the degree of infection is not too severe. Thinning in mistletoe-infested stands is a common forest management practice in western North America for mistletoe *control* but not *eradication*. Such thinning will improve growth, even in infected stands, as long as the residual trees are not too heavily infected. As a general rule, thinning should not be attempted in stands with dwarf mistletoe ratings of over 3.0 (Hawksworth, 1978b).

Other direct-control techniques are available for individual trees of high value, such as those in recreational or residential areas. Here, pruning of diseased branches is a practical means of control if at least the top half of the crown is disease-free (Lightle & Hawksworth, 1973). A recent major advance in management of dwarf mistletoe-infested stands has been the development of computer programs that predict yields from stands under various treatment strategies (Edminster, 1978).

Such programs are being used by the forest managers on millions of hectares in the western United States for comparing projected yields for stands under treatments, such as thinning to various stocking levels, various thinning frequencies, no treatment, etc. These comparisons allow for selecting the optimum control, given the particulars of the stand in question.

European quarantine regulations against *Arceuthobium* spp.

The European and Mediterranean Plant Protection Organization (EPPO) has established quarantine regulations to prevent the introduction of North American *Arceuthobium* (Smith, 1979). Dwarf mistletoe is placed on the 'A-1' list, meaning a dangerous organism absent from all EPPO countries and for which a zero tolerance is allowed. Although we feel that this quarantine is prudent, we suggest that chances of damage resulting from an introduction of *Arceuthobium* are minimal. Because of their limited host range and the slow rates of spread of dwarf mistletoes, an outbreak, if established, could be readily detected and eradicated before significant harm results. There have been a few cases in North America where dwarf mistletoes have become established outside their natural ranges; usually only a few trees were involved (Hawksworth & Wiens, 1972). If dwarf mistletoes were significantly dispersed by birds, their potential for spread after an introduction would be more significant, as in the case of the European mistletoe (*Viscum album*) in California (Scharpf & McCartney, 1975).

References

Aho P.E. (1982) Indicators of cull in western Oregon conifers. *USDA Forest Service, General Technical Report PNW-144*, 17 pp.

Alexander M.E. & Hawksworth F.G. (1975) Wildland fires and dwarf mistletoes: A literature review of ecology and prescribed burning. *USDA Forest Service General Technical Report RM-14*, 12 pp.

Andrade Escobar V. & Cibrian Tovar D. (1981) Evaluacion de effectos de muerdago enano, *Arceuthobium globosum* Hawks. y Wiens y *A. vaginatum* Willd., en rodales de *Pinus hartwegii* Lind. *Universidad Autonoma Chapingo, Mexico, Chapingo, Nueva Epoca* No **29–30**, pp. 9–15.

Andrews S.R. & Daniels J.P. (1960) A survey of dwarf mistletoes in Arizona and New Mexico. *U.S. Department of Agriculture, Forest Service, Rocky Mountain Forest and Range Experiment Station, Station Paper* **49**, 17 pp.

Baranyay J.A. & Smith R.B. (1972) Dwarf mistletoes in British Columbia and recommendations for their control. *Canadian Forestry Service Report BC-X-72*, 18 pp.

Bolsinger C.L. (1978) The extent of dwarf mistletoe in six principal softwoods in California, Oregon, and Washington, as determined from forest survey records. *USDA Forest Service General Technical Report PSW-31*, pp. 45–54.

Byler J.W. (1978) The pest damage inventory in California. *USDA Forest Service General Technical Report PSW-31*, 161–71.

Dobie J. & Britneff A.A. (1975) Lumber grades and volumes from lodgepole pine infected with dwarf mistletoe. *Wood and Fiber* **7**, 104–9.

Dooling, O.J. & Eder R.C. (1981) An assessment of dwarf mistletoes in Montana. *USDA Forest Service, Northern Region, Forest Pest Management Report 81–12*, 17 pp.

Drummond D.B. (1982) Timber loss estimates for the coniferous forests of the United States due to dwarf mistletoe. *USDA Forest Service, Forest Pest Management, Methods Application Group Report* No. **83–2**, 24 pp.

Edminster C.B. (1978) Development of growth and yield models for dwarf mistletoe-infested stands. *USDA Forest Service General Technical Report PSW-31*, pp. 172–7.

Etheridge D.E. (1973) Wound parasites causing tree decay in British Columbia. *Canadian Department of Environment, Forestry Service, Pacific Forestry Research Centre, Forest Pest Leaflet* **62**, 15 pp.

Hawksworth F.G. (1961) Dwarf mistletoe of ponderosa pine in the Southwest. *U.S. Department of Agriculture Technical Bulletin* **1246**, 112 pp.

Hawksworth F.G. (1975) Dwarf mistletoe and its role in lodgepole pine ecosystems. In: *Proceedings of the symposium on management of lodgepole ecosystems*, Vol. I. (Ed. by D.A. Baumgartner), pp. 342–58. Washington State University, Pullman.

Hawksworth F.G. (1977) The 6-class dwarf mistletoe rating system. *USDA Forest Service General Technical Report RM-48*, 7 pp.

Hawksworth F.G. (1978a) Biological factors of dwarf mistletoe in relation to control. *USDA Forest Service General Technical Report PSW-31*, pp. 5–15.

Hawksworth F.G. (1978b) Intermediate cuttings in mistletoe-infested lodgepole pine and southwestern ponderosa pine. *USDA Forest Service General Technical Report PSW-31*, pp. 86–92.

Hawksworth Frank G. & Graham Donald P. (1963) Spread and intensification of dwarf mistletoe in lodgepole pine reproduction. *Journal of Forestry* **61**, 587–91.

Hawksworth F.G. & Lusher A.A. (1956) Dwarf mistletoe survey and control on the Mescalero Apache Reservation, New Mexico. *Journal of Forestry* **54**, 384–90.

Hawksworth F.G. & Wiens D. (1972) Biology and classification of dwarf mistletoes (*Arceuthobium*). *U.S. Department of Agriculture Handbook* **401**, 234 pp.

Hawksworth F.G. & Wiens D. (1976) *Arceuthobium oxycedri* and its segregates *A. juniperi-procerae* and *A. azoricum* (Viscaceae). *Kew Bulletin* **31**, 71–80.

Hawksworth F.G. & Wiens D. (1977) *Arceuthobium* (Viscaceae) in Mexico and Guatemala: Additions and range extensions. *Brittonia* **29**, 411–18.

Hawksworth F.G. & Wiens D. (1980) A new species of *Arceuthobium* (Viscaceae) from central Mexico. *Brittonia* **32**, 348–52.

Hoffman J.T. & Hobbs L. (1979) Dwarf mistletoe loss assessment survey in Region 4, 1978. *USDA Forest Service, Intermountain Region, State and Private Forestry, Forest Insect and Disease Management Report R-4, 79–4*, 14 pp.

Hunt K. (1971) A comparison of Kraft pulping of wound and dwarf mistletoe-infected western hemlock wood. *Information Report VP-X-78*, 7 pp. Canadian Department of Fisheries and Forestry, Forest Products Lab., Vancouver.

Johnson D.W., Hawksworth F.G. & Drummond D.B. (1981) Yield loss of lodgepole pine stands to dwarf mistletoe in Colorado and Wyoming National Forests. *Plant Disease* **65**, 437–8.

Kiu H-X & Ren W. (1982) A new species of *Arceuthobium* from Xiazang. *Journal Yunnan Forestry College* No. **1**, pp. 42–5.

Knutson D.M. & Tinnin R. (1980) Dwarf mistletoe and host interaction in managed forests in the Pacific Northwest. *USDA Forest Service General Technical Report PNW-111*, 19 pp.

Korstian C.F., & Long W.H. (1922) The western yellow pine mistletoe: Effect on growth and suggestions for control. *U.S. Department of Agriculture Bulletin* **1112**, 35 pp.

Lightle P.C. & Hawksworth F.G. (1973) Control of dwarf mistletoe in a heavily used ponderosa pine recreation forest: Grand Canyon, Arizona. *USDA Forest Service Research Paper RM-106*, 22 pp.

McCambridge W.F., Hawksworth F.G., Edminster C.B. & Laut J.G. (1982) Ponderosa pine mortality resulting from a mountain pine beetle outbreak. *USDA Forest Service Research Paper RM-235*, 7 pp.

McGregor M.D. (1978) Management of mountain pine beetle in lodgepole pine stands in the Rocky Mountain area. In: *Proceedings of the symposium theory and practice of mountain pine beetle management in lodgepole pine forests*, pp. 129–39. *University of Idaho, Moscow.*

Munns E.N. (1919) Effect of fertilization on the seed of Jeffrey pine. *Plant World* **22**, 138–44.

Myers C.A. (1974) Multipurpose silviculture in ponderosa pine stands of the montane zone of central Colorado. *USDA Forest Service Research Paper RM-132*, 15 pp.

Piirto D.D., Crews D.L. & Troxell H.W. (1974) The effects of dwarf mistletoe on the wood properties of lodgepole pine. *Wood and Fiber* **6**, 26–35.

Schaffer B., Hawksworth F.G. & Jacobi W.R. (1983) Effects of comandra blister rust and dwarf mistletoe on cone and seed production of lodgepole pine. *Plant Disease* **67**, 125–7.

Scharpf R.F. & McCartney W. (1975) *Viscum album* in California — its introduction, establishment and spread. *Plant Disease Reporter* **59**, 257–62.

Scharpf R.F. & Parmeter J.R., Jr. (1978) Proceedings of the symposium on dwarf mistletoe control through forest management. *USDA General Technical Report PSW-31*, 190 pp.

Scharpf R.F. & Parmeter J.R., Jr. (1982) Population dynamics of dwarf mistletoe on young true firs in the central Sierra Nevada, California. *USDA Forest Service Research Paper PSW-161*, 9 pp.

Shaw C.G., III. (1982a) Mountain hemlock is occasional host for hemlock dwarf mistletoe in Alaska. *Plant Disease* **66**, 852–4.

Shaw C.G., III (1982b) Development of dwarf mistletoe in western hemlock regeneration in southeast Alaska. *Canadian Journal of Forestry Research* **12**, 482–8.

Smith I.M. (1979) EPPO: The work of a regional plant protection organization with particular reference to phytosanitary regulations. In: *Plant Health, the Scientific Basis for Administrative Control of Plant Diseases and Pests* (Ed. by D.L. Ebbels & J.E. King), pp. 17–22. Blackwell Scientific Publications, Oxford.

Tinnin R.O., Hawsworth F.G. & Knutson D.M. (1982) Witches' broom formation in conifers infected by *Arceuthobium* spp.: An example of parasitic impact upon community dynamics. *American Midland Naturalist* **107**, 351–9.

Weir J.R. (1916) Some suggestions on the control of mistletoe in the national forests of the Northwest. *Forest Quarterly* **14**, 567–77.

Wicker Ed F. & Leaphart C.D. (1976) Fire and dwarf mistletoe (*Arceuthobium* spp.) relationships in the northern Rocky Mountains. *Tall Timbers Fire Ecology Conference* **14**, 279–98.

Wicker E.F. & Shaw C.G. (1967) Target area as a klendusic factor in dwarf mistletoe infections. *Phytopathology* **57**, 1161–3.

Wilcox W.W., Pong W.Y. & Parmeter J.R. (1973) Effects of mistletoe and other defects on lumber quality in white fir. *Wood and Fiber* **4**, 272–7.

25 Market losses of fresh produce

R.A. NOON*

Imperial Chemical Industries PLC, Plant Protection Division, Jealott's Hill Research Station, Bracknell, Berkshire RG12 6EY, UK

Introduction

In view of the ever-increasing world population and the inadequacy of its food supply, it is essential to direct attention to both producing enough food and ensuring that it reaches the consumer in good condition with minimal loss. Unfortunately, it is well recognized that many millions of people, particularly those in the developing countries, suffer food shortages as a result of low crop yields and high post-harvest wastage. The plant pathologist has a major role to play in resolving both aspects of the problem. Particular attention needs to be directed to reducing marketing losses to ensure that the successes of modern agriculture benefit all consumers in the form of a plentiful supply of high-quality fresh produce at reasonable prices.

This chapter attempts to outline our knowledge of the extent and causes of market losses of fresh produce by reference to market-loss surveys conducted in the USA and Mexico. There follows a discussion of the factors which influence the post-harvest life of fresh produce and techniques which reduce wastage. Where possible, these are discussed in relation to the problems encountered in these two surveys.

Magnitude of losses

Few data are available on the extent of post-harvest losses, particularly in the developing countries. Loss assessment is difficult because of the complicated chain of events in the handling of a crop from the field to the consumer. Commercial market forces of supply and demand also combine to influence the amount of wastage and to determine the financial gain or loss on particular commodities.

There has been disagreement on methods for quantifying wastage and research workers have tended to avoid the problem. Past reviews have relied heavily upon quoting figures which usually have not been based on well-substantiated facts. When no figures have been available various estimates have been made for different markets.

In Latin America alone the total post-harvest losses of major crops have

* Formerly of Tropical Products Institute, 56–62 Gray's Inn Road, London WC1X 8LU, UK

Wood R.K.S. & Jellis G.J. (1984) *Plant Diseases: infection, damage and loss.* Blackwell Scientific Publications, Oxford.

been estimated to be of the order of 38 million tonnes annually. This is sufficient to feed more than 45 million people (Amezquita & La Gra, 1979).

Survey methods

Figures such as those quoted above only partially indicate the extent of wastage and fail to provide sufficient guidance to aid loss-reduction programmes. Detailed analyses of losses during marketing are essential in providing information on which recommendations for remedial action can be based. Absence of such data has frequently resulted in ineffective and inappropriate remedial programmes.

A few detailed surveys have been made, most notably in the USA (Wright & Billeter, 1969; Burton, 1970; Ceponis & Butterfield, 1974a, 1974b). Wright and Billeter (1969), working in Chicago, determined wholesale losses from the distribution centre of a chainstore and from an 'independent street broker'. Retail losses were estimated by measuring the wastage in four stores. Losses at the consumer level were obtained by simulated consumer tests on produce obtained from the retail displays of cooperating stores.

In a study in Mexico in 1978–79 by the author*, fresh fruit quality and losses were assessed in several representative wholesale warehouses in La Merced market, Mexico City. This market has been estimated (unpublished) by the Mexican National Fruit Commission to handle the distribution of 70% of all fresh produce formally marketed in the country. Details were obtained of the origin of fruit, production practices, field selection and transport and storage conditions by interviewing warehouse operators. Representative fruit packs from each consignment were sampled and individual fruits were graded for quality using commercial British, European, American or Canadian standards or experimental ones established specifically for the survey. Fruits which failed to meet the minimum characteristics were considered to be rejects and the reasons for rejection were recorded. Samples of fruit were taken to the laboratory when necessary to determine the cause of the defect resulting in rejection. Routine plant pathological laboratory procedures were followed to identify fruit diseases. Only those reject fruit which could not be sold were considered to constitute a total loss since fruit of low quality were often still considered marketable, albeit at lower prices.

In both studies, causes of total loss were grouped into three main categories: mechanical, non-parasitic and parasitic.

Loss statistics

Losses for the commodities surveyed in the Chicago market averaged 5.7% at the wholesale level, 3.0% at the retail level and 7.7% at the consumer level (Table 25.1).

* This survey was undertaken while the author was on secondment to the Mexican Government from the Tropical Products Institute, Overseas Development Administration, London.

Table 25.1. Marketing losses of selected fruits and vegetables in the Chicago area.

	% losses by weight		
Commodity (Cultivar)	Wholesale	Retail	Consumer
Apple (Red Delicious)	2.9	2.9	2.4
Lettuce (California Head)	5.7	1.7	13.9
Orange (California Navel)	2.0	1.6	2.2
Orange (Florida Valencia)	1.4	0.8	3.7
Peach	12.3	5.8	10.8
Potato (Red River Valley Red)	4.9	1.0	3.7
Potato (California Long White)	2.5	5.0	2.4
Strawberry (California & Southern)	13.5	5.5	22.2
Average	5.7	3.0	7.7

Data from Wright & Billeter (1969)

Mean losses at the wholesale level in the Mexico City market for the ten fruits surveyed were 10.4% (Table 25.2). Clearly, it is not only in developing countries that considerable post-harvest losses occur.

Financial implications

Comparison of wastage statistics combined with annual production figures for a particular crop allows estimates to be made of the total quantity of the commodity wasted. However, only very crude calculations may be made of the likely financial losses since prices paid are presumably modified by the market as a result of such wastage. Nevertheless crude calculations such as these may be worthwhile in demonstrating the potential size of the problem.

Comparing annual fruit production figures for Mexico with wastage statistics, it was demonstrated that losses of potential earnings on nine commodities at the wholesale level might have approached US$66 million during 1979 (Table 25.2). These figures ignore the added financial implications of low quality. By grading and valuing commodities according to quality, estimates may be made of the possible financial gains which might accrue if top quality commodities could be produced and the quality maintained during marketing. Reference to Table 25.3, where the quality of selected fruits in Mexico City market is tabulated, shows that 55.3% of apples were in the lower two classes. A further 33.5% were rejected and 18% were lost. Since lower quality apples were discounted by an average of 25% during the survey, low quality grading into classes 3 and 4 alone may have resulted in a loss of potential income approaching US$20 million. Consequently, economic losses from total wastage and low quality for apples may be crudely calculated to be of the order of US$50 million.

Table 25.2. Marketing losses during 1978–79 of selected fruits in La Merced market, Mexico City.

Commodity	% loss (by weight)	Estimated financial loss million Mexican pesos	(US$ million)
Apple	18.0	700	(30.4)
Avocado	2.3	103	(4.5)
Guava	10.0	77	(3.3)
Mango	4.4	15	(0.7)
Orange	5.0	no data	
Papaya	17.6	247	(10.7)
Peach	17.0	110	(4.8)
Pear	16.1	60	(2.5)
Pineapple	6.0	98	(4.3)
Watermelon	7.7	105	(4.6)
Average	10.4		
Total estimated financial loss*		1515	(65.9)

* excluding orange
Data from Noon (unpublished)

Table 25.3. Quality of selected fruits in La Merced market, Mexico City.

	% fruit in each quality class				
Commodity	1	2	3	4	Reject
Apple	0.2	11.0	29.6	25.7	33.5
Avocado	6.3	81.7	*	*	12.0
Mango	6.1	62.3	*	*	31.6
Orange	5.8	43.8	32.2	*	18.2
Pear	0.9	25.0	40.2	11.3	22.6

* These classes were not recognized
Data from Noon (unpublished)
Quality assessments were based on US standards for apple, orange and pear. Experimental standards were established for avocado and mango.

Causes of market losses

Quantitative and qualitative losses during marketing can be divided into those resulting from primary and from secondary causes. Bourne (1977) listed primary causes as biological and microbiological, chemical and biochemical, mechanical, physical, physiological, and psychological (this being the refusal to eat a food through dislike or for religious reasons).

Secondary causes of loss are those that lead to conditions in which the primary cause of loss can occur (Bourne, 1977). They include common deficiencies in the marketing chain such as those listed by Amezquita & La Gra (1979):

(a) poor organization of the marketing system;

(b) inadequate use of available technology during the various phases of production, harvesting, selection, packing, transportation, storage, processing and distribution;

(c) lack of knowledge and of remedial programmes;

(d) lack of services, guides or information about prices, markets, standards etc.;

(e) lack of a suitable infrastructure for the interchange and handling of agricultural products.

The post-harvest technologist must be able to interact effectively with specialists from such disciplines as engineering, agronomy, economics, sociology and politics since each has a rôle to play in correcting deficiencies and decreasing market losses. However, it is widely acknowledged that the application of methods to reduce waste tends only to accompany a general improvement in social conditions.

Most workers who have conducted market surveys have found it convenient to recognize only three broad primary causes of loss: parasitic, non-parasitic (or

Table 25.4. Causes of marketing losses of selected fruits and vegetables in Chicago area.

	% losses for each cause								
	Parasitic			Non-parasitic			Physical		
Commodity*	W†	R†	C†	W	R	C	W	R	C
Apple	0.1	0.2	0.2	0.5	0.2	1.0	2.3	2.5	1.2
Lettuce	0.5	0.7	0.6	1.6	0.4	3.9	3.6	0.6	9.4
Orange (California Navel)	1.5	1.2	2.1	0.3	0.2	0.1	0.2	0.2	T
Orange (Florida Valencia)	0.9	0.7	3.3	0.4	0.1	0.2	0.1	T	0.2
Peach	2.4	1.2	7.0	0.5	0.1	0.4	9.4	4.5	3.4
Potato (Red River Valley Red)	1.3	0.1	0.9	0.7	0.2	1.3	2.9	0.7	1.5
Potato (California Long White)	0.3	n.d.	0.3	0.5	n.d.	1.6	2.5	n.d.	2.4
Strawberry	2.7	1.5	3.9	2.1	0.1	1.3	8.7	3.9	17.0
Average	1.2	0.8	2.0	0.8	0.2	1.2	3.7	1.9	4.4

† W = Wholesale R = Retail C = Consumer

n.d. = no data

T = trace

* see Table 25.1 for full details

Data from Wright & Billeter (1969)

physiological) and physical (or mechanical). Wright & Billeter (1969) in their Chicago survey quantified loss into these three categories for each commodity at each level of marketing (Table 25.4). Clearly the importance of each primary cause of loss varies according to a wide range of factors such as commodity, variety, origin, growing environment and marketing conditions. The interactions between each type of loss are often so complex that it is difficult to assess the true importance of each factor. For example, most microorganisms which infect after harvest are wound pathogens requiring physical injury of tissue before they are able to infect. Those that do infect are often most aggressive when the host is nearing senescence. In these situations pathogens may merely increase tissue deterioration as secondary factors following primary physical and physiological changes which predispose the tissue to infection. A detailed appreciation of these aspects is necessary for the better control of primary causes of wastage in loss reduction programmes.

In the Chicago market survey (Table 25.4) physical damage was the major cause of direct loss and in most commodities was implicated in predisposing tissue to disease. In the analysis of causes of loss in Mexican Golden Delicious apples (Table 25.5) bruising and other mechanical injuries during post-harvest handling operations resulted in losses of 4.2%. Injuries at harvesting (0.2%) consisted mainly of finger-nail damage to the skins of fruits. Stem punctures accounted for 1.2% loss. Improved packaging would reduce mechanical damage in situations such as these where the value of the commodity is sufficiently high to warrant expenditure on packing materials. Hinsch (1966) demonstrated that apples packed in moulded trays showed 10 times less bruising than those loosely packed in polyethylene bags. In tight-fill packs, which are frequently used to minimize bruising in apples, fruit are placed in a box at random and then a pad and lid are placed on top of the pack. Correct loading

Table 25.5. Specific causes of marketing losses and low quality in Golden Delicious apples in Mexico City.

% number of fruits lost or rejected					
Parasitic		Non-parasitic		Physical	
Penicillium rots	4.0	Internal breakdown	4.8	Bruising	3.0
Alternaria rots	1.3	Bitter pit	1.1	Mechanical injury	1.2
Gloeosporium rots	0.6	Severe scald	0.7	Stem punctures	1.2
Rhizopus rots	0.4	Cracking	0.3	Harvest injury	0.2
Botrytis rots	0.1	Senescence	0.3	Below size	0.1
Other rots	0.1	Freezing damage	0.2		
Storage scab	0.1				
Insect damage	1.3				
Total	7.9	Total	7.4	Total	5.7

Data from Noon (unpublished)

of the tight-fill packs used in the Mexico City market would have minimized the incidence of stem punctures.

Important factors affecting market quality and losses

A wide range of factors influences the incidence of wastage and maintenance of quality. Storage temperature, for example, affects both the physiology and pathology of commodities while some packaging systems, may radically alter their rate of metabolism and disease spectrum. The important factors considered here are:
(a) handling systems;
(b) temperature;
(c) humidity;
(d) curing;
(e) controlled and modified atmospheres;
(f) chemical treatments.

Handling systems

Damage may be particularly severe during the harvesting and handling of root crops. In Britain in 1973, 23% of the harvested potato crop was shown to have sustained flesh damage by the time of loading into store (Anon., 1974). Clearly, improved harvesting and handling systems would lead to a lower rate of rejection.

Cassava, the most important carbohydrate staple root crop of the tropics, deteriorates rapidly following harvest injury, by a physiological process which is manifested by vascular discoloration (Noon & Booth, 1977). So rapid is its deterioration that cassava remains marketable for only about 2 days after harvest under typical tropical conditions.

Yams can be effectively stored and marketed over long periods provided mechanical damage is avoided. The internal metabolism of the tuber is disrupted by injury, leading to an increase in respiratory rate and a consequent loss of moisture and food reserve (Passam & Noon, 1977). Damaged tissue may also become infected by micro-organisms which cause rapid decay. Microbial decay is commonly regarded as the main cause of loss of yams during storage and marketing but usually develops as a result of primary mechanical damage during harvesting, transport and marketing (Noon, 1978).

Temperature

Correct refrigeration is essential during the marketing of many commodities and its absence in the marketing chain in many developing countries is a major cause of post-harvest losses. Commodities differ in optimum temperatures for storage and marketing (Lutz & Hardenburg, 1968). Commodities which tolerate low temperatures are usually best stored just above 0°C. However, many tropical commodites are subject to chilling injury at temperatures well above their freezing points. This form of low temperature injury causes direct loss and renders produce more susceptible to microbial decay. Noon & Colhoun (1981) demonstrated that for yams the critical temperature below which the disorder occurred was 10–12°C. As with other commodities, severity of injury increased as the temperature lowered or as exposure lengthened. Affected tubers showed rapid dehydration and weight loss. Subsequent deterioration was associated with saprophytic fungi, bacteria and yeasts rather than with the main post-harvest pathogens of yams.

Humidity

Generally, commodities are best held at high relative humidities to reduce dehydration. Those which may be held for long periods prior to sale e.g. grapes, apples, pears and cabbage are best stored at temperatures and humidities which avoid condensation on their surfaces but which minimize the drying capacity of the air. However, if free water is allowed to condense on produce, or under conditions of very high humidity, the incidence of disease may be greatly increased. Thus a balance is necessary between the physiological requirements of the commodity and those of potential pathogens.

The use of polyethylene film for consumer packs and box liners aids in the reduction of dehydration. Hardenburg & Anderson (1961) demonstrated that during storage for 6 months at 0°C, Golden Delicious apples lost less than 1% weight when box liners were used compared with 4 to 7% loss without liners. However, the incidence of fruit rots is usually increased with liners (McIntosh & Porritt, 1958) and it is advisable to treat with fungicides before packing (Hardenburg, 1974).

Curing

The exposure of several root and tuber crops to high temperature and humidities for a short period after harvest promotes the formation of suberin at wound sites, followed by the development of a periderm or cork layer. This process is termed 'curing' and the protective layer which forms prevents further infection by wound pathogens. The process has been described for potatoes by Artschwager (1927) sweet potatoes by Weimer & Harter (1921) and Lauritzen (1935) and for yams by Passam *et al.* (1976a, 1976b) and Noon (1976). In many countries potatoes are routinely cured as soon as they have been loaded into store or after initial drying. Ventilation is restricted to allow the temperature to rise to about 15°C and this is maintained for 10 days. This reduces infection of wounds by pathogens such as *Phoma exigua* (see Jellis & Boulton, this volume).

Controlled and modified atmospheres

Use of polyethylene films also serves to modify the atmosphere within the pack. Carbon dioxide or nitrogen enrichment and oxygen depletion slow metabolic processes, and such controlled or modified atmospheres allow the extended storage of produce, particularly climacteric commodities which ripen after harvest. The optimum concentrations of oxygen, carbon dioxide and nitrogen have been established for a wide range of commodities (Dewey, 1977). Methods have also been developed commercially for the removal of endogenous ethylene in order to delay ripening in several commodities, including bananas (Scott *et al.*, 1971).

Chemical treatments

Where infection occurs in the field, chemical control is usually best applied before harvest. Infections prior to harvest which remain latent until the commodity starts to ripen may be controlled either in the field or the packing house. Fewer chemicals are registered for post-harvest use than for field application because of the more stringent safety requirements for such products. They are normally applied to commodities by dipping in solutions or suspensions, as dusts or by volatile fumigation. Their action may eradicate existing infections, protect wounds from subsequent infections, or inhibit sporulation of pathogens which have already infected. These control methods are best illustrated with specific examples.

Common parasitic causes of low quality and loss in mango encountered in the market in Mexico City were caused by *Colletotrichum gloeosporioides* (anthracnose), *Elsinöe mangiferae* (scab), *Botryodiplodia theobromae* (stem-end rot), *Rhizopus stolonifer* and insect damage caused by infestation by fruit fly larvae (Table 25.6). Anthracnose may be controlled using protectant orchard sprays of dithiocarbamate or benzimidazole fungicides (McMillan, 1973). Alternatively, post-harvest eradicant treatment of green fruit with hot water between 50 and 55°C gives good control. Furthermore, the addition of benomyl or thiabendazole fungicides to the hot water allows the treatment temperature to be lowered and improves the degree of eradication (Muirhead, 1976). Trials on Mexican Manila mangos by the author confirmed these observations (Table 25.7). Scab is a superficial disorder which disfigures mango fruit surfaces. Since the disease originates in the orchard,

Table 25.6. Specific causes of marketing losses and low quality in Manila mangos in Mexico City.

% number of fruits lost or rejected					
Parasitic		Non-parasitic		Physical	
Anthracnose	25.3	Overmaturity)		Bruising)	
		)	8.6	)	
Scab	0.7	Dehydration)		Creasing)	
Rhizopus rots	0.5			)	4.9
Stem-end rots	0.4			Tearing)	
				)	
Insect damage	0.7			Harvest injury)	
Total	27.6	Total	8.6	Total	4.9

Data from Noon (unpublished)

Table 25.7. Effects of hot water/benzimidazole fungicide dip treatment on anthracnose and rate of ripening of Mexican Manila mangos up to 11 days after treatment.

	Ripeness Rating*			Anthracnose Rating*		
	days after treatment			days after treatment		
Treatment	4	8	11	4	8	11
Hot water (53°C)	2.3a	3.4a	3.9a	0a	0.5a	2.0a
Hot water + 500ppm Benomyl	2.2a	3.4a	3.8a	0a	0a	0b
Hot water + 500ppm Thiabendazole	2.4a	3.4a	4.0a	0a	0.1a	0.3b
Untreated — Control	2.3a	3.9b	4.0a	0.8b	2.6b	4.3c

Means in any one column flanked by a letter in common do not differ significantly (P = 0.05) using the Student-Newman-Keuls Range Test (Sokal and Rohlf, 1969)

* Ripeness and anthracnose were rated on scales from 0 (green fruit) to 4 (ripe fruit) and 0 (no disease) to 5 (greater than 50% disease)

Data from Noon (unpublished)

pre-harvest treatment is essential to prevent its development and protectant fungicides are used during flowering and fruit set (Ruehle and Ledin, 1955).

Eckert (1978) has reviewed the modern usage of chemicals in the citrus industry. An integrated approach is followed which combines pre- and post-harvest treatments. The stem-end rots (caused by *Diplodia* spp. and *Phomopsis* spp.) and penicillium decay may be controlled with orchard or post-harvest treatments with benzimidazole fungicides. However, modern citrus packing operations may exacerbate disease problems. Degreening using ethylene alters the susceptibility of fruits to disease so that fungicidal dips or 2-aminobutane fumigation are usually required prior to treatment. Injuries during the post-harvest handling of fruits may be disinfected with solutions of borax, sodium carbonate or sodium *o*-phenyl phenate. A second treatment with benzimidazole fungicides or imazalil inhibits the development of latent stem-end rot infections and those caused by *Pencillium* spp. Biphenyl-impregnated tissue paper fruit wraps, or pads at the bottom and top of boxes, are used during marketing to prevent the sporulation of *Penicillium* spp. on infected fruits and the resultant soiling of adjacent healthy fruits with spores of the pathogens.

Chemical control of *Botrytis cinerea* on grapes must normally commence in the vineyard at flowering, continue through the season until harvest and, for table grapes, be carried on during long-term storage before marketing. Until the pathogen developed resistance, benomyl was the preferred fungicide for use in the vineyard. It was then replaced in some markets by dicarboximide fungicides which were active against the benomyl-resistant strains of the fungus. During storage, however, use of these chemicals is not permitted and table grapes are fumigated instead with sulphur dioxide within 12 h of harvest and before cooling. This treatment surface-sterilizes the berries and inhibits wound infection. Delaying the treatment beyond

12 h allows the spores of *B. cinerea* to infect. For extended storage, further treatments are given with care since the phytotoxic properties of the gas are cumulative and the risk of injury increases with each treatment (Nelson, 1979).

Conclusions

The marketing of fresh produce is influenced by economic, sociological and political factors which may affect the quantity of losses and quality of commodities. It is necessary to identify the causes of loss and estimate their magnitude so that appropriate control programmes can be devised. Market surveys are therefore essential in indicating to the post-harvest technologist the priorities for remedial programmes. The few detailed surveys which have been attempted show that losses may be high, even in sophisticated marketing systems.

Improved handling and packing may reduce mechanical damage and thereby reduce direct loss; they have a further influence on indirect loss by decreasing disease resulting from infection by wound pathogens. Refrigeration throughout the marketing chain may greatly aid in the maintenance of quality by delaying adverse physiological changes and decay caused by pathogens. Relative humidity and storage atmospheres further interact and influence the market life of produce. The use of chemicals in the field and after harvest provides a means of maintaining quality during growth, harvesting and handling of the commodity. Appropriate application of these technologies to the marketing of produce in developing and developed countries should greatly reduce food losses and provide a better supply of high quality produce at reasonable prices to the world's growing population.

References

Amezquita R. & La Gra J. (1979) A methodological approach to identifying and reducing post-harvest losses. *Inter-American Institute of Agricultural Sciences, Office in the Dominican Republic, Miscellaneous Publication* No. **219**.

Anon. (1974) *Report on a National Damage Survey, 1973*. Potato Marketing Board, London, 37 pp.

Artschwager, E.F. (1927) Wound periderm formation in the potato as affected by temperature and humidity. *Journal of Agricultural Research* **35**, 995–1000.

Bourne M.C. (1977) The role of post-harvest food loss reduction in improving the economic and nutritional status of tropical populations. Text of a lecture given at a seminar on '*Reduction of Post Harvest Food Losses in the Caribbean and Central America*' in Santo Domingo, Dominican Republic, August 8, 1977. Unpublished.

Burton C.L. (1970) Diseases of tropical vegetables on the Chicago market. *Tropical Agriculture* (Trinidad) **47**, 303–13.

Ceponis M.J. & Butterfield J.E. (1974a) The nature and extent of retail and consumer loss in apples, oranges, lettuce, peaches, strawberries, and potatoes marketed in Greater New York. *United States Department of Agriculture, Marketing Research Report* No. **996**, 23 pp.

Ceponis M.J. & Butterfield J.E. (1974b) Retail and consumer losses in sweet potatoes marketed in metropolitan New York. *HortScience* **9**, 393–4.

Dewey D.H. (1977) Controlled atmospheres for the storage and transport of perishable agricultural commodities. *Michigan State University Horticulture Report* **28**, 301 pp.

Eckert J.W. (1978) Post-harvest diseases of citrus fruits. *Outlook on Agriculture* **9**, 225–32.

Hardenburg R.E. (1974) Benomyl and thiabendazole treatments reduce decay of apples in polyethylene consumer bags. *Journal of the American Society for Horticultural Science* **99**, 236–8.

Hardenburg R.E. & Anderson R.E. (1961) Polyethylene box liners for storage of Golden Delicious apples. *United States Department of Agriculture, Marketing Research Report* **461**, 36 pp.

Hinsch R.T. (1966) New overwrapped trays for McIntosh apples. *United States Department of Agriculture ARS 52–16*, 8 pp.

Lauritzen J.I. (1935) Factors affecting infection and decay of sweet potatoes by certain storage rot fungi. *Journal of Agricultural Research* **50**, 285–329.

Lutz J.M. & Hardenburg R.E. (1968) The commercial storage of fruits, vegetables, and florist and nursery stocks. *United States Department of Agriculture, Agriculture Handbook* No. **66**, 94 pp.

McIntosh D.L. & Porritt S.W. (1958) The effect of storing apples in polythene box liners on the development of Bull's Eye Rot, caused by *Neofabraea perennans. Plant Disease Reporter* **42**, 97–9.

McMillan R.T. (1973) Mango anthracnose control. *Proceedings of the Florida State Horticultural Society* **86**, 326–7.

Muirhead I.F. (1976) Post-harvest control of mango anthracnose with benomyl and hot water. *Australian Journal of Experimental Agriculture and Animal Husbandry* **16**, 600–3.

Nelson K.E. (1979) Harvesting and handling California table grapes for market. *Division of Agricultural Sciences, University of California, Publication* No. **4095**, 67 pp.

Noon R.A. (1976) The post-harvest pathology of tropical fruits vegetables and root crops in the United Kingdom. *PhD Thesis, University of Manchester.*

Noon R.A. (1978) Storage and market diseases of yams. *Tropical Science* **20**, 177–88.

Noon R.A. & Booth R.H. (1977) Nature of post-harvest deterioration of cassava roots. *Transactions of the British Mycological Society* **69**, 287–90.

Noon R.A. & Colhoun J. (1979) Market and storage diseases of yams imported into the United Kingdom. *Phytopathologische Zeitschrift* **94**, 289–302.

Noon R.A. & Colhoun J. (1981) Observations on the nature of chilling injury in yams. *Phytopathologische Zeitschrift* **100**, 10–8.

Passam H.C. & Noon R.A. (1977) Deterioration of yams and cassava during storage. *Annals of Applied Biology* **85**, 436–40.

Passam H.C., Read S.J. & Rickard J.E. (1976a) Wound repair in yam tubers: physiological processes during repair. *New Phytologist* **77**, 325–31.

Passam H.C., Read S.J. & Rickard J.E. (1976b) Wound repair in yam tubers: the dependence of storage procedures on the nature of the wound and its repair. *Tropical Science* **18**, 1–11.

Ruehle C.D. & Ledin R.B. (1955) Mango growing in Florida. *Florida Agricultural Experiment Station Bulletin* No. 174, **90** pp.

Scott K.J., Blake J.R., Strachan G., Tugwell B.L. & McGlasson W.B. (1971) Transport of bananas at ambient temperatures using polyethylene bags. *Tropical Agriculture* (Trinidad) **48**, 245–54.

Sokal R.R. & Rohlf F.J. (1969) *Biometry.* W.H. Freeman & Co., San Francisco.

Weimer J.L. & Harter L.L. (1921) Wound-cork formation in the sweet potato. *Journal of Agricultural Research* **21**, 637–47.

Wright W.R. & Billeter B.A. (1969) Marketing losses of selected fruits and vegetables at wholesale, retail and consumer levels in the Chicago area. *United States Department of Agriculture, Marketing Research Report No.* **1017**, 20 pp.

Table 26.2. Incidence of pathogens on samples of soybean in Bolivia.

	a	b	c
On leaves			
Peronospora manshurica	–	+	+
Septoria glycines	+	–	–
Diaporthe phaseolorum	+	+	–
Colletotrichum dermatium	+	–	–
Xanthomonas campestris pv *glycines*	+	–	–
Cercospora canescens	+	+	+
Cercospora kikuchii	+	+	+
Leptosphaerulina trifolii	–	+	+
On stems			
Diaporthe phaseolorum	+	+	–
Colletotrichum dermatium	+	–	–
Pyrenochaeta glycines	+	–	–

a, b, c = soybean plants sampled from different areas

and foliage as well as bolls. They have been largely controlled in many countries by the use of resistant cultivars and clean seed but significant boll rot damage is still being caused by other minor foliar pathogens such as *Diplodia gossypina* and *Ascochyta gossypii* and by soil-borne fungi such as *Fusarium* spp. and *Rhizoctonia solani* in wet years.

Root pathogens present the most difficult problems of disease assessment, particularly as they commonly occur together and are influenced strongly by environmental interactions. Moisture and nutrient stresses so common in tropical soils predispose roots of annual crops to a range of pathogens. *Fusarium*, *Rhizoctonia*, *Pythium* and *Macrophomina* together with nematodes are the main constitutuents of these complexes. Secondary fungal rotting following primary nematode damage is a common feature on many tropical crops, e.g. Booth & Stover (1974), and synergistic reactions occur with other fungi, e.g. between *Heterodera glycines* and *Phytophthora megasperma* on soybean (Adeniji et al., 1975) and between *Meloidogyne incognita* and *Rhizoctonia solani* (Reddy *et al.*, 1979). Studies on soil-borne pathogen complexes on beans (Pieczarka & Abawi, 1978) and groundnuts (Garcia & Mitchell, 1975a, 1975b), indicated that *Pythium* species tend to be the major component with *Fusarium* and *Meloidogyne* having a strongly synergistic role. Even so, Krikun & Frank (1982) could control groundnut pod rot using metham sodium which apparently had little effect on *Pythium* but eliminated other organisms interacting synergistically with it. *Rhizoctonia* and *Macrophomina* seem to be mainly secondary although these may well be the most obvious fungi seen on badly diseased roots, an effect which led Garrett (1956) to describe *M. phaseolina* as an 'imposter'.

Apportioning blame between root and shoot disease on the same plant is notoriously difficult even where one does not obviously affect susceptibility to the other. Usually the most damage is attributed to the more obvious disease (i.e. that on the aerial plant parts and therefore more readily seen) but root disease will reduce the plant's capacity for compensatory growth — an aspect which seems to have received very little attention.

Complex or confused etiology

As well as making disease assessment more difficult, pathogen interactions may produce diseases of complex etiology which require a substantial research effort to unravel the cause of the damage before a start can be made on preventing losses. Alomae disease of taro (*Colocasia esculenta*) has caused severe and prolonged damage to taro cultivation in the Solomon Isles and threatens the crop in neighbouring SE Asia/Pacific areas. It is now known to be caused by the synergistic interaction of two bacilliform virus particles (Gollifer, *et al.*, 1978) but adequate control measures are confounded by lack of resistant germplasm and by traditional agricultural practices.

Many diseases of apparent complex or confused etiology still await resolution as they continue to constrain the productivity of some crops. Citrus dieback provides a typical example. This problem has bedevilled citrus production for years in Africa and in Asia especially on the Indian sub-continent where it appears to be caused by a pathogen complex (Raychaudhuri *et al.*, 1979). Citrus decline and dieback are symptoms that can be caused by many different pathogens and their interactions. *Botryodiplodia theobromae* is commonly associated with citrus dieback; 'diplodia' gummosis dieback was previously thought to be a major disease of citrus in many parts of Africa and Asia. It is now recognized that *B. theobromae* is a weak pathogen at best and is a typical example of an imposter, being a common secondary invader of moribund stem tissue on many tropical crops. However, the citrus nematode *Tylenchulus semipenetrans* is known to be a major cause of citrus decline in many areas; it predisposes citrus roots to infection by *Fusarium solani* (Van Gundy & Tsao, 1963) which exacerbates the condition leading to a dieback in which *B. diplodia* is involved. Other components of this decline-dieback complex include *Glomerella cingulata* and *Diaporthe citri* which can hasten leaf abcission, cause twig lesions and further reduce the physiological capacity of the tree. The situation is yet further complicated by citrus greening disease (caused by a mycoplasma-like organism) and by tristeza virus, both of which are prevalent in some areas of Asia and Africa and can be particularly virulent on some genotypes. These pathogens occur in all possible combinations on citrus, greatly confusing the etiology of the disease and the amount of damage caused which varies according to location, cultivar, and cultural practice.

Another dieback disease causing problems in many parts of Africa is that of mango. *Botryosphaeria ribis* and *Hendersonula toruloidea* are involved. This combination also occurs in almond dieback (England *et al.*, 1975) when *H. toruloidea* follows *B. ribis* and increases disease severity but only when moisture stress and overheating of branches occur.

Several diseases of confused etiology occur on palms, e.g. Kerala wilt of coconuts in India and lethal spear rot of oil palm in South America. Because a variety of potential pathogens have been isolated from such diseases, interacting pathogen complexes have often been blamed. Until a primary pathogen has been demonstrated and control can be initiated these diseases will continue to cause losses and to constrain coconut and oil palm development. Oil palm 'blast', a major nursery problem which hampered development of this crop in West Africa in the middle of this century, was shown to be due to a synergistic interaction of *Pythium splendens* and *Rhizoctonia lamellifera* which restricted root development and thus made plants unduly susceptible to moisture stress (Robertson, 1959) but recently other possible causes of blast have been investigated including insect transmitted agents (Turner, 1981).

Cryptic pathogens

Pathogens (and pests) can be obscured by other, more aggressive, competitors and their potential for damage held in check by negative interactions. Removal of competitors or sources of negative interaction allows these cryptic pathogens to emerge. Although not widely recognized, this effect is often responsible for the failure of disease control measures to achieve their full potential, control of one pathogen allowing its substitution by another. This can lead to the discovery of a succession of problems as each one is overcome, as commonly occurs when attempting to establish vegetable cultivation in recently cleared bush. Controlling insect pests which devastate young plants enables them to produce sufficient foliage to select fungal pathogens from the native microflora; controlling these in turn reveals a complex of viruses which then became predominant; plants also became infested with root-knot nematode and other soil-borne pathogens. As each problem is overcome, another pest or pathogen emerges from the natural ecosystem to become significant on the crop.

Some examples of competitive interactions among foliar pathogens of tropical crops have been studied. *Mycosphaerella berkelii* and *M. arachidis* are major pathogens causing Tikka leaf spot of groundnuts, but they suppress the potential damage that groundnut rust (*Puccinia arachidis*) can cause. Where benomyl was used to control *Mycosphaerella* spp., rust, which is unaffected by this fungicide, became dominant and could be as damaging as the *Mycosphaerella* spp. unless a fungicide was used which controlled both types of pathogen (Harrison, 1973). In Africa, web blotch caused by *Phoma arachidicola* is suppressed by the *Mycosphaerella* spp. and only causes significant damage where Tikka leaf spot is mild. Leaves infected with the *Mycosphaerella* spp. have been shown to produce a substance which prevents growth of the *Phoma* sp. (Cole, 1982). On banana, the more aggressive black leaf streak pathogen, *Mycosphaerella fijiensis*, has superseded *M. musicola*, the cause of Sigatoka disease, in many areas. Ascospores are more important in the epidemiology of *M. fijiensis* than of *M. musicola* and the use of fungicides in oil may have influenced this competitive interaction by favouring ascospore infection. Where both are controlled, other foliar pathogens such as rust (*Uromyces musae*), *Haplobasidion musae* and *Cordana musae* become more evident (Firman, 1972). Negative interactions between root and shoot diseases can also suppress the incidence of one or both pathogens; *Uromyces phaseoli* and *Meloidogyne incognita* are both reduced when present together on *Phaseolus vulgaris* (Bookbinder & Bloom, 1982).

Cocoa pod pathogens provide a more complex situation. There are basically three primary pathogens of cocoa pods: *Moniliophthora roreri* (frosty pod), which occurs in the northern parts of South America, *Crinipellis perniciosa* (witches' broom), which exists over much of South America and *Phytophthora* spp. (black pod), which as a group occur world wide. The three diseases occur together in the northern parts of South America where multiple infections are known to occur, especially with *Crinipellis* and *Moniliophthora* (H.C. Evans, personal communication) which are the major pod pathogens in this area. *Moniliophthora* sporulates only on diseased pods and its epidemic potential is dependent on this source of inoculum. However, *Crinipellis* inoculum is derived from sporophores produced on witches' brooms and diseased pods appear to have no epidemiological role. Thus competition between the two pathogens to colonize pods has little effect on the inoculum level and epidemic potential of *Crinipellis* but reduces those of *Moniliophthora*. In areas such as Colombia where *Crinipellis* has increased, *Moniliophthora* has declined and controlling *Crinipellis* alone may well enhance *Moniliophthora* pod rot losses. If the incidence of pod infections by both pathogens was reduced, the incidence of *Phytophthora* pod rot could increase so that this would then become the major pod pathogen, as it is in the Bahia region of Brazil and in West Africa where neither *Moniliophthora* nor *Crinipellis* occur.

Interactions between the *Phytophthora* spp. causing black pod also occur. *P. palmivora* exists throughout West Africa, but *P. megakarya* only occurs in eastern areas such as Nigeria where it is dominant because of its greater pathological potential and somewhat different epidemiology (Gregory & Maddison, 1981). Losses attributable solely to *P. palmivora* in these areas are small compared to those occurring elsewhere. However measures which reduce the incidence of *P. megakarya* alone would enable *P. palmivora* damage to increase.

Conclusion

Multiple infection and damage from many different pathogens is normal on most tropical crops. Interactions between the components of multiple infection increase crop damage and reduce the efficiency of many disease control measures. Often a substantial research effort is needed to unravel the components of etiologically complex diseases before a start can be made on reducing the damage they are causing. Obtaining the necessary support for such efforts, already difficult to undertake on a continuing basis in many developing countries, is often thwarted by the difficulties of attributing and assessing losses caused by multiple pathogen damage. Many diseases ascribed to interacting pathogen complexes may well be caused by primary pathogens or have separate causes under different circumstances. Because control of well established diseases appears to pave the way for cryptic pathogens to emerge, a continuing capacity to investigate new crop protection problems in developing countries is essential. The study of pathogen (and pest) interactions is still fairly new but it is becoming increasingly evident that substantial advances in preventing crop loss in tropical areas can only be achieved by close attention to the problems created by these interactions.

References

Adeniji M.O., Edwards D.I., Sinclair J.B. & Malek R.B. (1975) Interrelationship of *Heterodera glycines* and *Phytophthora megasperma* var *sojae*. *Phytopathology* **65**, 722–5.

Bhowmik T.P. & Singh A. (1977) Combined effect of *Rhizoctonia* root rot and *Alternaria* leaf blight on sunflower. *Indian Phytopathology* **30**, 195–7.

Bookbinder M.G. & Bloom J.P. (1982) Interaction of *Uromyces phaseoli* and *Meloidogyne incognita* on bean (*Phaseolus vulgaris*). *Journal of Nematology* **12**, 177–82.

Booth C.B. & Stover R.H. (1974) *Cylindrocarpon musae* sp. nov., commonly associated with burrowing nematode (*Radopholus similis*) lesions on bananas. *Transactions of the British Mycological Society* **63**, 503–7.

Cole Desirée L. (1982) Interactions between *Cercospora arachidicola* and *Phoma arachidicola* and their effects on defoliation and kernel yield of groundnuts. *Plant Pathology* **31**, 355–62.

Dickinson C.H. (1979) External synergisms among organisms inducing disease. *Plant disease, an advanced treatise*, Vol. IV. (Ed. by J.G. Horsfall & E.B. Cowling), pp. 97–111. Academic Press, New York.

English H. Davis J.R. & Devay J.E. (1975) Relationship of *Botryosphaeria dothidae* & *Hendersonula toruloidea* to a canker disease of Almond. *Phytopathology* **65**, 114–22.

Firman I.D. (1972) Black leaf streak of bananas in Fiji. *Annals of Applied Biology* **70**, 19–24.

Garcia R. & Mitchell D.J. (1975a) Synergistic interactions of *Pythium myriotylum* with *Fusarium solani* & *Meloidogyne arenaria* in pod rot of peanut. *Phytopathology* **65**, 832–3.

Garcia R. & Mitchell D.J. (1975b) Interactions of *Pythium myriotylum* with *Fusarium solani*, *Rhizoctonia solani* and *Meloidogyne arenaria* in pre-emergence damping-off of peanut. *Plant Disease Reporter* **59**, 665–9.

Garrett S.D. (1956) *Biology of root infecting fungi*. Cambridge University Press, Cambridge.

Golden J.K. & Van Gundy S.D. (1975) A disease complex of okra and tomato involving the nematode *Meloidogyne incognita* and the soil inhabiting fungus *Rhizoctonia solani*. *Phytopathology* **65**, 265–73.

Gollifer D.E., Jackson G.V.H., Dabek A.J. and Plumb R.T. (1978) Incidence, and effects on yield of virus diseases of taro (*Colocasia esculenta*) in the Solomon Islands. *Annals of Applied Biology* **88**, 131–5.

Gregory P.H. & Maddison A.C. (eds.) (1981) Epidemiology of *Phytophthora* on cocoa in Nigeria: Final report of the International Cocoa Research Project. *Phytopathological Paper* No. **25**, Commonwealth Agricultural Bureaux, Farnham Royal.

Hamilton R.I. (1980) Defenses triggered by previous invaders: viruses. *Plant disease: an advanced treatise*, Vol. V. (Ed. by J.G. Horsefall & E.B. Cowling), pp. 279–303. Academic Press, New York.

Harrison A.L. (1973) Control of peanut leaf rust alone or in combination with *Cercospora* leaf spot. *Phytopathology* **63**, 668–73.

Hooker W.J. & Fronek F.R. (1960) The influence of virus Y infection on early blight susceptibility in potato. *Proceedings of the Conference on Potato Virus Diseases* No. **4** 76–81.

Hyer A.H., Jorgenson E.C., Garber R.H. & Smith S. (1979) Resistance to root knot nematode in control of root knot nematode — *Fusarium* wilt disease complex in cotton. *Crop Science* **19**, 898–907.

Kleineke-Borchers A. (1982) Auswirkung der Interaktion zwischen *Meloidogyne incognita* und *Fusarium oxysporum* f.sp. *lycopersici* auf den Phytohormonhaushalt in Tomatenpflanzen. *Zeitschrift fur Pflanzenkrankheiten und Pflanzenschutz* **89**, 132–43.

Krikun J. & Frank Z.R. (1982) Metham sodium applied by sprinkler irrigation to control pod rot and *Verticillium* wilt of peanut *Plant Disease* **66**, 128–30.

Nitzany F.E., Joffee A.Z. & Palti J. (1973) Synergism between *Fusarium* spp. and cucumber mosaic virus. *Phytopathologische Zeitschrift* **76**, 314–8.

Pieczarka D.J. & Abawi G.S. (1978) Effect of interactions between *Fusarium, Pythium* and *Rhizoctonia* on severity of bean root rot, *Phytopathology* **68**, 403–8.

Powell N.T. (1979) Internal synergisms among organisms inducing disease. *Plant disease, an advanced treatise*, Vol. IV. (Ed. by J.G. Horsfall & E.B. Cowling), pp. 113–33. Academic Press, New York.

Raychaudhuri S.D., Nariani T.K. & Ahlawat Y.S. (1979) Dieback of citrus in India. *Proceedings of the International Society of Citriculture* **3**, 914–8.

Reddy P.P., Singh D.B. & Sharma S.R. (1979) Interaction of *Meloidogyne incognita* and *Rhizoctonia solani* in a root rot disease complex of French bean. *Indian Phytopathology* **32**, 651–2.

Richardson D.E. & Doling D.A. (1957) Potato blight and leaf-roll virus. *Nature* (London) **180**, 866–7.

Robertson J.S. (1959) Co-infection by a species of *Pythium* and *Rhizoctonia lamellifera* Small in blast disease of oil palm seedlings. *Transactions of the British Mycological Society* **42**, 401–5.

Sidhu G.S. & Webster J.M. (1977) The use of amino acid fungal auxotrophs to study the predisposition phenomena in the root knot/wilt fungus disease complex of tomato. *Physiological Plant Pathology* **11**, 117–27.

Simkin M.B. & Wheeler B.E.J. (1974) Effects of dual infection of *Puccinia hordei* and *Erysiphe graminis* on barley, cv. Zephyr. *Annals of Applied Biology* **78**, 237–50.

Suzuki H. (1980) Defenses triggered by previous invaders: fungi. *Plant disease, an advanced treatise*, Vol. V. (Ed. by J.G. Horsefall & E.B. Cowling), pp. 319–32. Academic Press, New York.

Turner P.D. (1981) *Oil palm diseases and disorders*. Oxford University Press, Oxford.

Van Gundy S.D. & Tsao P.J. (1963) Growth reduction of citrus seedlings by *Fusarium solani* as influenced by the citrus nematode and other soil factors. *Phytopathology* **53**, 488–9.

Index